Generative Artificial Intelligence

Exploring the Power and Potential of Generative AI

Shivam R Solanki
Drupad K Khublani

Apress®

Generative Artificial Intelligence: Exploring the Power and Potential of Generative AI

Shivam R Solanki
Dallas, TX, USA

Drupad K Khublani
Salt Lake City, UT, USA

ISBN-13 (pbk): 979-8-8688-0402-1
https://doi.org/10.1007/979-8-8688-0403-8

ISBN-13 (electronic): 979-8-8688-0403-8

Copyright © 2024 by Shivam R Solanki, Drupad K Khublani

This work is subject to copyright. All rights are reserved by the Publisher, whether the whole or part of the material is concerned, specifically the rights of translation, reprinting, reuse of illustrations, recitation, broadcasting, reproduction on microfilms or in any other physical way, and transmission or information storage and retrieval, electronic adaptation, computer software, or by similar or dissimilar methodology now known or hereafter developed.

Trademarked names, logos, and images may appear in this book. Rather than use a trademark symbol with every occurrence of a trademarked name, logo, or image we use the names, logos, and images only in an editorial fashion and to the benefit of the trademark owner, with no intention of infringement of the trademark.

The use in this publication of trade names, trademarks, service marks, and similar terms, even if they are not identified as such, is not to be taken as an expression of opinion as to whether or not they are subject to proprietary rights.

While the advice and information in this book are believed to be true and accurate at the date of publication, neither the authors nor the editors nor the publisher can accept any legal responsibility for any errors or omissions that may be made. The publisher makes no warranty, express or implied, with respect to the material contained herein.

>Managing Director, Apress Media LLC: Welmoed Spahr
>Acquisitions Editor: Celestin Suresh John
>Development Editor: Laura Berendson
>Coordinating Editor: Kripa Joseph
>Copy Editor: Kim Wimpsett

Cover designed by eStudioCalamar

Cover image by kjpargeter on freepik (www.freepik.com)

Distributed to the book trade worldwide by Apress Media, LLC, 1 New York Plaza, New York, NY 10004, U.S.A. Phone 1-800-SPRINGER, fax (201) 348-4505, e-mail orders-ny@springer-sbm.com, or visit www.springeronline.com. Apress Media, LLC is a California LLC and the sole member (owner) is Springer Science + Business Media Finance Inc (SSBM Finance Inc). SSBM Finance Inc is a **Delaware** corporation.

For information on translations, please e-mail booktranslations@springernature.com; for reprint, paperback, or audio rights, please e-mail bookpermissions@springernature.com.

Apress titles may be purchased in bulk for academic, corporate, or promotional use. eBook versions and licenses are also available for most titles. For more information, reference our Print and eBook Bulk Sales web page at http://www.apress.com/bulk-sales.

Any source code or other supplementary material referenced by the author in this book is available to readers on GitHub (https://github.com/Apress). For more detailed information, please visit https://www.apress.com/gp/services/source-code.

If disposing of this product, please recycle the paper

To my mother, whose strength and love have guided me; to my wife, my rock and inspiration; and to my family, who have always believed in me. This book is a tribute to your unwavering support and belief in my dreams, with all my love and gratitude.

Shivam R Solanki

To my beloved family—Mummy and Papa, whose unwavering faith and love have been my guiding light; to Didi and Jiju, whose encouragement never faltered; and to my wife, Suman, my inspiration and support. This book stands as a testament to your belief and love, dedicated with all my heart and gratitude.

Drupad K Khublani

Reality is unpredictable only as long as we see it without the lens of statistics. Statistics' potential to collapse reality to a handful of possibilities is what drew us to this field. We want to pay our respects to Alan Turing for initiating humanity's endeavor toward training Turing machines (what we call computers today), which paved the way for artificial intelligence.

The authors

Table of Contents

About the Authors xi

About the Technical Reviewer xiii

Introduction xv

Chapter 1: Introduction to Generative AI 1

    Unveiling the Magic of Generative AI 1

    The Genesis of Generative AI 2

        Milestones Along the Way.. 4

    Fundamentals of Generative Models.. 5

        Neural Networks: The Backbone of Generative AI 6

        Understanding the Difference: Generative vs. Discriminative Models........................... 8

        Understanding the Core: Types and Techniques 9

        Diffusion Models... 10

        Generative Adversarial Networks 10

        Variational Autoencoders... 11

        Restricted Boltzmann Machines.. 11

        Pixel Recurrent Neural Networks 12

    Generative Models in Society and Technology... 13

        Real-World Applications and Advantages of Generative AI.. 13

        Ethical and Technical Challenges of Generative AI 15

        Impact of Generative Models in Data Science... 18

    The Diverse Domains of Generative AI 20

        Visuals: From Pixel to Palette 20

        Audio: Symphonies of AI.. 21

        Text: Weaving Words into Worlds 22

        The Future of Generative AI: A Symphony of Possibilities 22

Table of Contents

Setting Up the Development Environment 23
    Setting Up a Google Colab Environment 23
    Hugging Face Access and Token Key Generation 30
    OpenAI Access Account and Token Key Generation 32
    Troubleshooting Common Issues 33
Summary 35

Chapter 2: Text-to-Image Generation 37

Introduction 37
Bridging the Gap Between Text and Image Data 39
    Understanding the Fundamentals of Image Data 40
    Correlation Between Image and Text Data Using CLIP Model 43
    Diffusion Model 49
Text-to-Image Generation 67
    Using a Pre-trained Model 68
    Fine-Tuning Text-to-Image Models 71
Conclusion 79

Chapter 3: From Script to Screen: Unveiling Text-to-Video Generation 81

Introduction 81
Understanding Video Data 84
    Challenges in Working with Video Data 87
    The Synergy of Video and Textual Data 91
Hands-On: Demonstrating a Pre-Trained Model 93
    Step 1: Installing Libraries 94
    Step 2: Model Inference 95
Fine-Tuning for Custom Applications 96
    Step 1: Installing Libraries 99
    Step 2: Data Loading and Preprocessing 100
    Step 3: Model Training (Fine-Tuning) 103
    Step 4: Model Inference 107
Conclusion 111

Chapter 4: Bridging Text and Audio in Generative AI 113
Brief History 113
Fundamentals and Challenges 115
    Understanding Audio Data 115
    Challenges in Working with Audio Data 118
    Mitigating Challenges in Audio Data Processing 119
Bridging Text and Audio: The CLAP Model Implementation 120
    Step 1: Installing Libraries and Data Loading 122
    Step 2: Model Inference 123
Understanding AI-Driven Text and Audio Conversion Models 125
    Understanding CTC Architectures 125
    Understanding Seq2Seq Architectures 128
Implementation AI-Driven Text and Audio Conversion Modes 130
    Speech to Text 130
    Text to Speech 149
Conclusion 170

Chapter 5: Large Language Models 173
Introduction 173
    Phases of Training and Adoption of Large Language Models 175
Types of Language Transformers Models 179
    Encoder Models 183
    Fine-Tuning BERT 188
    Decoder-Only Models (Generative Pre-trained Transformer) 219
    Encoder-Decoder Models 222
A Glimpse into the LLM Horizon: Where Do We Go from Here? 226
Summary 228

Chapter 6: Generative Large Language Models 229
Introduction 229
NLP Tasks Using LLMs 230
    Sentiment Analysis 231

TABLE OF CONTENTS

- Entity Extraction 236
- Topic Modeling 239
- Natural Language Generation Tasks Using LLMs 241
  - Creative Writing 241
  - Text Summarization 244
  - Dialogue Generation 247
- Advanced Prompting Techniques 250
  - Few-Shot Prompting 251
  - Chain-of-Thought 253
  - Prompting vs. Fine-Tuning 255
- Fine-Tuning LLMs 258
  - Case Study: Fine-Tuning an LLM for Sentiment Analysis 260
  - Parameter Efficient Fine-Tuning 261
  - Fine-Tuning LLM for Question Answering 263
- Summary 295

Chapter 7: Advanced Techniques for Large Language Models 297

- Introduction 297
- Fine-Tuning LLMs for Abstractive Summarization 298
  - Fine-Tuning an Encoder-Decoder Model 299
  - Abstractive Summarization Using a Decoder-Only Model 311
- Guidelines on Fine-Tuning a Large Language Model 322
  - Types of SFT (Supervised Fine-Tuning) 323
  - Memory Consumption During SFT 324
- Reinforcement Learning from Human Feedback 324
  - What Is RLHF? 325
  - How Does RLHF Work? 325
  - Reward Model Implementation 328
  - Controlled Review Generation 330
  - RLHF Summary 347
- Summary 348

Chapter 8: Building Demo Applications Using LLMs 349

Making Sense of Website Content 349
    Data Scraping 351
    Question-answering 353
    Summarization 357
    User Interface/Application 360

Uncovering Insights and Gaining a Quick Understanding of PDF Documents 368
    Question-Answering for PDF 369
    PDF Summarization 375

Extracting Insights from Video Transcripts 383
    Video Caption Summarization and Q&A 384
    Video Transcript Analysis Using Langchain and OpenAPI 394

Summary 398

Chapter 9: Building Enterprise-Grade Applications Using LLMs 401

Retrieval-Augmented Question-Answering Chatbot 402
    Real-World Use Cases of Retrieval Augmentation Generation 405
    RAG Architecture 406
    Creating a Knowledge Base 408
    Setting Up a Retrieval System 412
    Neural Reranker 418
    Generative LLM 422
    User Interface 426
    Suggested Improvements in the RAG Pipeline for Generative Q&A 436

Summary 438

Conclusion: Generative AI Journey 440

References 443

Index 449

About the Authors

Shivam R Solanki is an accomplished senior advisory data scientist leading an AI team in solving challenging problems using artificial intelligence (AI) in a worldwide partner ecosystem. Shivam holds a master's degree from Texas A&M University with major coursework in applied statistics. Throughout his career, he has delved into various AI fields, including machine learning (ML), deep learning (DL), and natural language processing (NLP). His expertise extends to Generative AI, where his practical experience and in-depth knowledge empower him to navigate its intricacies. As a researcher in AI, Shivam has filed two patents for ML and NLP, co-authored a book on DL, and published a paper on Generative AI.

Drupad K Khublani is a skilled senior data scientist and part of the revenue management team in a real estate company. His leadership in partnering with teams across marketing, call center operations, product management, customer experience, and operations has cultivated a wealth of experience, empowering him to extract actionable insights and co-create innovative solutions. Drupad completed graduate and postgraduate programs at the Indian Institute of Technology (Indian School of Mines) and Texas A&M University. Collaborating with Dr. Jean-Francois Chamberland on the development of technology to identify obstacles and gauge distances using only a monocular camera highlights Drupad's inventive approach and dedication to real-world applications, alongside his accomplishments in both the commercial and academic arenas.

About the Technical Reviewer

Durgesh Gurnani is a key influencer in Generative AI, earning a master's degree in the United States and currently residing in Delhi, India. He's shared his deep knowledge on TV and at international events. Universities around the world invite him for special lectures and AI bootcamps. In addition to his collaborations with multinational companies, Durgesh conducts online classes every Sunday. Discover his insights at https://gurnaninotes.com. Join the community and explore the world of Generative AI with Durgesh.

Introduction

This book explains the field of generative artificial intelligence (Generative AI), focusing on its potential and applications, and aims to provide you with an understanding of the underlying principles, techniques, and practical use cases of Generative AI models.

The book begins with an introduction to the foundations of Generative AI, including an overview of the field, its evolution, and its significance in today's AI landscape. Next it focuses on generative visual models, exploring the exciting field of transforming text into images and videos. Then it covers text-to-video generation and provides insights into synthesizing videos from textual descriptions, opening new possibilities for creative content generation. The next chapter covers generative audio models and prompt-to-audio synthesis using text-to-speech (TTS) techniques. Then it switched gears, diving into the realm of generative text models and exploring the concepts of large language models (LLMs), natural language generation (NLG), fine-tuning, prompt tuning, and reinforcement learning. The chapters explore techniques for fixing LLMs and making them grounded and instructible, along with practical applications in enterprise-grade applications such as question answering, summarization, and knowledge base generation.

After reading this book, you will understand generative text, audio, and visual models and have the knowledge and tools necessary to harness the creative and transformative capabilities of Generative AI.

CHAPTER 1

Introduction to Generative AI

Unveiling the Magic of Generative AI

Imagine a world where the lines between imagination and reality blur. *Generative AI* refers to the subset of artificial intelligence focused on creating new content—from text to images, music, and beyond—based on learning from vast amounts of data. A few words whispered into a machine can blossom into a breathtaking landscape painting, and a simple melody hummed can transform into a hauntingly beautiful symphony. This isn't the stuff of science fiction but the exciting reality of Generative AI. You've likely encountered its early forms in autocomplete features in email or text editors, where it predicts the end of your sentences in surprisingly accurate ways. This transformative technology isn't just about analyzing data; it's about breathing life into entirely new creations, pushing the boundaries of what we thought machines could achieve.

Gone are the days of static, preprogrammed responses. Generative AI models learn and adapt, mimicking humans' ability to observe, understand, and create. These models decipher the underlying patterns and relationships defining each domain by analyzing massive images, text, audio, and more datasets. Armed with this knowledge, they can then transcend mere imitation, generating entirely new content that feels fresh, original, and often eerily similar to its real-world counterparts.

This isn't just about novelty, however. Generative AI holds immense potential to revolutionize various industries and reshape our daily lives. Imagine the following:

Designers: Creating unique and personalized product concepts based on user preferences.

Musicians: Composing original soundtracks tailored to specific emotions or moods.

Writers: Generating creative content formats such as poems, scripts, or entire novels.

Educators: Personalizing learning experiences with AI-generated practice problems and interactive narratives.

Scientists: Accelerating drug discovery by simulating complex molecules and predicting their properties.

From smart assistants crafting detailed travel itineraries to sophisticated photo editing tools that can alter the time of day in a photograph, Generative AI is weaving its magic into the fabric of our everyday experiences.

The possibilities are endless, and Generative AI's magic lies in its versatility. It can be used for artistic expression, entertainment, education, scientific discovery, and countless other applications. But what makes this technology truly remarkable is its ability to collaborate with humans, pushing the boundaries of creativity and innovation in ways we never thought possible.

So, as you begin your journey into the world of Generative AI, remember this: it's not just about the technology itself but about the potential it holds to unlock our creativity and imagination. With each new model developed and each new application explored, we inch closer to a future where the line between human and machine-generated creation becomes increasingly blurred, and the possibilities for what we can achieve together become genuinely limitless.

The Genesis of Generative AI

The saga of Generative AI unfolds like a tapestry woven from the early threads of artificial intelligence, evolving through decades of innovation to become the powerhouse of creativity and problem-solving we see today. From its inception in the 1960s to the flourishing ecosystem of today's technology, Generative AI has traced a path of remarkable growth and transformation.

The Initial Spark (1960s): The odyssey commenced with the development of ELIZA, a simple chatbot devised to simulate human conversation. Despite its rudimentary capabilities, ELIZA ignited the imaginations of many, sowing the seeds for future advancements in natural language processing (NLP) and beyond, laying a foundational stone for the intricate developments that would follow.

The Era of Deep Learning Emergence (1980s–2000s): The concept of neural networks and deep learning was not new, but it lay dormant, constrained by the era's computational limitations. It wasn't until the turn of the millennium that a confluence of enhanced computational power and burgeoning data availability set the stage for significant breakthroughs, signaling a renaissance in AI research and development.

Breakthrough with Generative Adversarial Networks (2014): The introduction of generative adversarial networks (GANs) by Ian Goodfellow marked a watershed moment for Generative AI. This innovative framework, consisting of dueling networks—one generating content and the other evaluating it—ushered in a new era of image generation, propelling the field toward the creation of ever more lifelike and complex outputs.

A Period of Rapid Expansion (2010s–present): The landscape of Generative AI blossomed post-2010, driven by GANs and advancements in deep learning technologies. This period saw the diversification of generative models, including convolutional neural networks (CNNs) and recurrent neural networks (RNNs) for text and video generation, alongside the emergence of variational autoencoders and diffusion models for image synthesis. The development of large language models (LLMs), starting with GPT-1, demonstrated unprecedented text generation capabilities, marking a significant leap in the field.

Mainstream Adoption and Ethical Debates (2022): The advent of user-friendly text-to-image models like Midjourney and DALL-E 2, coupled with the popularity of OpenAI's ChatGPT, catapulted Generative AI into the limelight, making it a household name. However, this surge in accessibility and utility also brought to the forefront critical discussions on copyright issues, the potential displacement of creative professions, and the ethical use of AI technology, emphasizing the importance of mindful development and application.

CHAPTER 1 INTRODUCTION TO GENERATIVE AI

Milestones Along the Way

The evolution of Generative AI (see Figure 1-1) has been punctuated by several key milestones that have significantly shaped its trajectory, pushing the boundaries of what's possible and setting new standards for innovation in the field.

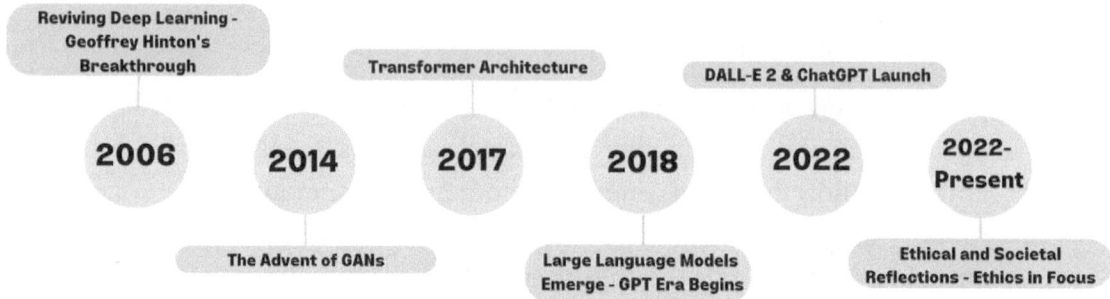

Figure 1-1. Generative AI evolution timeline

Reviving Deep Learning (2006): A pivotal moment in the resurgence of neural networks came with Geoffrey Hinton's groundbreaking paper, "A Fast Learning Algorithm for Deep Belief Nets." This work reinvigorated interest in restricted Boltzmann machines (RBMs) and deep learning, laying the groundwork for future advancements in Generative AI.

The Advent of GANs (2014): Ian Goodfellow and his colleagues introduced GANs, a novel concept that employs two neural networks in a form of competitive training. This innovation not only revolutionized the generation of realistic images but also opened new avenues for research in unsupervised learning.

Transformer Architecture (2017): The "Attention Is All You Need" paper by Vaswani et al. introduced the transformer architecture, fundamentally changing the landscape of NLP. This architecture, which relies on self-attention mechanisms, has since become the backbone of LLMs, enabling more efficient and coherent text generation.

Large Language Models Emerge (2018–Present): The introduction of GPT by OpenAI marked the beginning of the era of large language models. These models, with their vast capacity for understanding and generating human-like text, have drastically expanded the applications of Generative AI, from writing assistance to conversational AI.

Mainstream Breakthroughs (2022): The release of models like DALL-E 2 for text-to-image generation and ChatGPT for conversational AI brought Generative AI into mainstream awareness. These tools demonstrated the technology's potential to the public, showcasing its ability to generate creative, engaging, and sometimes startlingly lifelike content.

Ethical and Societal Reflections (2022–Present): With greater visibility came increased scrutiny. The widespread adoption of Generative AI technologies sparked important conversations around copyright, ethics, and the impact on creative professions. This period has highlighted the need for thoughtful consideration of how these powerful tools are developed and used.

These milestones underscore the rapid pace of advancement in Generative AI, illustrating a journey of innovation that has transformed the landscape of artificial intelligence. Each landmark not only represents a leap forward in capabilities but also sets the stage for the next wave of discoveries, challenging us to envision a future where AI's creative potential is harnessed for the greater good while navigating the ethical complexities it brings.

Fundamentals of Generative Models

With their ability to "dream up" new data, generative models have become a cornerstone of AI, reshaping how we interact with technology, create content, and solve problems. This section delves deeper into their inner workings, applications, and limitations, equipping you to harness their power responsibly.

CHAPTER 1   INTRODUCTION TO GENERATIVE AI

Neural Networks: The Backbone of Generative AI

Neural networks form the foundation of Generative AI, enabling machines to generate new data instances that mimic the distribution of real data. At their core, neural networks learn from vast amounts of data, identifying patterns, structures, and correlations that are not immediately apparent. This learning capability allows them to produce novel content, from realistic images and music to sophisticated text and beyond. The versatility and power of neural networks in Generative AI have opened new frontiers in creativity, automation, and problem-solving, fundamentally changing our approach to content creation and data analysis.

Key Neural Network Architectures Relevant to Generative AI

Generative AI has been propelled forward by several key neural network architectures, each bringing unique strengths to the table in terms of learning patterns, processing sequences, and generating content.

Convolutional Neural Networks

Convolutional neural networks are specialized in processing structured grid data such as images, making them a cornerstone in visual data analysis and generation. By automatically and adaptively learning spatial hierarchies of features, CNNs can generate new images or modify existing ones with remarkable detail and realism. This capability has been pivotal in advancing fields such as computer vision, where CNNs are used to create realistic artworks, enhance photos, and even generate entirely new visual content that is indistinguishable from real-world images. DeepDream, developed by Google, is an iconic example of CNNs in action. It enhances and modifies images in surreal, dreamlike ways, showcasing CNNs' ability to interpret and transform visual data creatively.

Recurrent Neural Networks

Recurrent neural networks excel in handling sequential data, making them ideal for tasks that involve time series, speech, or text. RNNs can remember information for long durations, and their ability to process sequences of inputs makes them perfect for generating coherent and contextually relevant text or music. This architecture has revolutionized natural language processing and generation, enabling the creation of sophisticated AI chatbots, automated writing assistants, and dynamic music composition software. Google's Magenta project utilizes RNNs to create new pieces of music, demonstrating RNNs' prowess in understanding and generating complex sequences, such as musical compositions, by learning from vast datasets of existing music.

Generative Adversarial Networks

Generative adversarial networks consist of two neural networks—the generator and the discriminator—competing in a zero-sum game framework. This innovative structure allows GANs to generate highly realistic and detailed images, videos, and even sound. The competitive nature of GANs pushes them to continually improve, leading to the generation of content that can often be indistinguishable from real-world data. Their application ranges from creating photorealistic images and deepfakes to advancing drug discovery and material design. StyleGAN, developed by NVIDIA, exemplifies GANs' capabilities by generating highly realistic human faces and objects. This technology has been used in fashion and design to visualize new products and styles in stunning detail.

Transformers

Transformers have revolutionized the way machines understand and generate human language, thanks to their ability to process words in relation to all other words in a sentence, simultaneously. This architecture underpins some of the most advanced language models like Generative Pre-trained Transformer (GPT), enabling a wide range of applications from generating coherent and contextually relevant text to translating languages and summarizing documents. Their unparalleled efficiency in handling sequential data has made them the model of choice for tasks requiring a deep understanding of language and context. OpenAI's GPT-3 showcases the power of transformer architectures through its ability to generate human-like text across a variety of applications, from writing articles and poems to coding assistance, illustrating the model's deep understanding of language and context.

Transitioning from these architectures, it's essential to appreciate the distinction between generative and discriminative models in AI. While the former focuses on generating new data instances, the latter is concerned with categorizing or predicting outcomes based on input data. Understanding this difference is crucial for leveraging the right model for the task at hand, ensuring the effective and responsible use of AI technologies.

CHAPTER 1　INTRODUCTION TO GENERATIVE AI

Understanding the Difference: Generative vs. Discriminative Models

The world of AI models can be vast and complex, but two key approaches stand out: generative and discriminative models. Though they deal with data and learning, their goals and functionalities differ significantly.

Generative models, the creative minds of AI, focus on understanding the underlying patterns and distributions within data. Imagine them as artists studying various styles and techniques. They analyze the data, learn the "rules" of its creation, and then use that knowledge to generate entirely new content. This could be anything from realistic portraits to captivating melodies to even novel text formats.

Discriminative models, on the other hand, function more like meticulous detectives. Their focus lies on identifying and classifying different types of data. They draw clear boundaries between categories, enabling them to excel at tasks like image recognition or spam filtering. While they can recognize a cat from a dog, they can't create a new image of either animal on their own.

Here's an analogy to further illustrate the distinction:

- Imagine you're learning a new language. A generative model would immerse itself in the language, analyzing grammar, vocabulary, and sentence structures. It would then use this knowledge to write original stories or poems.

- A discriminative model would instead focus on understanding the differences between different languages. It could then identify which language a text belongs to but couldn't compose its own creative text in that language.

Table 1-1 summarizes the differences.

Table 1-1. Generative and Discriminative Comparison

Aspect	Generative Models	Discriminative Models
Primary focus	Understanding and learning the distribution of data to generate new instances	Identifying and classifying data into categories
Functionality	Generates new data samples similar to the input data	Classifies input data into predefined categories
Learning approach	Analyzes and learns the "rules" or patterns of data creation	Learns the decision boundary between different classes or categories of data
Key characteristics	Creative and productive; can create something new based on learned patterns	Analytical and selective; focuses on distinguishing between existing categories
Applications	Image and text generation (e.g., DALL-E, GPT-3); music composition (e.g., Google's Magenta); drug discovery and design	Spam email filtering; image recognition (e.g., identifying objects in photos); fraud detection
Examples	Creating realistic images from textual descriptions; composing original music; writing poems or stories	Categorizing emails as spam or not spam; recognizing faces in images; predicting customer churn
Real-world example	GPT-3 by OpenAI: uses generative modeling to produce human-like text	Google Photos: uses discriminative algorithms to categorize and label photos by faces, places, or things

In essence, generative models are the dreamers, conjuring up new possibilities, while discriminative models are the analysts, expertly classifying and categorizing existing data. Both play crucial roles in various fields, and understanding their differences is essential for choosing the right tool for the right job.

Understanding the Core: Types and Techniques

Generative models are a fascinating and versatile group of algorithms used across a wide range of applications in artificial intelligence and machine learning. Each model has its own strengths and is suited to particular types of tasks. Here's an expanded view of each generative model mentioned, along with examples of their real-life use cases:

CHAPTER 1   INTRODUCTION TO GENERATIVE AI

Diffusion Models

Diffusion models gradually transform data from a simple distribution into a complex one and have revolutionized digital art and content creation. They generate realistic images and animations from textual descriptions and are also applied in enhancing image resolution, including medical imaging, where they can generate detailed images for research and training purposes. While Chapter 2 will delve into diffusion models, let's build a foundational understanding with some pseudocode first.

```
import torch
from torch import nn

class DiffusionModel(nn.Module):
    def __init__(self, channels):
        super().__init__()
        # ... (layers for diffusion process)

    def forward(self, x, t):
        # ... (diffusion steps based on time step t)
        return x
```

Generative Adversarial Networks

GANs consist of two neural networks—the generator and the discriminator—engaged in a competitive training process. This innovative approach has found widespread application in creating photorealistic images, deepfake videos, and virtual environments for video games, as well as in fashion, where designers visualize new clothing on virtual models before production. To gain a clearer picture of the model's implementation, let's examine the pseudocode.

```
import torch
from torch import nn

class Generator(nn.Module):
    # ... (generator architecture)
```

```
class Discriminator(nn.Module):
  # ... (discriminator architecture)

# Train the GAN
# ... (training loop for generator and discriminator)
```

Variational Autoencoders

Variational autoencoders (VAEs) are renowned for their ability to compress and reconstruct data, making them ideal for image denoising tasks where they clean up noisy images. Furthermore, in the pharmaceutical industry, VAEs are utilized to generate new molecular structures for drug discovery, demonstrating their capacity for innovation in both digital and physical realms. Let's delve into the pseudocode to unravel the implementation specifics.

```
import torch
from torch import nn

class VAE(nn.Module):
  def __init__(self, input_dim, latent_dim):
    super().__init__()
    self.encoder = nn.Sequential(
        # ... (encoder layers)
    )
    self.decoder = nn.Sequential(
        # ... (decoder layers)
    )
  def forward(self, x):
    z = self.encoder(x)
    reconstruction = self.decoder(z)
    return reconstruction, z
```

Restricted Boltzmann Machines

Restricted Boltzmann machines learn probability distributions over their inputs, making them instrumental in recommendation systems. By predicting user preferences for items like movies or products, RBMs personalize recommendations, enhancing user experience by leveraging learned user-item interaction patterns. By reviewing the pseudocode, we can better comprehend the practical implementation of this model.

CHAPTER 1 INTRODUCTION TO GENERATIVE AI

```python
import numpy as np

class RBM:
    def __init__(self, visible_size, hidden_size):
        self.weights = np.random.rand(visible_size, hidden_size)
        self.visible_bias = np.zeros(visible_size)
        self.hidden_bias = np.zeros(hidden_size)

    def sample_hidden(self, v):
        # ... (calculate hidden layer probabilities based on visible layer)
        return hidden_states

    def sample_visible(self, h):
        # ... (calculate visible layer probabilities based on hidden layer)
        return visible_states

    def train(self, data, epochs):
        # ... (training loop for weight and bias updates)
```

Pixel Recurrent Neural Networks

Pixel Recurrent Neural Networks (PixelRNNs) generate coherent and detailed images pixel by pixel, considering the arrangement of previously generated pixels. This capability is crucial for generating textures in virtual reality environments or for photo editing applications where filling in missing parts of images with coherent detail is required. A walkthrough of the pseudocode will help us grasp the model's implementation structure.

```python
import torch
from torch import nn

class PixelRNN(nn.Module):
    def __init__(self, input_dim):
        super().__init__()
        self.rnn = nn.LSTM(input_dim, input_dim)

    def forward(self, x):
        # ... (iterate through pixels, feeding previous output to RNN)
        return generated_image
```

Generative Models in Society and Technology

As we embark on the exploration of generative models, we delve into a domain where artificial intelligence not only mirrors the complexities of human creativity but also propels it into new dimensions. These models stand at the confluence of technology and society, offering groundbreaking solutions, enhancing creative endeavors, and presenting new challenges. Their integration into various sectors underscores a transformative era in AI application, where the potential for innovation is boundless yet accompanied by the imperative of ethical stewardship.

Real-World Applications and Advantages of Generative AI

Generative models are not just about creating new data; their advantages span a wide array of applications, significantly impacting various facets of human civilization. Their transformative effects can be seen in the following areas, ordered by their potential to reshape industries and improve lives:

> **Healthcare and Medical Research:** Generative models are a boon to healthcare, especially in data-limited areas. They can synthesize medical data for research, facilitating the development of diagnostic tools and personalized medicine. This ability to augment datasets is pivotal for training robust AI systems that can predict diseases and recommend treatments, potentially saving lives and improving healthcare outcomes worldwide.
>
> **Security and Fraud Detection:** In the financial sector, generative models enhance security by identifying anomalous patterns indicative of fraudulent transactions. Their capacity to understand and model normal transactional behavior enables them to pinpoint outliers with high accuracy, safeguarding financial assets and consumer trust in banking systems.
>
> **Design and Creativity:** The impact of generative models in design and creative industries is profound. They foster innovation by generating novel concepts in architecture, product design, and even fashion, challenging traditional boundaries and inspiring

new trends. This not only accelerates the design process but also introduces a new era of creativity that blends human ingenuity with computational design.

Content Personalization: By tailoring content to individual preferences, generative models enhance user experiences across digital platforms. Whether it's personalizing music playlists, curating movie recommendations, or customizing news feeds, these models ensure that content resonates more deeply with users, elevating engagement and satisfaction.

Cost Reduction and Process Efficiency: In manufacturing and entertainment, among other industries, generative models streamline operations by automating the creation of content, designs, and solutions. This automation translates into significant cost savings and operational efficiencies, enabling businesses to allocate resources more effectively and focus on innovation.

Adaptability Across Learning Scenarios: The flexibility of generative models to function in unsupervised, semi-supervised, and supervised learning environments underscores their versatility. This adaptability makes them invaluable tools across a broad spectrum of applications, from language translation to generating synthetic training data for machine learning models.

Educational Tools and Simulations: Expanding on their applications, generative models offer innovative ways to create educational content and simulations. They can generate interactive learning materials that adapt to the student's learning pace and style, making education more engaging and personalized. This has the potential to revolutionize teaching methodologies and make learning more accessible to diverse learner populations.

Generative models stand at the vanguard of technological innovation, their influence transcending mere data creation to catalyze advancements across multiple domains. Elon Musk, reflecting on this transformative power, has stated, "Generative AI is the most powerful tool for creativity that has ever been created. It has the potential to unleash a new era of human innovation." This echoes the sentiment that Generative AI's capacity to drive healthcare innovations, bolster security, ignite creativity, customize content, and streamline industry operations marks a significant societal shift. Furthermore, Bill Gates captures the expansive potential of these technologies, noting, "Generative AI has the potential to change the world in ways that we can't even imagine. It has the power to create new ideas, products, and services that will make our lives easier, more productive, and more creative. It also has the potential to solve some of the world's biggest problems, such as climate change, poverty, and disease. The future of Generative AI is bright, and I'm excited to see what it will bring." As generative models continue to evolve and mature, their influence on the fabric of human civilization is set to deepen, highlighting the critical need for responsible harnessing of their potential.

Ethical and Technical Challenges of Generative AI

Generative AI, despite its transformative potential, is accompanied by a spectrum of ethical and technical challenges that necessitate careful consideration and management. These challenges, ordered by their potential impact on society, highlight the delicate balance between innovation and responsibility.

> **Ethical Dilemmas and Misuse:** At the forefront are the ethical concerns associated with the creation of hyper-realistic content. The potential for misuse in generating deepfakes, propagating misinformation, or infringing on copyright and privacy rights poses significant societal risks. Navigating these ethical minefields requires stringent guidelines and ethical frameworks to ensure that the power of Generative AI serves to benefit rather than harm society.
>
> **Bias and Fairness:** The issue of bias in AI outputs, rooted in biased training datasets, is a critical challenge. Without careful curation and oversight, generative models can perpetuate or even amplify existing societal biases, leading to unfair or discriminatory content. Addressing this requires a concerted effort toward ethical data collection, model training, and continuous monitoring to ensure fairness and inclusivity.

Data Privacy and Security: The reliance on vast amounts of data for training generative models raises concerns around data privacy and security. Ensuring that data is sourced ethically, with respect to individual privacy rights, and secured against breaches is paramount to maintaining trust in AI technologies.

Quality Control and Realism: Guaranteeing the quality and realism of generated outputs, while avoiding subtle anomalies, is a technical hurdle. These anomalies, if unnoticed, could pose risks, especially in sensitive applications such as medical diagnosis or legal documentation. Implementing rigorous quality control measures and validation processes is essential to mitigate these risks.

Interpretability and Transparency: The "black box" nature of some Generative AI models, particularly those based on deep learning, complicates efforts to understand their decision-making processes. This lack of interpretability is especially concerning in critical applications where understanding AI's rationale is crucial. Advancing toward more transparent AI models is a necessary step to ensure accountability and trust.

Training Complexity and Resource Requirements: The sophisticated nature of generative models means they require extensive computational resources and expertise to train, presenting barriers to entry and sustainability concerns. Efforts to optimize model efficiency and reduce computational demands are ongoing challenges in making Generative AI more accessible and environmentally sustainable.

Overfitting and Lack of Diversity: The tendency of models to overfit to their training data, resulting in outputs that lack diversity or creativity, is a technical challenge. This can limit the generality and applicability of AI-generated content. Developing techniques to encourage diversity and novelty in AI outputs is key to unlocking the full creative potential of generative models.

> **Mode Collapse in GANs:** A specific challenge for GANs is mode collapse, where the model generates a limited variety of outputs, undermining the diversity and richness of generated content. Addressing mode collapse through improved model architectures and training methodologies is crucial for realizing the vast creative possibilities of GANs.

As we continue to harness the capabilities of Generative AI, addressing these challenges and considerations with a mindful approach to ethics, fairness, and sustainability will be critical in shaping a future where Generative AI technologies contribute positively to human civilization.

DeepMind's Approach to Data Privacy and Security

DeepMind, a pioneer in artificial intelligence research, has been at the forefront of developing advanced AI models, including generative models that require access to large datasets. Recognizing the critical importance of data privacy and security, DeepMind has implemented robust measures to address these concerns, showcasing a commitment to ethical AI development.

The development of Generative AI models necessitates the collection and analysis of vast amounts of data, raising significant concerns regarding privacy and the potential for data misuse. DeepMind's challenge was to ensure that its research and development practices not only complied with data protection laws but also set a benchmark for ethical AI research.

DeepMind's approach to navigating data privacy and security challenges involves several key strategies.

> **Ethical Data Sourcing:** DeepMind adheres to stringent guidelines for data collection, ensuring that data is sourced ethically and with explicit consent from individuals. This includes anonymizing data to protect personal information and reduce the risk of identification.

> **Data Access Controls:** DeepMind implements strict access controls and encryption to safeguard data integrity and confidentiality. Access to sensitive or personal data is tightly regulated, with protocols in place to prevent unauthorized access.

> **Transparency and Accountability:** DeepMind fosters a culture of transparency, regularly publishing research findings and methodologies. To ensure accountability, the company engages with external ethical review boards and seeks feedback from the broader AI community.
>
> **Collaboration on Data Security Standards:** By collaborating with industry partners, academic institutions, and regulatory bodies, DeepMind contributes to the development of global standards for data privacy and security in AI. This collaborative approach helps advance the field while promoting best practices for data protection.

DeepMind's proactive measures in addressing data privacy and security have not only enhanced trust in its AI technologies but also served as a model for responsible AI development. By prioritizing ethical considerations and implementing robust security measures, DeepMind demonstrates that advancing AI research can be balanced with protecting individual privacy rights.

Impact of Generative Models in Data Science

In the rapidly evolving landscape of data science, generative models like GPT-4 are at the forefront of innovation, offering unparalleled tools that extend well beyond the initial stages of data exploration. Their application is reshaping industries, enhancing decision-making processes, and fostering new forms of creativity. Here's an expanded look at the pivotal areas where generative models are making their mark, arranged by their significance and potential for societal impact:

> **Natural Language Processing:** Generative models have revolutionized the way we interact with language, automating content creation, enabling real-time translation, and refining communication systems to be more intuitive and interactive. This transformation extends across various sectors, from customer service enhancements to accessibility improvements, making information more universally accessible and fostering global connections.

Predictive Analysis: The ability of generative models to sift through extensive historical data and predict future trends and outcomes is transforming critical decision-making processes. In finance, healthcare, and environmental studies, these predictions inform strategic planning, risk management, and preventive measures, contributing to more informed, data-driven decisions that can save lives, optimize operations, and protect resources.

Data Exploration: Generative models are redefining data exploration by quickly summarizing complex datasets into natural language descriptions of key statistics, trends, and anomalies. This not only accelerates the analytical process but also democratizes data analysis, making it accessible to nonexperts and facilitating cross-disciplinary collaboration and innovation.

Customization and Personalization: Expanding their influence, generative models offer sophisticated customization and personalization options in products, services, and content delivery. From personalized shopping experiences to customized learning modules, these models are enhancing user engagement and satisfaction by tailoring offerings to individual preferences and behaviors.

Ethical and Responsible Use: As the capabilities of generative models expand, so does the need for ethical considerations and responsible use. Ensuring that these powerful tools are used to benefit society, protect privacy, and promote fairness requires ongoing vigilance, transparent practices, and a commitment to ethical principles in their development and deployment.

Incorporating these models into data science practices not only necessitates a technical understanding of their mechanisms but also a thoughtful approach to their potential impact on society. By prioritizing areas that offer the greatest benefits while addressing ethical considerations, the data science community can harness the power of generative models to drive positive change and innovation.

After exploring the vast realm of Generative AI's applications—from revolutionizing healthcare and transforming creative industries to enhancing security measures and personalizing digital experiences—how do we envision the future trajectory of these technologies in a way that prioritizes human welfare and societal progress? Reflect on the potential long-term impacts of integrating Generative AI into everyday life; the ethical frameworks that should accompany such integration to address challenges like privacy, bias, and control; and how individuals and communities can contribute to a future where the benefits of Generative AI are accessible and equitable for all.

The Diverse Domains of Generative AI

The creative potential of generative models extends far beyond a single domain, painting a vibrant landscape of possibilities across visual, audio, and textual realms. In subsequent chapters (see Figure 1-2), we will explore each of these visual, audio, and textual domains. Let's explore how these models are revolutionizing diverse fields.

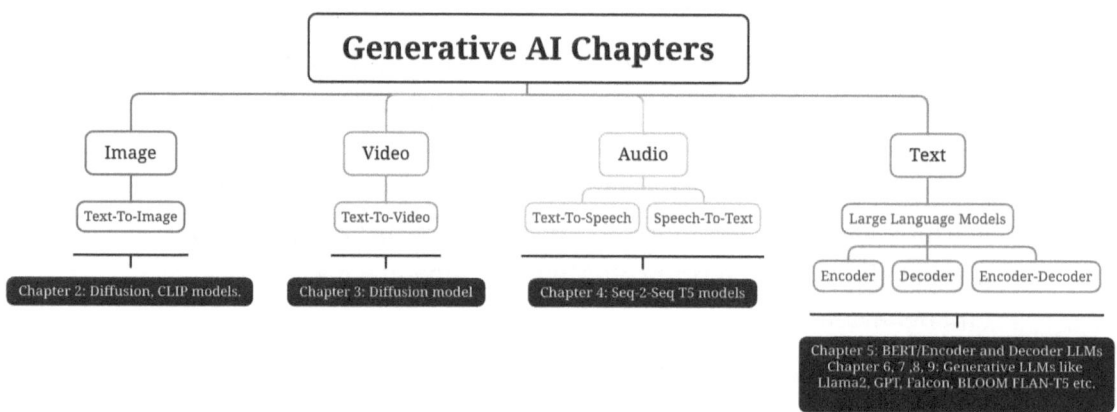

Figure 1-2. Overview of chapters in this book

Visuals: From Pixel to Palette

Bridging the gap between imagination and reality, Generative AI models offer unparalleled capabilities in transforming simple inputs into complex, mesmerizing outputs across visual, audio, and textual landscapes. As we delve into the visual domain, we uncover the transformative power of Generative AI in redefining the essence of image generation, video synthesis, and 3D design, marking a new epoch in digital expression and innovation.

Image Generation: From photorealistic portraits to whimsical landscapes, generative models push the boundaries of image creation. Tools like DALL-E 2 and Midjourney allow artists to explore new styles and generate unique concepts while researchers leverage them to study visual perception and develop advanced medical imaging techniques.

Video Synthesis: Imagine generating realistic videos from just a text description. This is the promise of generative models in video creation. Imagine-A-Video and Imagen Video models pave the way for personalized video experiences, revolutionizing advertising, entertainment, and even education.

3D Design: Sculpting virtual worlds with generative models is no longer science fiction. DreamFusion and Magic 3D empower designers to create intricate 3D models from simple sketches, accelerating product design and animation workflows.

Audio: Symphonies of AI

The realm of audio is undergoing a transformative renaissance, thanks to the advent of Generative AI. From the harmonious intricacies of music composition to the vibrant textures of sound design and the personal touch in voice synthesis, these models are not just creating audio; they're crafting experiences. As we explore the symphonies of AI, we unveil how these tools are harmonizing technology and creativity, offering new dimensions of auditory expression that were once unimaginable.

Music Composition: From composing original melodies to mimicking specific genres, generative models are changing the way music is created. Tools like Jukebox and MuseNet allow musicians to collaborate with AI co-creators while researchers explore the potential for personalized music experiences and music therapy applications.

Sound Design: Imagine creating realistic sound effects for games or movies with just a text description. Generative models like SoundStream and AudioLM are making this a reality, enabling sound designers to work faster and explore new sonic possibilities.

Voice Synthesis: From creating realistic audiobook voices to personalizing voice assistants, generative models transform how we interact with audio. Tools like Tacotron 2 and MelNet make synthetic voices more natural and expressive, opening doors for new applications in education, accessibility, and entertainment.

Text: Weaving Words into Worlds

In the domain of text, Generative AI is like a master weaver, turning the threads of language into rich tapestries of meaning. Whether it's spinning narratives, bridging languages, or coding the future, these models are redefining the art of the written word. Through the lens of AI, we explore how text generation, translation, and code generation are expanding the horizons of communication, creativity, and technological innovation, making every word a world to discover.

Text Generation: From writing creative fiction to generating marketing copy, generative models are changing how we produce text. GPT-3 and LaMDA are pushing the boundaries of natural language processing, enabling writers to overcome writer's block and businesses to personalize content for their audience.

Translation: Imagine breaking down language barriers in real time with AI-powered translation. Generative models like T5 and Marian are making strides in machine translation, facilitating cross-cultural communication and understanding.

Code Generation: Automating repetitive coding tasks and generating code from natural language descriptions is now possible with generative models like Codex and GitHub Copilot. This is revolutionizing software development by boosting programmer productivity and fostering innovation.

The Future of Generative AI: A Symphony of Possibilities

Generative models are revolutionizing the way we approach creativity and problem-solving in the visual, auditory, and textual realms. As we stand on the brink of new technological breakthroughs, the promise of Generative AI extends into a horizon filled with unparalleled innovation. The advancements in these models herald a future where the generation of new, original content is not only more efficient but also increasingly sophisticated. Yet, as we navigate this exciting landscape, it's imperative to anchor our pursuits in ethical development and the responsible use of technology. By doing so, we ensure that the advancements in Generative AI not only spur creative expression but also enrich society in meaningful ways. The journey ahead for Generative AI is one of exploration and discovery, where the synergy between human creativity and artificial intelligence opens a realm of possibilities previously unimagined.

Setting Up the Development Environment

You can use any available resource for training/fine-tuning and inferencing the models in the subsequent sections. Google Colab is one of the many options you can use to get started. In this section, we will walk through the steps of setting up the environment. These steps will help you get started for the subsequent chapters. You can always refer to this section when working on the application section of any of the chapters.

Setting Up a Google Colab Environment

Google Colab provides a fantastic platform to explore Generative AI due to its free access to graphic processing units (GPUs) and computational resources. You can switch to a Colab Pro premium subscription (see Figure 1-3) if you need more GPU compute hours.

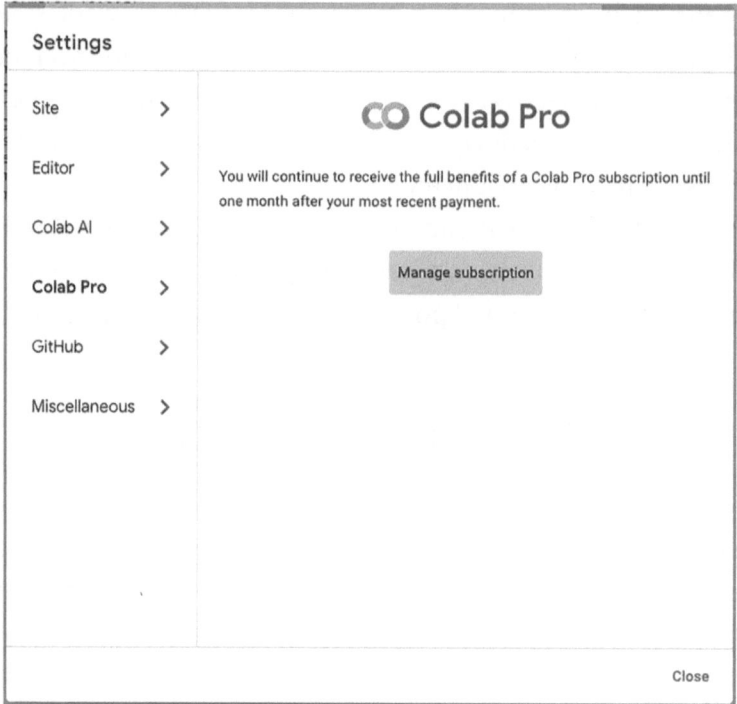

Figure 1-3. *Colab Pro subscription*

Here's a step-by-step guide set up your Colab environment:

1. **Choose a Colab runtime:**

    Go to **Runtime ➤ Change runtime type**.

    Select a **GPU runtime** based on your model's requirements and available resources (see Figure 1-4). Consider factors such as model size, training complexity, and desired processing speed.

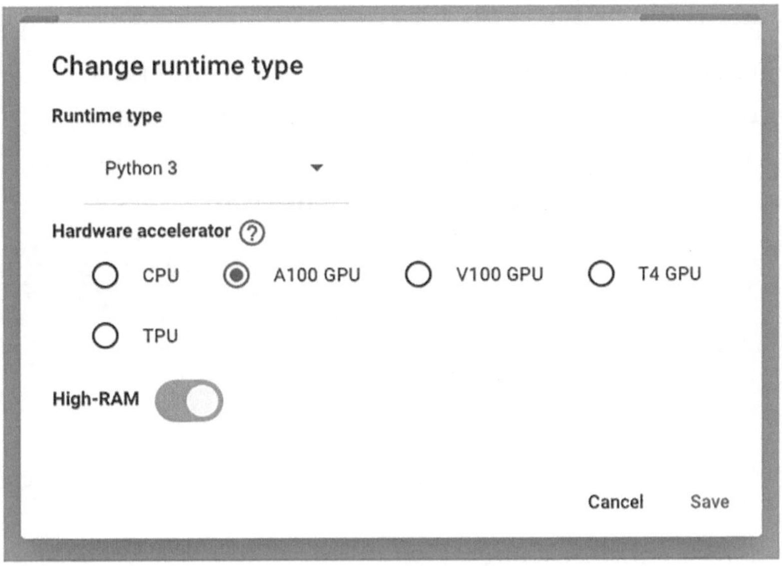

CHAPTER 1   INTRODUCTION TO GENERATIVE AI

Figure 1-4. *Colab runtime type*

2. **Install the necessary libraries:**

    Use !pip install commands to install libraries such as PyTorch, Hugging Face Transformers, and other dependencies specific to your model (details can be found in the subsequent chapters); see Figure 1-5.

Figure 1-5. *Colab install libraries*

25

3. **Mount Google Drive:**

    If you want to save/load models from your Drive, use `from google.colab import drive` and follow the authentication steps. Click the link provided and follow the on-screen instructions to grant Colab access to your Drive (see Figure 1-6).

Figure 1-6. Colab mounting Google Drive

- Once authenticated, a new directory called `content` will appear in the left sidebar (see Figure 1-7). This represents your mounted Google Drive.

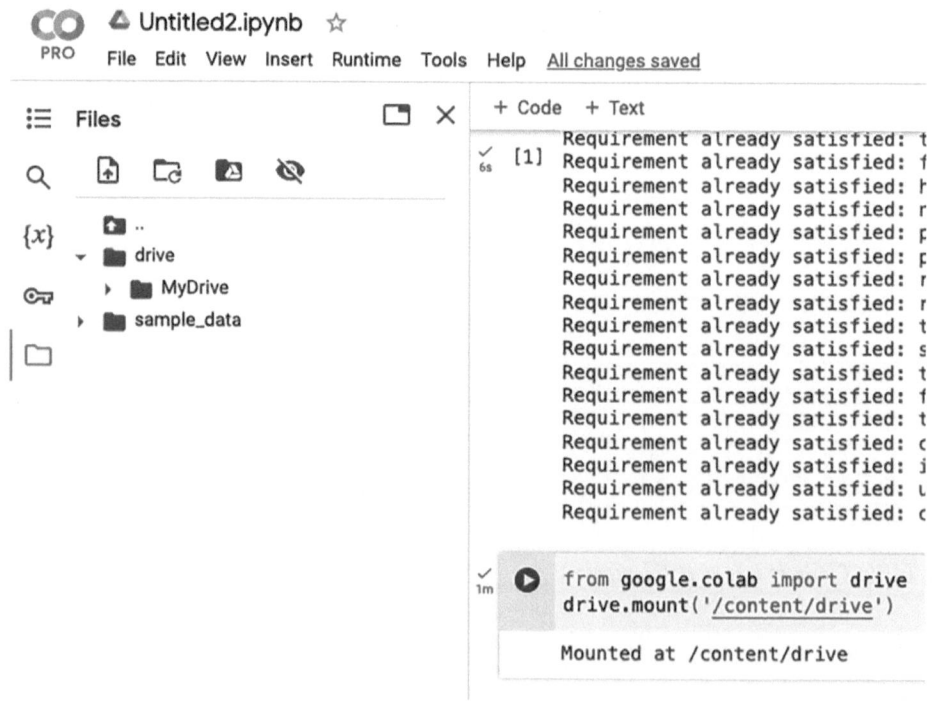

Figure 1-7. Colab new content

CHAPTER 1   INTRODUCTION TO GENERATIVE AI

4. **Clone or upload your model code:**

   - Option 1: Clone from a repository:

     Use the !git clone command followed by the repository URL to clone your code directly into Colab (see Figure 1-8).

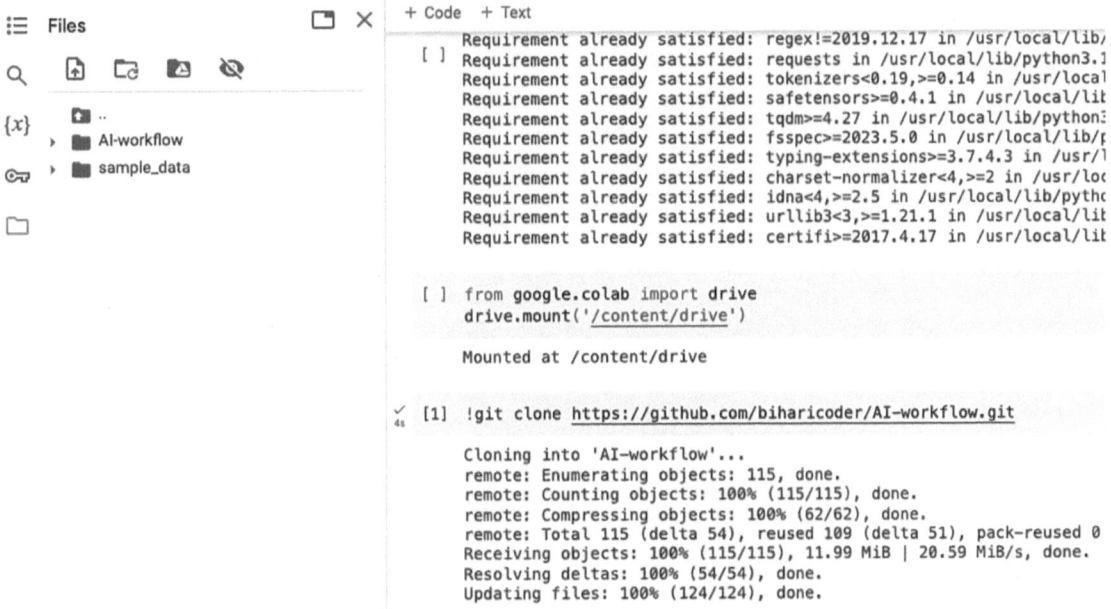

Figure 1-8. Colab clone repository

   - Option 2: Uploading manually:

     Click the **Files** tab and then the **Upload** button to manually upload your notebook from your local machine (see Figure 1-9).

CHAPTER 1   INTRODUCTION TO GENERATIVE AI

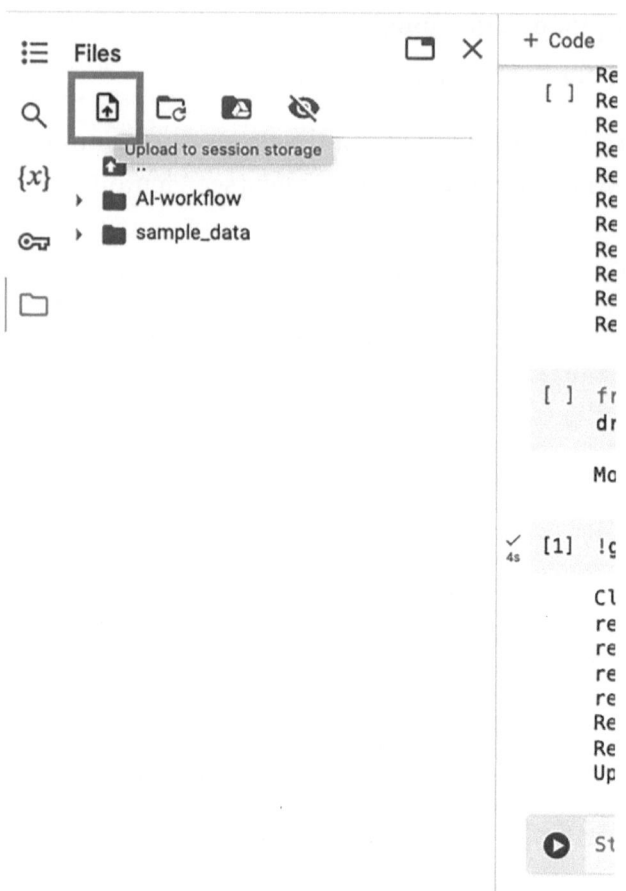

Figure 1-9. Colab upload manually

5. **Save/download/reuse the model:**

   - Saving locally:

     Use model.save("my_model") or equivalent for your framework to save the model locally in the Colab environment.

   - Saving to Drive:

     If you mounted your Drive, use model.save("/content/drive/MyDrive/my_model") to save the model directly to your Google Drive.

CHAPTER 1   INTRODUCTION TO GENERATIVE AI

- Downloading:

    Use the Files tab to browse the saved model file and/or the notebook and click the download arrow to download it to your local machine (see Figure 1-10).

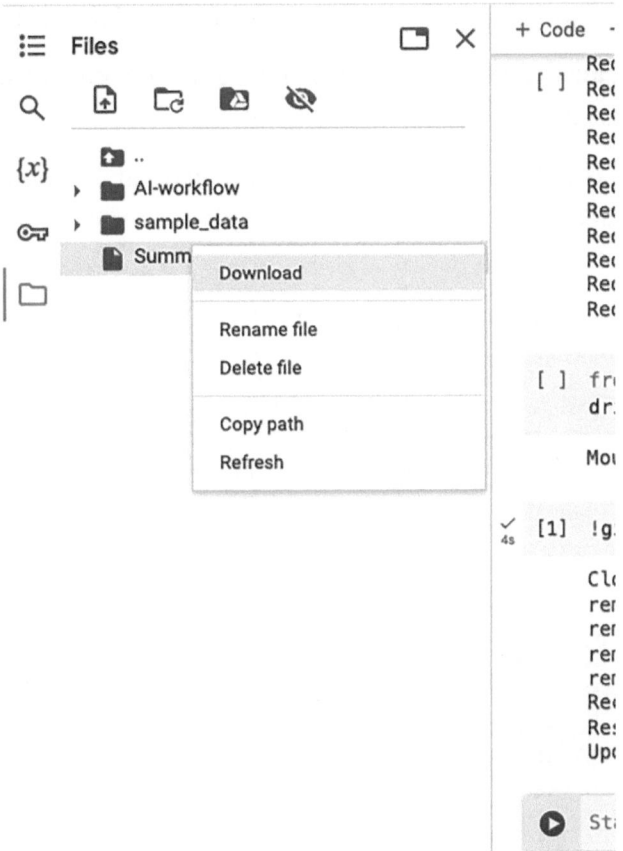

Figure 1-10.   Colab download file

- Reusing:

    Load the saved model using model.load_state_dict(torch.load("my_model.pt")) or equivalently to use it for further training or inference.

CHAPTER 1   INTRODUCTION TO GENERATIVE AI

Hugging Face Access and Token Key Generation

We will use the Hugging Face library to access the models, datasets, and Python classes extensively throughout all the chapters in the book. So, you need to set up your Hugging Face account before diving into the application section of the subsequent chapters.

Follow these steps to set up a free account on Hugging Face and an API token:

1. **Create an account:** Go to `https://huggingface.co` and sign up for a free account.

2. **Generate an API token:** Click your profile icon at the top right and go to Settings (see Figure 1-11).

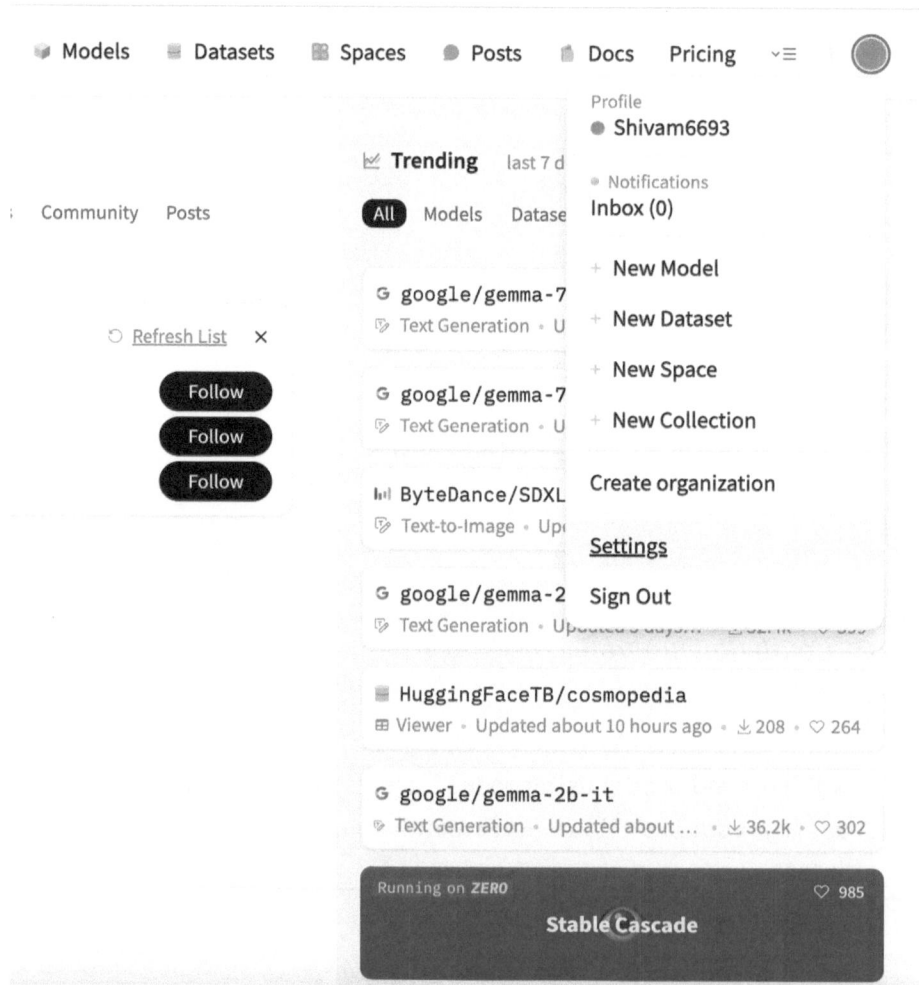

Figure 1-11. Huggingface account

CHAPTER 1   INTRODUCTION TO GENERATIVE AI

3. Select *Access Tokens* from the left nav bar. Click the *New token* button. Enter a suitable name and select the write permissions (read permissions also work if you are not pushing anything to the Hugging Face hub); see Figure 1-12.

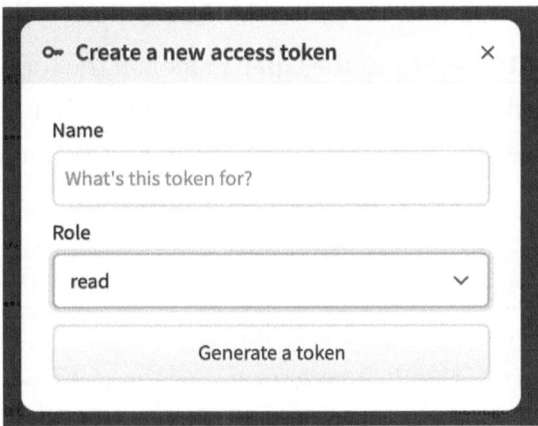

Figure 1-12. *Huggingface token*

4. **Store your token securely:** Never share your API token publicly and consider using environment variables or secure storage methods.

5. **Integrating the token in your Colab notebook:**

   a. You can set the environment variable:

   ```
   os.environ["HF_TOKEN"] = "your_token"
   ```

   b. Or you can use it directly in your code:

   ```
   from transformers import AutoModelForSeq2SeqLM, AutoTokenizer;
   model = AutoModelForSeq2SeqLM.from_pretrained("bigscience/bloom", auth_token="your_token")
   ```

CHAPTER 1 INTRODUCTION TO GENERATIVE AI

OpenAI Access Account and Token Key Generation

Similarly, for OpenAI resources, you'll need an account and an API key:

1. **Create an account:** Go to https://platform.openai.com/ and sign up for a free or paid account, depending on your needs.

2. **Generate an API key:** Click the OpenAI icon at the top left, click **API keys**, and then click **Create new secret key**. Then, enter a suitable name and click the **Create secret key** button (see Figure 1-13).

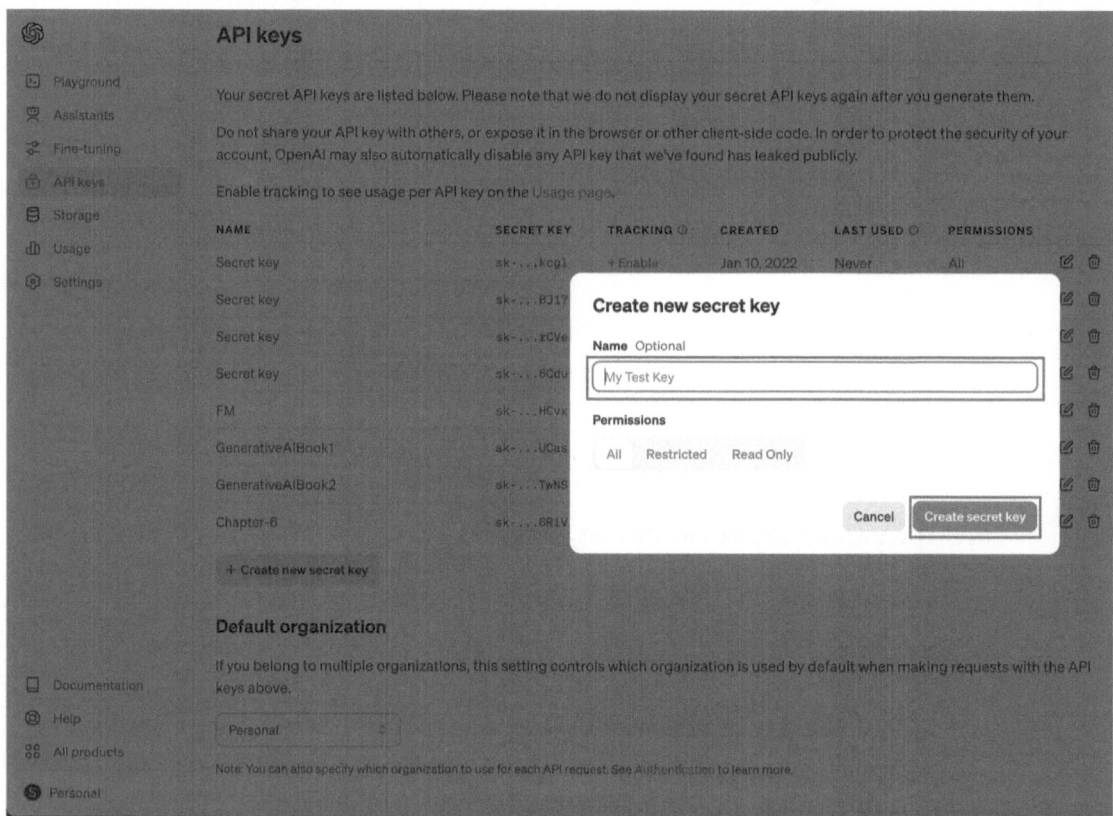

Figure 1-13. OpenAI token

3. **Store your key securely:** Like with Hugging Face, keep your API key confidential and use secure storage methods.

4. **Integrating the key in your Colab notebook:**

    a. You can set the environment variable:

    ```
    os.environ["OPENAI_API_KEY"] = "your_key"
    ```

    b. Or you can use it directly in your code:

    ```
    import openai; openai.api_key = "your_key"
    ```

Remember, both platforms have usage limits and specific terms of service, so be sure to familiarize yourself with them before proceeding.

Troubleshooting Common Issues

After setting up your development environment and accessing various APIs, you might encounter some common issues. Here are troubleshooting tips to help you navigate and resolve these challenges efficiently:

GPU Runtime Limitations in Google Colab

- **Issue:** Google Colab may restrict access to GPU resources after extensive use.
- **Solution:** Consider using Colab Pro for extended access to GPUs or alternate between different free resources like Kaggle Kernels.

Library Installation Errors

- **Issue:** Errors during the installation of libraries using `!pip install`.
- **Solution:** Ensure you're using the correct version of the library compatible with your runtime. Use specific library versions (e.g., `!pip install tensorflow==2.3.0`) to avoid compatibility issues.

Google Drive Mounting Issues

- **Issue:** Difficulty in mounting Google Drive or accessing files from it.
- **Solution:** Double-check your authentication process and ensure you've copied the authentication code correctly. If the issue persists, try reconnecting your Google account or restarting your Colab session.

CHAPTER 1   INTRODUCTION TO GENERATIVE AI

API Rate Limit Exceeded for Hugging Face or OpenAI

- **Issue:** Receiving an error indicating that you've exceeded the API rate limit.

- **Solution:** For Hugging Face, consider downloading datasets or models directly and accessing them locally. For OpenAI, review the current limits and pricing plans to adjust your usage or upgrade your plan.

Model Loading or Saving Errors

- **Issue:** Errors when trying to load or save models using specific paths.

- **Solution:** Verify the paths used for loading or saving models, especially when using mounted Google Drive. Ensure the path exists, or correct the path syntax.

OpenAI API Key Security Concerns

- **Issue:** Risk of exposing your OpenAI API key when sharing notebooks.

- **Solution:** Always use environment variables to store API keys (`os.environ["OPENAI_API_KEY"] = "your_key"`) and remove or obfuscate these values before sharing notebooks publicly.

Connectivity Issues with External APIs

- **Issue:** Sometimes, you might face connectivity issues when your notebook tries to access external APIs.

- **Solution:** Check your Internet connection and ensure that there are no firewall or network settings blocking the requests. If the issue is intermittent, try rerunning the cell after a brief wait.

Dealing with Deprecated Functions or APIs

- **Issue:** Encountering errors due to deprecated or updated functions in libraries or APIs.

- **Solution:** Refer to the official documentation for the libraries or APIs in question for updated methods or functions. Consider using alternative functions that offer the same or similar functionality.

Summary

This chapter introduced you to Generative AI, from its origin story to its practical applications. You explored the key differences between generative and discriminative models, delving into their fundamental principles and the diverse techniques they employ.

You witnessed the real-world impact of Generative AI across various domains, seeing the benefits it offers while also learning the challenges and ethical considerations that arise with such powerful tools.

As we move forward, remember that the journey doesn't end here. The possibilities of Generative AI are constantly evolving, pushing the boundaries of creativity and innovation. The following chapters will delve deeper into the diverse domains where Generative AI shines, showcasing its potential in visuals, audio, and text generation.

We also covered the environment setup that will help you get prepared with the right tools and technologies so that you can start building amazing things with Generative AI models in the subsequent chapters.

Stay tuned as we explore the specific applications and unleash the magic of Generative AI in these fascinating areas.

CHAPTER 2

Text-to-Image Generation

Introduction

In this chapter, we will dive into how amazing technologies can turn words into stunning images. Text-to-image generation, a significant advancement in Generative AI, blends creativity with technology, opening new doors in the art world and artificial intelligence. This fascinating intersection of language understanding and visual creativity allows us to generate detailed and coherent images from textual descriptions, showcasing the incredible potential of AI to augment and participate in creative processes.

Real-life use cases of text-to-image generation technology extend far beyond traditional business applications, permeating various facets of our daily lives and industries. In the realm of content creation and digital media, this technology empowers creators with the ability to instantaneously bring their visions to life. Imagine bloggers, writers, and social media influencers crafting unique, tailor-made images to accompany their posts with just a few keystrokes, enhancing reader engagement without the need for extensive graphic design skills. Similarly, advertising and marketing professionals can leverage text-to-image generation to produce visually compelling campaigns that perfectly align with their narrative, significantly reducing the time and cost associated with traditional photography and graphic design. This allows for rapid prototyping of ideas and concepts, enabling teams to visualize and iterate on creative projects with unprecedented speed and flexibility.

Furthermore, the impact of text-to-image generation extends into education and research, offering innovative methods to aid learning and exploration. Educational content developers can use this technology to create custom illustrations for textbooks and online courses, making complex subjects more accessible and engaging for students. In scientific research, especially in fields like biology and astronomy, researchers can generate visual representations of theoretical concepts or distant celestial bodies, facilitating a deeper understanding of phenomena that are difficult to observe directly. Art and entertainment industries also stand to benefit immensely; filmmakers and game developers can generate detailed concept art and backgrounds, streamlining the creative process from ideation to production. Additionally, the technology opens up new avenues for personalized entertainment, allowing consumers to create custom avatars, scenes, and visual stories based on their own descriptions, fostering a more interactive and engaging user experience.

These real-life applications underscore the transformative potential of text-to-image generation technology, bridging the gap between imagination and visual representation. By democratizing access to high-quality visual content creation, it enables individuals and professionals across various sectors to innovate, educate, and entertain in ways that were previously inconceivable. As this technology continues to evolve, its integration into our daily lives promises to redefine creativity, making the act of bringing ideas to visual fruition as simple as describing them in words.

The journey toward text-to-image generation is a hallmark of the broader evolution within artificial intelligence and deep learning. In its infancy, the ambition to transform textual descriptions into visual representations grappled with the limitations of early neural network designs. Emerging in the late 1990s and early 2000s, initial models employed basic forms of neural-style transfer and direct concatenative methods, attempting to blend textual and visual information. Despite their pioneering nature, these early attempts often lacked the sophistication needed to fully bridge the gap between the complex expressiveness of human language and the precise visual accuracy required for coherent image creation. The resultant images, while groundbreaking, underscored the vast divide between nascent AI capabilities and the depth of human creativity.

A pivotal shift in this landscape was heralded by the development of generative adversarial networks (GANs), introduced by Ian Goodfellow and his colleagues in 2014. GANs brought a novel competitive framework into play, where two networks, the generator and the discriminator, engage in a dynamic contest. This adversarial approach led to the generation of images that were significantly more detailed and realistic, propelling the capabilities of text-to-image models forward. The subsequent integration of transformer models, initially designed for tasks like translation in natural language processing (NLP), further revolutionized the field. Transformers, such as Google's BERT or OpenAI's GPT series, showcased an unparalleled proficiency in parsing and understanding complex textual inputs, setting the stage for more sophisticated text-to-image conversions.

Among these advancements, the introduction of OpenAI's Contrastive Language–Image Pretraining (CLIP) model marked a zenith in the journey of text-to-image generation. CLIP embodies a harmonious blend of linguistic and visual comprehension, trained across diverse datasets to master the subtle art of matching text with corresponding images. This model not only signifies a leap in the AI's ability to generate visually coherent outputs from textual descriptions but also symbolizes a paradigm shift toward creating AI that mirrors human-like understanding and creativity.

This chapter is structured to guide you through the fascinating landscape of text-to-image generation. We begin with an exploration of the CLIP model, delving into its architecture, functioning, and how it can be implemented to bridge the gap between text and image. Following this, we will introduce the concept of diffusion models, starting with a hands-on approach to build a diffusion model from scratch, progressing to the implementation of the stable diffusion model using Hugging Face, and concluding with insights on fine-tuning a pre-trained model. Through this journey, we aim to equip you with both a theoretical understanding and practical skills in the latest advancements of text-to-image generation, preparing you to contribute to this exciting field or leverage its capabilities in your projects.

By the end of this chapter, you will not only grasp the technical workings behind these models but also appreciate their potential to transform creative and business endeavors. Join us as we explore the cutting-edge of Generative AI, where words become the brush and canvas for creating stunning imagery.

Bridging the Gap Between Text and Image Data

As we delve into this section, we embark on an insightful journey that begins with the fundamentals of image data. This foundational understanding is crucial for appreciating the complex interplay between textual descriptions and visual representations. We explore the intricate correlation between text and image data, highlighting how this relationship forms the backbone of innovative AI models. Transitioning smoothly from the conceptual groundwork laid by CLIP, we further amplify our exploration into the realm of creativity with diffusion models. These models, which we explain and show how to build from scratch, stand at the cutting edge of generating highly creative and visually compelling images from textual descriptions. Diffusion models represent a significant leap in the AI's ability to generate images that are not just representations of text but are imbued with elements of creativity and imagination. On the other hand, large language models (LLMs) like Falcon and LLaMA, which are types of pre-trained transformer models, are initially developed to anticipate subsequent text tokens based on provided input. With billions of parameters and training on trillions of tokens over extensive periods, these models gain remarkable power and flexibility. They can address various NLP tasks immediately through user prompts crafted in everyday language.

By combining the understanding of text and image correlation with the creative capabilities afforded by diffusion models, we set the stage for a comprehensive approach to text-to-image generation. This approach not only encapsulates the theoretical and practical aspects of bridging the text-image gap but also emphasizes the progression toward enhancing creativity in image generation, thereby enriching the reader's expertise in the fascinating domain of AI-driven artistry.

Understanding the Fundamentals of Image Data

This section delves into the foundational elements that form the backbone of digital images, offering insights into how these visual representations are not just seen but understood and manipulated by computers. This exploration delves into the essentials of digital imagery, from the pixels that form our screen visuals to the distinction between vector and raster images and their specific uses. Color models like RGB and CMYK play a critical role in accurately capturing and reproducing colors across various platforms, while the principles of image resolution and quality highlight the significance of precision in digital visuals. Additionally, understanding the array of image file formats, including JPEG and RAW, is key to effectively storing and managing visual data. Collectively, these elements reveal the complex interplay of technology and art in digital imagery, significantly enriching our digital experiences and creative expressions.

- **Digital Images as Arrays of Pixels:** Digital images are fundamentally composed of arrays of pixels (Figure 2-1), where each pixel represents the smallest unit of visual information. These pixels act like a mosaic, with each tiny square contributing a specific color to the overall image. The arrangement and color value of these pixels determine the image's appearance, detail, and color depth. This pixel-based structure allows digital images to be displayed on electronic devices, manipulated in editing software, and compressed for storage and transmission, making them versatile tools in digital communication and media. A generative model like DALL-E utilizes the knowledge of pixel arrays to transform textual descriptions into detailed images by meticulously arranging each pixel's color value to match the described scene.

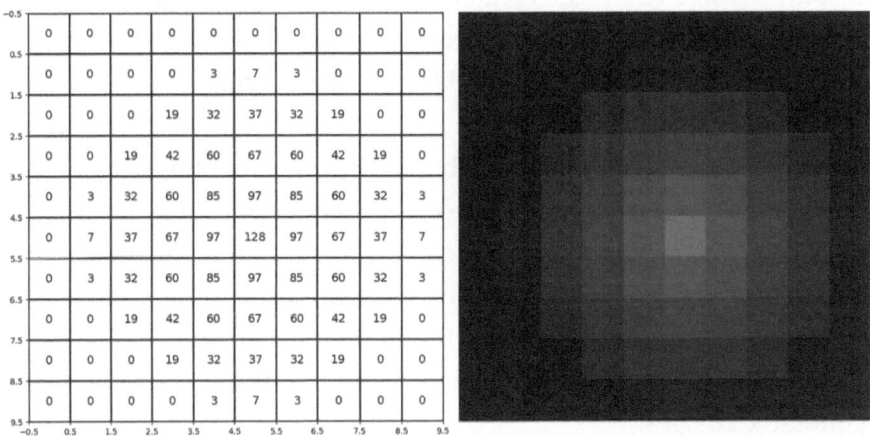

Figure 2-1. *Digital image pixel array and grayscale rendering (source: https://neubias.github.io/training-resources/pixels/index.html)*

- **Vector and Raster Images:** The digital imaging world is broadly divided into two categories: vector and raster images. Vector images are made up of paths defined by mathematical equations, allowing them to be scaled infinitely without any loss of quality. This makes them ideal for logos, text, and simple illustrations that require clean lines and scalability. Raster images, on the other hand, consist of a fixed grid of pixels, making them better suited for complex and detailed photographs. However, resizing raster images can result in a loss of clarity and detail, highlighting the fundamental differences in how these two image types are used and manipulated. Generative AI leverages vector models for scalable graphics and raster models for detailed, photorealistic images, highlighting the importance of choosing the right synthesis approach.

CHAPTER 2 TEXT-TO-IMAGE GENERATION

- **Pixels and Color Models:** Pixels serve as the foundational building blocks of digital images, and color models dictate how these pixels combine to produce the spectrum of colors we see. The Red, Green, Blue (RGB) color model (Figure 2-2) is predominant in electronic displays, where colors are created through the additive mixing of light in these three hues. In contrast, the Cyan, Magenta, Yellow, Key/Black (CMYK) model is used in printing, relying on subtractive mixing to absorb light and create colors. Additionally, the Grayscale model represents images using shades of gray, providing a spectrum from black to white. Each model serves distinct purposes, from on-screen visualizations to physical printing, influencing the choice of color representation in digital imaging projects. In Generative AI, the choice between RGB for vibrant digital displays and CMYK for accurate printed artworks crucially impacts the visual quality of generated images.

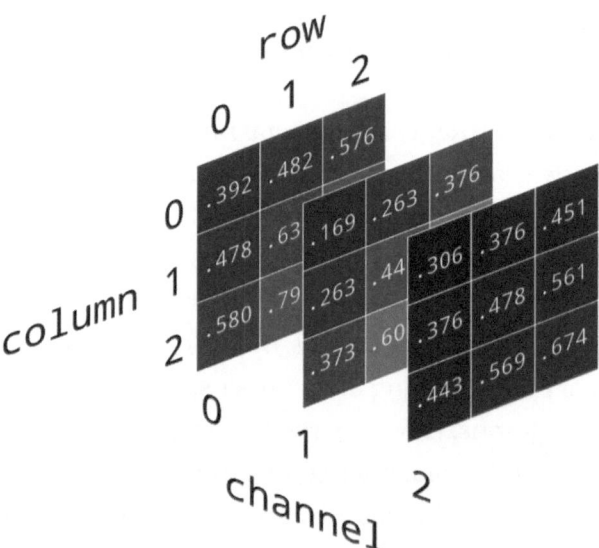

Figure 2-2. Visual representation of a three-dimensional tensor (source: https://e2eml.school/convert_rgb_to_grayscale)

- **Image Resolution and Quality:** Image resolution, typically measured in pixels per inch (PPI), plays a crucial role in defining the quality and clarity of digital images. High-resolution images contain more pixels, offering finer detail and allowing for larger print sizes without losing visual fidelity. Conversely, low-resolution images may appear blurry or pixelated, especially when enlarged. The resolution impacts not only the aesthetic quality of an image but also its file size and suitability for various applications, from web graphics, which require lower resolution, to high-quality print materials that demand higher resolution settings. For high-resolution artworks, Generative AI models require training on high-quality images to ensure the produced outputs maintain clarity and detail, crucial for digital art where visual quality affects viewer experience.

- **Image File Formats:** Digital images can be saved in various file formats, each with its own advantages and disadvantages. Popular formats include JPEG, known for its efficient compression and wide compatibility, making it suitable for web images where file size is a concern. PNG offers lossless compression, supporting transparency and making it ideal for web graphics. GIF is favored for simple animations. BMP retains image quality at the cost of larger file sizes, and RAW files preserve all data directly from a camera's sensor, offering the highest quality and flexibility in post-processing. Choosing the right format is crucial for balancing image quality, file size, and compatibility needs across different platforms and uses. In Generative AI, selecting the right file format, like JPEG for online galleries or lossless PNG/RAW for archival quality, is crucial to balance image quality, size, and detail preservation.

Correlation Between Image and Text Data Using CLIP Model

Before we dive into the realm of text-to-image translation, it's essential to grasp how machines determine the degree of similarity or connection between a given text and image pair. This foundational knowledge is critical for appreciating the complex mechanics that allow a system to discern and measure the relevance of visual content to linguistic descriptors. It's the underpinning science that informs the later stages of image generation, ensuring that the resulting visuals are not just random creations but are intricately linked to their textual prompts.

Among the various algorithms designed to bridge the gap between textual descriptions and visual imagery, CLIP[1] stands out as one of the most proficient. CLIP was developed by OpenAI, which is a multimodal, zero-shot model. This approach redefines how machines understand and correlate the contents of an image with the semantics of text. By leveraging CLIP's capabilities, we examine how the model processes and aligns the nuances of visual data with corresponding textual information, creating a multimodal understanding that paves the way for advanced applications in the field of artificial intelligence.

Architecture and Functioning

Let's begin by exploring the fundamental architecture that underpins the design of CLIP. These are the two main parts of CLIP:

- **Image Encoder:** CLIP uses a vision transformer or convolutional neural network as an image encoder. It divides an image into equal-sized patches, linearly embeds each of them, and then processes them through a multiple layer. Therefore, the model can consider global information from the entire input image and not just local features.

- **Text Encoder:** CLIP leverages a transformer-based model for a text encoder. It processes text data into a sequence of tokens and then applies self-attention mechanisms to understand relationships between different words in a sentence.

CLIP's training is aligning the embedding spaces of the image and text encoders. The model maps both images and text into a shared high-dimensional space (Figure 2-3). The objective is to learn a space where semantically related images and texts are close to each other despite originating from different modalities. CLIP uses a contrastive loss function that encourages the model to project the image and its correct text description close together in the embedding space while pushing the nonmatching pairs apart.

(1) Contrastive pre-training

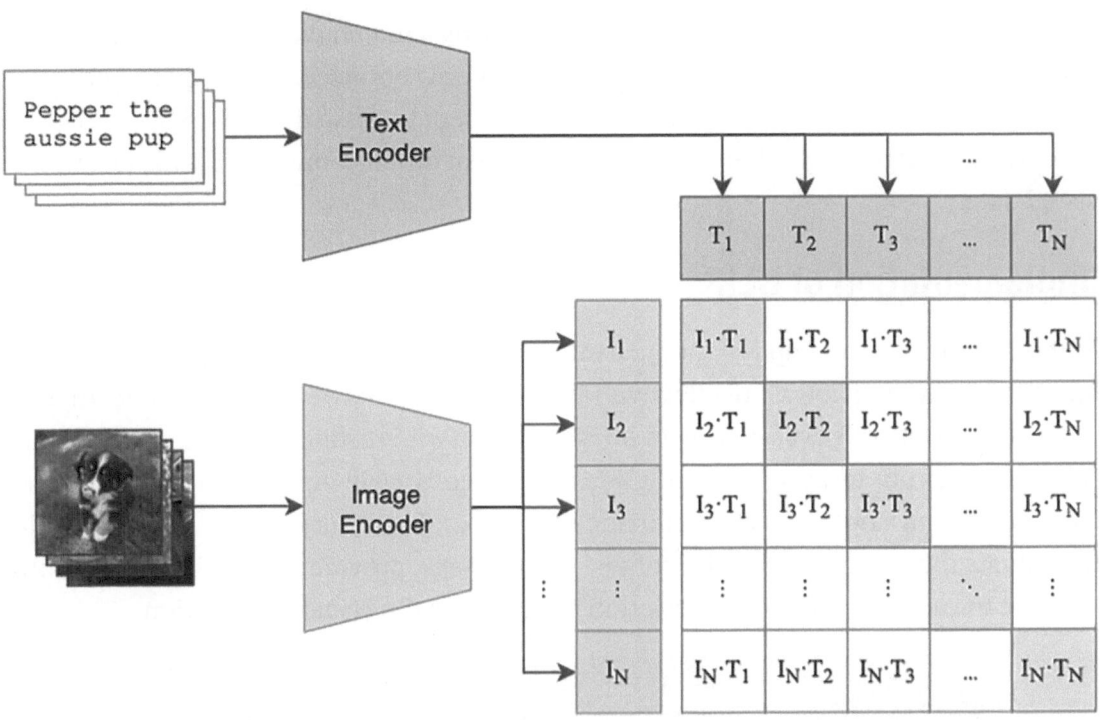

Figure 2-3. CLIP pre-training architecture (source: https://openai.com/research/clip)

Now that we have explored the fundamental architecture and how CLIP aligns the nuances of visual data with textual information, let's examine a real-world application that showcases the model's practicality and potential beyond theoretical uses.

CLIP Case Study

After understanding the architecture and functioning of CLIP, let's explore a real-world application that demonstrates its practicality and innovation. A notable example is the use of CLIP in enhancing visual search engines for e-commerce platforms. These platforms face the challenge of understanding and matching user queries with relevant product images from extensive catalogs. By leveraging CLIP, an e-commerce platform can significantly improve the accuracy and relevance of search results. For instance, when a user searches for "vintage leather backpack," CLIP helps the platform's search engine interpret the textual query and find product images that not only match the

CHAPTER 2 TEXT-TO-IMAGE GENERATION

description but also align with the nuanced style and quality implied by "vintage." This is accomplished by CLIP's ability to understand the semantic content of both the search terms and the images in the catalog, ensuring a match that is both visually and contextually appropriate. Such an application not only enhances user experience by making product discovery more intuitive and efficient but also demonstrates CLIP's potential to bridge the gap between complex textual descriptions and a wide array of visual data.

Implementation of CLIP

In this section, we will implement the CLIP model in Google Colab Notebook. We will focus on unraveling how CLIP effectively bridges the gap between visual and textual data. To illustrate this, we will use an image of a dog with distinctive black and white fur (Figure 2-4) as our test subject. Alongside this image, we will input a series of sentences, each describing a potential characteristic of the dog. The beauty of CLIP lies in its ability to evaluate these sentences in the context of the image, providing probability scores that indicate the accuracy of each description in matching the visual information.

Figure 2-4. *Beautiful dog featuring an elegant blend of black and white fur*

Step 1: Installing Libraries and Data Loading

Kickstarting our journey with CLIP, the first step involves setting up the necessary libraries and loading the data required for the model to function.

```
pip install transformers
```

The previous command installs the `transformers` library developed by Hugging Face. This library offers a wide range of pre-trained models to perform tasks on different modalities.

```
from PIL import Image
import os
from transformers import CLIPProcessor, CLIPModel
import pandas as pd
import torch
```

You have now imported all the essential libraries required for handling images and utilizing the CLIP model in this implementation. Next, we will connect to Google Drive to access the image that will be our input.

```
from google.colab import drive
drive.mount('/content/drive')
```

After connecting to Google Drive, we will specify the path where the image is located and store the image's name in a variable named image_path.

```
os.chdir("/content/drive/My Drive/Colab Notebooks")
image_path = 'benali_image.jpg'
Image.open(image_path)
```

Step 2: Data Preprocessing

Moving onto step 2, we delve into data preprocessing.

```
model = CLIPModel.from_pretrained("openai/clip-vit-base-patch32")
```

Next, we create an instance of the CLIP model using the `clip-vit-base-patch32` variant. This variant employs the ViT-B/32 transformer architecture for image encoding and a masked self-attention transformer as the text encoder.

```
processor = CLIPProcessor.from_pretrained("openai/clip-vit-base-patch32")
inputs = processor(text=["a photo of a cat", "a photo of a dog with black
and white fur", "a photo of a boy", "baseball field with a lot of people
cheering"], images=image, return_tensors="pt", padding=True)
```

As a next step, we use the processor from the CLIP model to prepare a batch of text and image data for input into a neural network. The processor is responsible for tokenizing text and processing images. It processes a list of textual descriptions of a photo of a cat, a photo of a dog with black and white fur, and a few other descriptions, along with an image (`images=image`). The processor converts these inputs into PyTorch tensors (`return_tensors="pt"`) and applies padding to ensure that all text inputs are of uniform length (`padding=True`). This preprocessing step is essential for the inputs to be correctly processed by the CLIP model.

Step 3: Model Inference

In step 3, we enter the model inference stage, where we apply the CLIP model to our processed data, enabling the extraction of insights and correlations between text and images.

```
outputs = model(**inputs)
logits_per_image = outputs.logits_per_image
probs = logits_per_image.softmax(dim=1)
```

The previous code starts with feeding preprocessed inputs, including text descriptions and an image, into the CLIP model for evaluation. The model then computes how closely each text description matches the image, yielding image-text similarity scores. These scores are then transformed using the `softmax` function to obtain probabilities. The resulting probabilities indicate the likelihood of each text description being an accurate depiction of the image.

```
probs_percentage = probs.detach().numpy() * 100
text_inputs = ["a photo of a cat", "a photo of a dog with black and white
fur", "a photo of a boy", "baseball field with a lot of people cheering"]
```

```
df = pd.DataFrame({
    'text_input': text_inputs,
    'similarity_with_image (%)': probs_percentage[0]
})
print(df)
```

The previous command prints the probability between each text and image pair. The output from the previous command is as follows:

```
   text_input                                    similarity_with_image (%)
0  a photo of a cat                                               1.893156
1  a photo of a dog with black and white fur                     62.819950
2  a photo of a boy                                              35.285812
3  baseball field with a lot of people cheering                   0.001074
```

From the output, it's evident that the sentence `a photo of a dog with black and white fur` has the highest similarity, accurately matching our input image. Impressively, CLIP achieved this result in just a few seconds.

Diffusion Model

Diffusion models are a transformative innovation in the world of Generative AI, marking a new frontier in how artificial intelligence creates complex data like images, audio, or text. At their core, these models operate on a fascinating concept: they start by introducing randomness or noise into a data sample and then methodically learn to reverse this process. This dance of noise addition and subtraction unfolds through a two-stage process. In the forward stage, the model progressively corrupts the original data with noise, while in the reverse stage, it meticulously works backward, reconstructing the original data from its noisy state. This mechanism enables diffusion models to generate highly detailed and realistic outputs.

Chapter 2 Text-to-Image Generation

Figure 2-5. *Diffusion models: gradually adding Gaussian noise and then reversing (source: `https://lilianweng.github.io/posts/2021-07-11-diffusion-models/`)*

Implement Diffusion Model from Scratch

In this section, we will delve into implementing a diffusion model from scratch. We will gradually improve a fully noise image into a clear one. To achieve this, our deep learning model will rely on two crucial inputs: the input image, which is the noisy image that needs processing, and the timestamp, which informs the model about the current noise status. The timestamp plays a key role in guiding the model's learning process, making it easier for the model to understand and reverse the noise addition at each stage. This approach allows us to create a model that enhances image quality and provides insights into the dynamic process of image transformation in a diffusion model.

Step 1: Installing Libraries

In step 1, we embark on the implementation of the diffusion model by installing the necessary libraries and loading the data, establishing the initial setting for our generative journey.

```
import numpy as np
import torch
import torch.nn as nn
import torch.nn.functional as F
import torchvision
from torchvision import transforms
from tqdm.auto import trange, tqdm
import matplotlib.pyplot as plt
```

First, we import all the essential libraries. We will be utilizing Pytorch for our implementation. We will be using the MNIST dataset from the `torchvision` library for our implementation. `tqdm` is a library for displaying progress bars for loops.

```
image_size = 32
batch_size = 64
number_of_timesteps = 30
time_scale_array = 1 - np.linspace(0, 1.0, number_of_timesteps + 1)
```

Next, we define a set of variables. We have scaled MNIST images to 32x32 pixels, so we have set `image_size` to 32. The batch size for training has been set as 64. `number_of_timesteps` denotes the number of steps for transforming a noisy image to a clear one. `time_scale_array` creates a linearly spaced array from 1 to 0. `time_scale_array` guides how much noise should be present in the data at each step of the process.

```
device = torch.device('cuda')
```

The previous command in PyTorch is used to define a GPU as the device for tensor computations, leveraging its parallel processing power for the efficient execution of deep learning models.

Step 2: Data Preprocessing

Step 2 is dedicated to data preprocessing, a critical stage where we tailor our datasets for compatibility with the diffusion model.

```
mnist_image_transform = transforms.Compose([
    transforms.Resize((32, 32)),
    transforms.ToTensor(),
    transforms.Lambda(lambda x: x.repeat(3, 1, 1)),
    transforms.Normalize(mean=[0.5, 0.5, 0.5], std=[0.5, 0.5, 0.5])
])

dataset = torchvision.datasets.MNIST(root='./data', train=True,
download=True, transform=mnist_image_transform)
```

Now we import the MNIST dataset using PyTorch's `torchvision` library. `train=True` indicates that the training set of the MNIST dataset should be loaded. If set to False, the test set would be loaded instead. `download=True` tells the function to download the dataset to the specified directory if it's not already there. `transform=mnist_image_ transform` applies the specified transformations to the dataset. We are applying transformations on the MNIST dataset as follows:

- `transforms.Resize((32, 32))`: This function resizes the input images to 32x32 pixels.
- `transforms.ToTensor()`: This converts the PIL image or `numpy. ndarray` to a `torch.FloatTensor`.
- `transforms.Lambda(lambda x: x.repeat(3, 1, 1))`: This is a custom transformation using Lambda. It repeats the single grayscale channel of the MNIST images three times to create a three-channel (RGB-like) image. This would enable us to implement the code on a colored image dataset.
- `transforms.Normalize(mean=[0.5, 0.5, 0.5], std=[0.5, 0.5, 0.5])`: This normalizes the images using the specified mean and standard deviation for each channel (RGB). We apply the same mean and standard deviation to all three channels, scaling the pixel values accordingly.

```
index_for_7 = [i for i, (img, label) in enumerate(dataset) if
label == 7]
subset_training = torch.utils.data.Subset(dataset, index_for_7)

traindata_loader = torch.utils.data.DataLoader(subset_training,
batch_size=batch_size, shuffle=True, num_workers=4)
```

We will just be training our model on the subset of images that are labeled as 7. `index_for_7` contains the indices of all images in the `dataset` that are labeled as 7. `enumerate(dataset)` iterates over the dataset, which contains pairs of images (`img`) and their corresponding labels (`label`). For each pair, if the label is 7, that images index (`i`) is added to the list. `subset_training` is a subset of the original MNIST dataset containing images labeled as 7. `traindata_loader` sets up a `DataLoader` to iterate over this subset for training purposes, with specified batch size, shuffling, and parallel loading settings.

Step 3: Model Training

In step 3, we dive into the heart of model training, where the diffusion model learns from our prepared data.

```
def add_stepwise_forward_noise(x_img, t_step):
    p = time_scale_array[t_step]
    q = time_scale_array[t_step + 1]
    noise = np.random.normal(size=x_img.shape)
    p = p.reshape((-1, 1, 1, 1))
    q = q.reshape((-1, 1, 1, 1))
    img_p = x_img * (1 - p) + noise * p
    img_q = x_img * (1 - q) + noise * q
    return img_p, img_q
```

The `add_stepwise_forward_noise` function is designed to add controlled amounts of noise to an image at two specific stages (timesteps) in a diffusion process. The function takes two arguments: an image `x_img` and a timestep `t_step`. p and q are obtained from the `time_scale_array` at indices `t_step` and `t_step + 1`, respectively. These values are scaling factors for how much noise to add to the image. `noise = np.random.normal(size=x_img.shape)` generates a noise array with the same shape as the input image. This noise is drawn from a normal distribution. p and q are reshaped to have four dimensions to match the shape of the image data (`x_img`). This is necessary for element-wise multiplication. Then, the function outputs noisy images (`img_p` and `img_q`). The function's purpose is to simulate the forward diffusion process by corrupting the original image (`x_img`) with noise in a gradual, step-wise manner.

```
def generate_random_timesteps(num):
    return np.random.randint(0, number_of_timesteps, size=num)
```

The previous function generates an array of random integers representing a set of random timesteps, which are critical in the context of simulating or training a diffusion-based generative model.

```
def display_image_grid(image):
    image = image.permute([0, 2, 3, 1])
    image = image - image.min()
    image = (image/image.max())
    return image.numpy().astype(np.float32)
```

CHAPTER 2 TEXT-TO-IMAGE GENERATION

The previous function `display_image_grid` displays a grid of images. It first converts the images to a displayable format using the `cvtImg` function and then creates a 5x5 grid of subplots to display the first 25 images.

```
def normalize_and_convert(x):
    plt.figure(figsize=(10, 10))
    imgs = display_image_grid(x)
    for i in range(25):
        plt.subplot(5, 5, i+1)
        plt.imshow(imgs[i])
        plt.axis('off')
```

The previous function is essential for preprocessing image tensors for display. It adjusts the tensor dimensions for compatibility with `matplotlib` (changing channel order), normalizes the image pixel values to the [0, 1] range, and converts them to NumPy arrays.

```
rand_tsteps = generate_random_timesteps(25)
x, _ = next(iter(traindata_loader))
p, q = add_stepwise_forward_noise(x[:25], rand_tsteps)
normalize_and_convert(p)
```

In the previous block of code, first we generate 25 random timestep values using `rand_tsteps = generate_random_timesteps(25)`. Then, `x, _ = next(iter(traindata_loader))` retrieves a batch of images from `traindata_loader`. The function `add_stepwise_forward_noise(x[:25], rand_tsteps)` is applied to the first 25 images in this batch, adding noise based on the randomly generated timesteps, resulting in two sets of noisy images, p and q. Finally, `normalize_and_convert(p)` normalizes and converts the set p of these noisy images into a format suitable for display and shows the output (Figure 2-6).

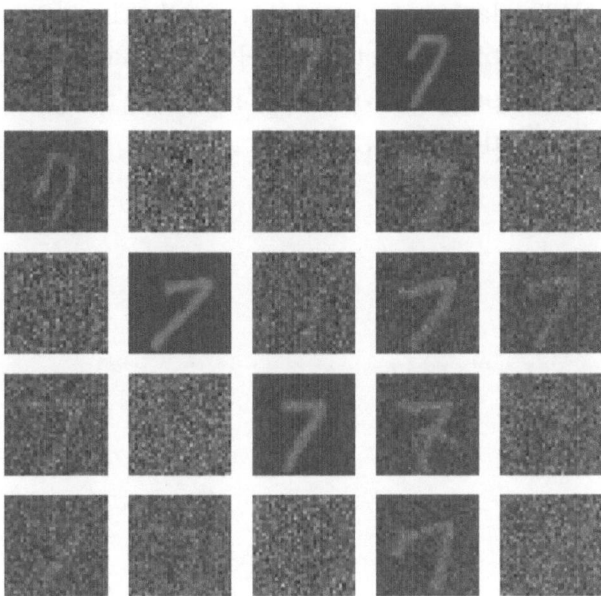

Figure 2-6. *Stages of digit 7 image denoising*

Before moving to the next part, let's get familiar with the U-Net architecture. This will help us lay the foundation for the next part of implementing the diffusion model. The U-Net model is a type of convolutional neural network that is specifically designed for the task of image segmentation. Its architecture is characterized by a symmetric "U" shape (Figure 2-7), which is made up of a contracting path to capture the context and an expansive path that enables precise localization.

- **Contracting Path:** This part consists of a series of convolutional and max pooling layers. The convolutional layers help in extracting and learning features from the image, while the max pooling layers reduce the spatial dimensions of the image, thereby increasing the depth of the feature maps. This process is crucial for understanding the context of the image.

- **Expansive Path:** Following the contraction, the expansive path uses transposed convolutions to upsample the feature maps, gradually increasing their spatial resolution to reconstruct the segmentation map. What makes U-Net particularly effective is its use of skip connections, which connect the layers of the contracting path with

CHAPTER 2   TEXT-TO-IMAGE GENERATION

the corresponding layers of the expansive path. These connections are pivotal for combining the high-level contextual information and the low-level spatial information, facilitating the generation of accurate segmentation maps.

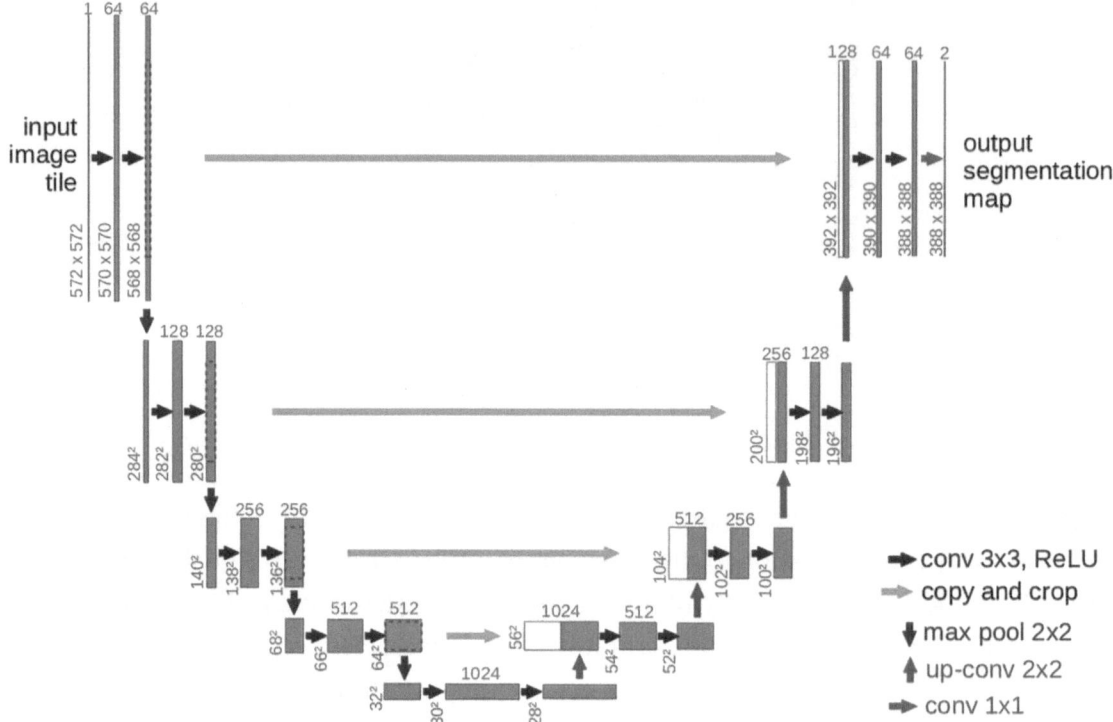

Figure 2-7. Unet architecture

Originally designed for biomedical image segmentation, U-Net's ability to efficiently learn from a limited amount of data, with the help of extensive data augmentation, has made it a popular choice for various image-related tasks. We are now going to establish a fundamental building block for our diffusion model. Following this, we will use this key component as the basis for constructing our U-Net model.

```
class building_block(nn.Module):
    def __init__(self, in_channels=128, size=32):
        super(building_block, self).__init__()
        self.image_feature_conv = nn.Conv2d(in_channels=in_channels,
        out_channels=128, kernel_size=3, padding=1)
```

```python
        self.output_conv = nn.Conv2d(in_channels=in_channels, out_
        channels=128, kernel_size=3, padding=1)
        self.timestep_dense = nn.Linear(192, 128)
        self.normalized_layer = nn.LayerNorm([128, size, size])

    def forward(self, image, timestep):
        image_feature = F.relu(self.image_feature_conv(image))
        timestep_feature = F.relu(self.timestep_dense(timestep))
        timestep_feature = timestep_feature.view(-1, 128, 1, 1)
        conditioned_image_feature = image_feature * timestep_feature
        final_output = self.output_conv(image)
        final_output = final_output + conditioned_image_feature
        final_output = F.relu(self.normalized_layer(final_output))
        return final_output
```

In the initialization method (`__init__`), the class defines several neural network layers to process input data. `self.image_feature_conv` and `self.output_conv` are convolutional layers, each taking an input with `in_channels` channels and outputting 128 channels, with a kernel size of 3 and padding of 1. These layers would extract and process features from the input image. `self.timestep_dense` is a linear (fully connected) layer designed to process the time-step information, transforming a 192-dimensional input into a 128-dimensional output. `self.normalized_layer` is a layer normalization that standardizes the outputs of the convolutional layers, regularizing the data flow in the network. It's configured to normalize data across channels of the specified `size`, which matches the spatial dimensions of the convolutional layer outputs.

The `forward` method (Figure 2-8) of the class defines the computation performed on the input data. It takes two inputs: an `image` and a `timestep`. The image is first processed through the `image_feature_conv` layer, followed by a ReLU activation function, extracting features and applying nonlinearity. Simultaneously, the timestep data is processed through the `timestep_dense` layer and reshaped to match the image feature dimensions. This time-conditioned feature is then element-wise multiplied with the image feature, creating a conditioned image feature. This integration suggests a role in modifying image processing based on the specific stage of the diffusion process. The `final_output` is obtained by adding this conditioned feature to the output of the `output_conv` layer and passing it through another ReLU activation after normalization by `normalized_layer`.

CHAPTER 2   TEXT-TO-IMAGE GENERATION

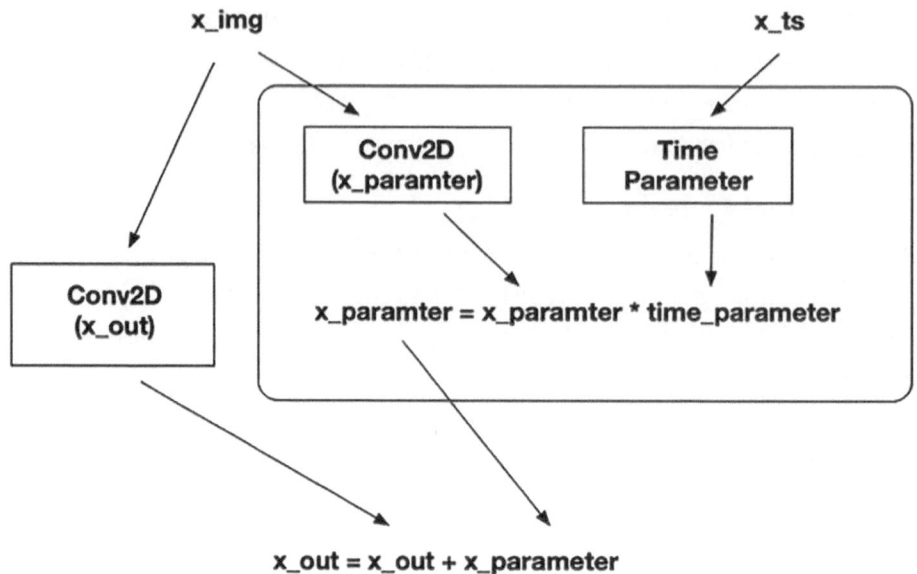

the flow of block

Figure 2-8. *Unet architecture*

```
class TimeStepUNet(nn.Module):
    def __init__(self):
        super(TimeStepUNet, self).__init__()
        self.timestep_processor = nn.Sequential(
            nn.Linear(1, 192),
            nn.LayerNorm([192]),
            nn.ReLU(),
        )
        self.downsample_x32 = building_block(in_channels=3, size=32)
        self.downsample_x16 = building_block(size=16)
        self.downsample_x8 = building_block(size=8)
        self.downsample_x4 = building_block(size=4)
        self.combined_feature_processor = nn.Sequential(
            nn.Linear(2240, 128),
            nn.LayerNorm([128]),
            nn.ReLU(),
            nn.Linear(128, 32 * 4 * 4),
```

```python
        nn.LayerNorm([32 * 4 * 4]),
        nn.ReLU(),
    )
    self.upsample_x4 = building_block(in_channels=32 + 128, size=4)
    self.upsample_x8 = building_block(in_channels=256, size=8)
    self.upsample_x16 = building_block(in_channels=256, size=16)
    self.upsample_x32 = building_block(in_channels=256, size=32)
    self.cnn_output = nn.Conv2d(in_channels=128, out_channels=3,
    kernel_size=1, padding=0)
    self.opt = torch.optim.Adam(self.parameters(), lr=0.0008)
def forward(self, x, x_ts):
    x_ts = self.timestep_processor(x_ts)
    blocks = [self.downsample_x32,
        self.downsample_x16,
        self.downsample_x8,
        self.downsample_x4,]
    downsampled_layers = []
    for i, block in enumerate(blocks):
        x = block(x, x_ts)
        downsampled_layers.append(x)
        if i < len(blocks) - 1:
            x = F.max_pool2d(x, 2)

    x = x.view(-1, 128 * 4 * 4)
    x = torch.cat([x, x_ts], dim=1)
    x = self.combined_feature_processor(x)
    x = x.view(-1, 32, 4, 4)

    blocks = [self.upsample_x4,
        self.upsample_x8,
        self.upsample_x16,
        self.upsample_x32,]

    for i, block in enumerate(blocks):
        # cat left
        x_left = downsampled_layers[len(blocks) - i - 1]
        x = torch.cat([x, x_left], dim=1)
```

```
            x = block(x, x_ts)
            if i < len(blocks) - 1:
                x = F.interpolate(x, scale_factor=2, mode='bilinear')
        x = self.cnn_output(x)
    return x
```

In the previous code block, we create `TimeStepUNet` class, which consists of the model and follows U-Net architecture, known for its effectiveness in various image transformation tasks. `TimeStepUNet` has two methods: `__init__` and `forward`.

In the `__init__` method, the model initializes its components.

- `self.timestep_processor` is a sequential module to process the time-step data, consisting of a linear layer, layer normalization, and ReLU activation.

- `self.downsample_x32`, `self.downsample_x16`, `self.downsample_x8`, and `self.downsample_x4` are instances of `building_block`, designed to progressively downsample the input image to smaller spatial dimensions.

- `self.combined_feature_processor` is a multilayer perceptron (MLP) consisting of linear layers and normalizations. This is used for processing the combined image and time-step features.

- `self.upsample_x4`, `self.upsample_x8`, `self.upsample_x16`, and `self.upsample_x32` are again instances of `building_block` but used for upsampling the image back to higher resolutions.

- `self.cnn_output` is a convolutional layer for producing the final output of the model.

- `self.opt` is an Adam optimizer with a specified learning rate for training the model.

The `forward` method defines how the model processes input data:

- Time-step data `x_ts` is first processed through `self.timestep_processor`.

- The model performs a series of downsampling operations using the defined blocks, storing each layer's output in `downsampled_layers`. Max pooling is applied between these blocks to reduce dimensions.

- The last downsampled output is reshaped and combined with the processed time-step data and then passed through the MLP (self.combined_feature_processor).

- The model performs upsampling using the upsampling blocks, concatenating corresponding layers from the downsampling path at each step, which is a characteristic feature of U-Net architectures.

- The final output is obtained by passing the data through the self.cnn_output layer.

In summary, TimeStepUNet is a neural network model that combines U-Net's powerful image-processing capabilities with the integration of time-step information.

The previous line of code initializes the TimeStepUNet model and ensures that it is placed on the GPU for efficient computation. This step is essential in preparing the TimeStepUNet model for subsequent training or evaluation tasks.

```
unet_model = TimeStepUNet().to(device)
```

The previous line of code initializes the TimeStepUNet model and ensures that it is placed on the GPU for efficient computation. This step is essential in preparing the TimeStepUNet model for subsequent training or evaluation tasks.

```
def generate_images():
    generated_images = torch.randn(32, 3, image_size, image_size).
    to(device)
    with torch.no_grad():
        for time_step in trange(number_of_timesteps):
            current_time_step = time_step
            generated_images = unet_model(generated_images, torch.full([32,
            1], current_time_step, dtype=torch.float, device=device))

    normalize_and_convert(generated_images.cpu())

generate_images()
```

The function generate_images, which is used to create images using the unet_model. Line generated_images = torch.randn(32, 3, image_size, image_size).to(device) creates a tensor of random values with the specified shape, and image_size is the height and width of the images. This tensor is then moved to a GPU specified by the device. With `torch.no_grad()`, we disable gradient calculations, reducing memory consumption and speeding up computations. This is done because we would use this function to only generate images from u_net and not for backpropagation.

The `for` loop iterates over `number_of_timesteps`. Each iteration represents a step in the generative process. Inside the loop, `current_time_step` is set to the loop's index `time_step`. `generated_images` are updated in each iteration by passing it through the `unet_model` along with the current time step information, formatted as a tensor.

After the loop, `normalize_and_convert(generated_images.cpu())` is called, which normalizes and converts the generated images for visualization. The `.cpu()` method moves the images from the GPU back to the CPU. Finally, `generate_images()` calls the function to perform the image generation process.

Since the model hasn't been trained yet, the output produced by the previous code is from an untrained model (see Figure 2-9). From the visuals, it's evident that the results are not particularly useful.

Figure 2-9. *Result from untrained model*

```
def train_batch(diffusion_images):
    timestep_indices = generate_random_timesteps(len(diffusion_images))
    noisy_images, target_images = add_stepwise_forward_noise(diffusion_
    images, timestep_indices)

    timestep_indices = torch.from_numpy(timestep_indices).view(-1,
    1).float().to(device)
    noisy_images = noisy_images.float().to(device)
    target_images = target_images.float().to(device)

    predicted_images = unet_model(noisy_images, timestep_indices)
    diffusion_loss = torch.mean(torch.abs(predicted_images - target_
    images))
    unet_model.opt.zero_grad()
    diffusion_loss.backward()
    unet_model.opt.step()

    return diffusion_loss.item()
```

The function `train_batch` takes a batch of images and trains the `unet_model` on this batch. We will go through the functionality of this function step-by-step:

- `timestep_indices = generate_random_timesteps(len(diffusion_images))` generates random timestep indices for the diffusion process based on the batch size.

- `noisy_images, target_images = add_stepwise_forward_noise(diffusion_images, timestep_indices)` applies noise to the images based on the generated timesteps. This step simulates the forward process of a diffusion model, producing noisy images and their corresponding targets.

- The next three lines of code convert timestep indices, noisy images, and target images into PyTorch tensors. This prepares the data for processing by the `unet_model`.

- `predicted_images = unet_model(noisy_images, timestep_indices)` passes the noisy images and timestep indices through the model, which outputs its predictions.

- diffusion_loss = torch.mean(torch.abs(predicted_images - target_images)) calculates the mean absolute error between the model's predictions and the target images, quantifying how well the model is performing.

- The next three steps are for backpropagation and optimization. unet_model.opt.zero_grad() clears existing gradients to avoid accumulation. diffusion_loss.backward() computes the gradient of the loss. unet_model.opt.step() updates the model parameters based on the calculated gradients.

- return diffusion_loss.item() returns the loss value as a Python number.

train_batch involves generating noisy versions of images, passing them through the model, calculating loss, and performing backpropagation and optimization steps. This function is a key part of the training loop in a diffusion-based generative modeling approach.

```
def execute_training_epochs(num_epochs=20):
    progress_bar = trange(num_epochs)
    total_batches = len(traindata_loader)
    for epoch in progress_bar:
        for batch_index, (batch_images, _) in enumerate(traindata_loader):
            batch_loss = train_batch(batch_images)
            progress_percentage = (batch_index / total_batches) * 100
            if batch_index % 5 == 0:
                progress_bar.set_description(f'Epoch {epoch+1}, Batch {batch_index+1}/{total_batches}, Loss: {batch_loss:.5f}, Progress: {progress_percentage:.2f}%')
```

The previous function execute_training_epochs is designed to manage and execute the training process over a specified number of epochs. progress_bar = trange(num_epochs) initializes a progress bar for the training process. total_batches = len(traindata_loader determines the total number of batches in the traindata_loader. The outer for loop iterates over each training epoch. An epoch is a complete pass through the entire training dataset.

CHAPTER 2 TEXT-TO-IMAGE GENERATION

The inner for loop iterates over each batch of images in the `traindata_loader`. The loader provides tuples of (`batch_images, labels`), but since the labels are not needed for the training step, they are ignored with _.

- `batch_loss = train_batch(batch_images)`: Calls the train_batch function for each batch of images.

- `progress_percentage = (batch_index / total_batches) * 100`: Calculates the progress percentage for the current epoch. The `if` condition checks if the current batch index is a multiple of 5. If so, the progress bar description is updated. `progress_bar.set_description` updates the progress bar with information about the current epoch, batch index, total number of batches, loss for the current batch, and progress percentage.

As a next step, train the model for 20 epochs and then visualize the results using the generate_images function (see Figure 2-10).

```
execute_training_epochs()
generate_images()
```

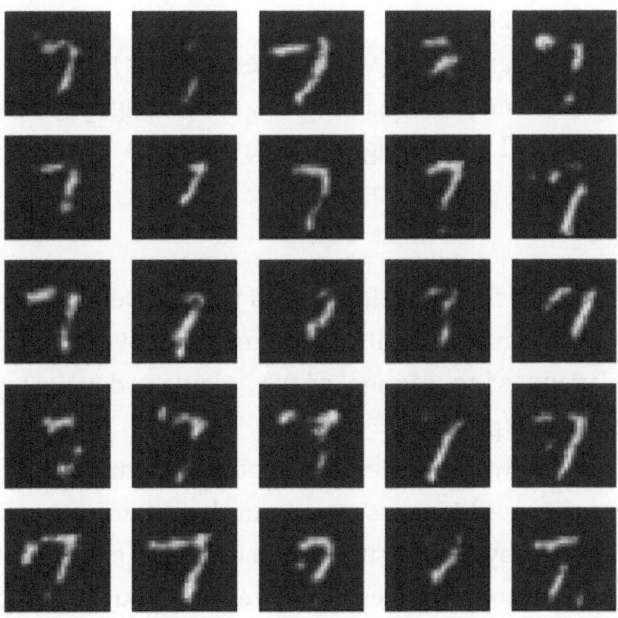

Figure 2-10. Result from trained model after 20 epochs

The output indicates that with just 20 epochs, the model has effectively learned to transform a noisy image into a clear one. Now, we will proceed to visually demonstrate the gradual improvement in image quality, step-by-step.

```python
def visualize_diffusion_steps():
    diffusion_sequence = []
    evolving_image = torch.randn(size=(8, 3, image_size, image_size),
    device=device)

    with torch.no_grad():
        for timestep in trange(number_of_timesteps):
            current_timestep = timestep
            evolving_image = unet_model(evolving_image, torch.full([8, 1],
            current_timestep, dtype=torch.float, device=device))
            if timestep % 2 == 0:
                diffusion_sequence.append(evolving_image[0].cpu())
    diffusion_sequence = torch.stack(diffusion_sequence, dim=0)
    diffusion_sequence = torch.clip(diffusion_sequence, -1, 1)
    visualized_images = display_image_grid(diffusion_sequence)

    plt.figure(figsize=(20, 2))
    for image_index in range(len(visualized_images)):
        plt.subplot(1, len(visualized_images), image_index+1)
        plt.imshow(visualized_images[image_index])
        plt.title(f'Step {image_index}')
        plt.axis('off')
```

In the previous function, first we initialize `diffusion_sequence = []` with an empty list to store the sequence of images at different timesteps. `evolving_image = torch.randn(size=(8, 3, image_size, image_size), device=device)` generates a tensor of random noise as the starting point.

In the next few lines, we would generate images over timesteps. `for timestep in trange(number_of_timesteps)` iterates over a number of timesteps. In each iteration, `evolving_image` is updated by passing it through the `unet_model` along with the current timestep, formatted as a tensor. This step simulates the diffusion process, where the initial noise is progressively transformed into more structured images. `if timestep % 2 == 0` checks if the current timestep is even. If so, it appends the first image of the current batch.

Now we process the data for visualization. `diffusion_sequence = torch.stack(diffusion_sequence, dim=0)` stacks the collected images into a single tensor for easier handling. `diffusion_sequence = torch.clip(diffusion_sequence, -1, 1)` ensures that the values in the image tensor are clipped to the range [-1, 1]. `visualized_images = display_image_grid(diffusion_sequence)` converts the sequence of tensors into a grid format suitable for visualization. `plt.figure(figsize=(20, 2))` sets up a Matplotlib figure with a specific size and `for` loop and then iterates over each image in `visualized_images`, displaying it in a subplot with the corresponding timestep as the title. As a next step, we run the `visualize_diffusion_steps` function and visualize the results (Figure 2-11).

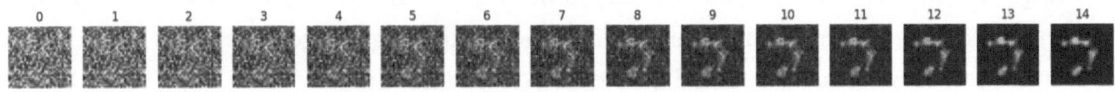

Figure 2-11. *Step-by-step improvement in noisy image*

The displayed image demonstrates the capability of our model to refine a clear image from an initially noisy one, effectively concluding our implementation of the diffusion model. We have now gained both an understanding of how diffusion models operate and hands-on experience in developing a basic version of such a model. Moving forward, the next section will involve importing and deploying a more advanced, pre-trained diffusion model from Hugging Face.

Text-to-Image Generation

In this section, we'll demonstrate the implementation of a pre-trained stable diffusion model—a cutting-edge tool that translates descriptive text into compelling imagery with remarkable accuracy. Additionally, we'll navigate through the fine-tuning process, which tailors the model to specific creative needs, enhancing its ability to produce customized and detailed images that resonate with the nuances of the input text.

Using a Pre-trained Model

In the previous section, we explored how the diffusion model can methodically remove noise from an image. Building upon what we've learned so far, this section takes a step further. Here, we'll focus on generating images directly from textual inputs, employing a pre-trained stable diffusion model provided by Hugging Face.

For our purposes, we'll be working with the `stable-diffusion-xl-base-1.0` model from Hugging Face. This model operates through a two-stage pipeline. Initially, it utilizes the base model to create latents corresponding to the target output size. Following this, a specialized high-resolution model is employed in the second stage. This model applies a technique known as SDEdit to the initially generated latents, using the same text prompt. Essentially, this model is adept at generating and modifying images based on text prompts and incorporates two fixed, pre-trained text encoders.

Stochastic Differential Equation editing (SDEdit, or `img2img`) is an innovative technique in the realm of generative models that allows for refining and modifying generated images. At its core, SDEdit leverages the principles of stochastic differential equations to transition smoothly between different visual representations, enabling enhancements in image quality or alterations in content while maintaining a high degree of fidelity to the original concept. By applying SDEdit to latent representations of images generated by a base model, it is possible to iteratively adjust and improve these images based on specific prompts or desired outcomes. This process, while slightly more time-consuming due to the additional computational steps involved, offers a powerful tool for creating highly detailed and contextually accurate visual content, opening up new possibilities for image generation and editing in fields ranging from digital art to automated content creation (see Figure 2-12).

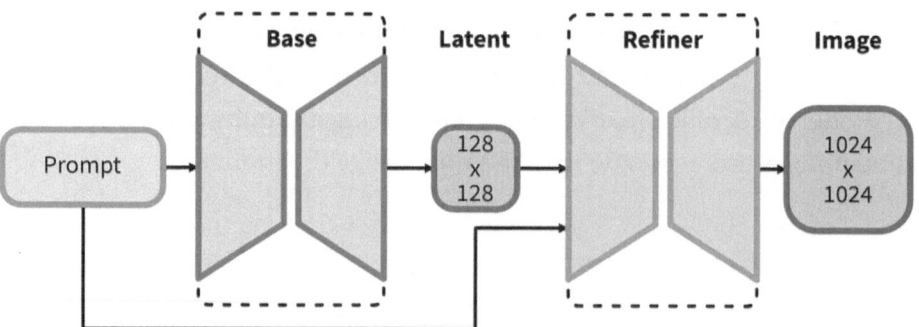

Figure 2-12. *Flow of SDEdit (source: `https://huggingface.co/stabilityai/stable-diffusion-xl-base-1.0`)*

Step 1: Installing Libraries

In step 1, we set the stage for utilizing a pre-trained diffusion model by installing the requisite libraries, equipping our toolkit with the necessary components for image synthesis.

```
!pip install diffusers
```

The previous command instructs the package manager `pip` to install the `diffusers` library.

```
from diffusers import DiffusionPipeline
import torch
```

The first line in the previous code block imports the `DiffusionPipeline` class from the `diffusers` library. `DiffusionPipeline` provides tools and functionalities to work with diffusion-based generative models.

The second line, `import torch`, imports PyTorch, a machine learning framework that offers a wide range of functions and classes for building and training neural networks. We will use it with diffusion models for image generation.

Step 2: Model Inference

Step 2 focuses on model inference, where we leverage the pre-trained diffusion model to generate images, translating textual descriptions into visual representations.

```
pipe = DiffusionPipeline.from_pretrained("stabilityai/stable-diffusion-xl-base-1.0", torch_dtype=torch.float16, use_safetensors=True, variant="fp16")
```

The previous line of code initializes a diffusion pipeline using a pre-trained model from the `diffusers` library in Python. It creates an instance of `DiffusionPipeline` using a `stable-diffusion-xl-base-1.0`[2] model pre-trained by Stability AI.

`torch_dtype=torch.float16` specifies the data type for PyTorch tensors as float16. It can lead to faster computation and reduced memory usage compared to the default float32, although it might result in slightly less precision in the calculations.

`use_safetensors=True` indicates the use of `SafeTensors` in the pipeline. `SafeTensors` provide additional safety measures, such as preventing accidental exposure to explicit or unsafe content.

variant="fp16" specifies the pipeline should use the fp16 variant of the model. This variant is optimized for float16 precision, aligning with the earlier torch data type specification.

```
pipe.to("cuda")
```

The previous command moves pipe onto the GPU for processing. This will enable us to leverage the power of the GPU for efficient performance in image generation tasks.

```
prompt = "a pixel art character with square dark green glasses, a film strip-shaped head and a yellow-colored body on a warm background"
images = pipe(prompt=prompt).images[0]
images
```

In the first line of code, we are storing a text prompt that describes the scene or subject to be generated by the model in the variable prompt.

In the second line, we pass the text prompt to the diffusion model pipeline (pipe) to generate an image. The .images[0] part accesses the first image in the generated image list.

In the last line, we display the generated image, and the output image looks like Figure 2-13.

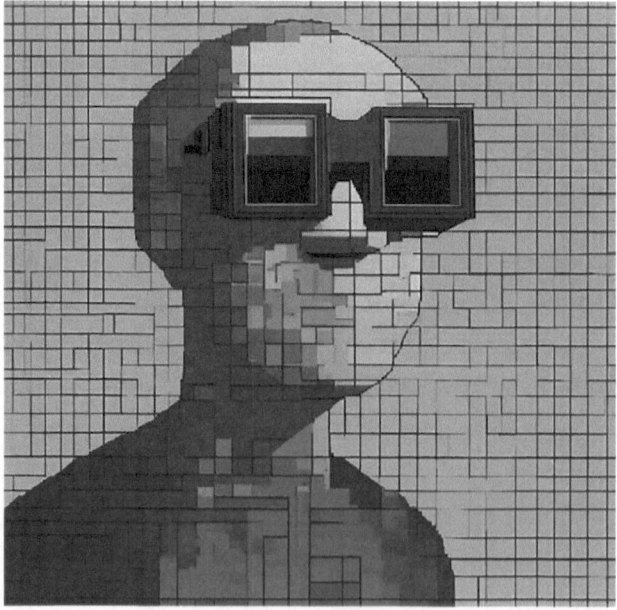

Figure 2-13. *Image generated from pre-trained model*

While the output image quality from the pre-trained model appears impressive, it falls short of matching the specific styling I'm aiming for. Acknowledging this gap, the next segment of our chapter will guide us through the process of fine-tuning the model on the m1guelpf/nouns dataset, available on Hugging Face. This strategic step is designed to align the model's generative capabilities more closely with the distinctive styling present in this dataset. By customizing the model to learn from the nuances and aesthetic qualities of the m1guelpf/nouns collection, we'll pave the way for generating images that not only boast high quality but also resonate with the desired stylistic attributes. This approach underscores the transformative potential of fine-tuning in achieving tailored visual outputs that meet our exact creative standards.

Fine-Tuning Text-to-Image Models

Fine-tuning a pre-trained stable diffusion model opens up a realm of possibilities for businesses looking to leverage customized visual content tailored to their unique branding and communication needs. By adjusting the model to specific styles or themes, companies can generate distinctive imagery for marketing campaigns, product designs, and digital content that stands out in a crowded marketplace. This process not only enhances the relevance and appeal of visual assets but also significantly reduces the time and resources spent on traditional content creation methods. With the m1guelpf/nouns dataset available on Hugging Face, we have a rich source of stylistic guidance to refine the model's output. Let's embark on this exciting journey to fine-tune the stable diffusion model, harnessing its potential to produce images that are not just visually striking but perfectly aligned with our desired styling. This step is not just about tweaking a model; it's about unlocking a new level of creative expression and operational efficiency for businesses ready to innovate in their visual storytelling.

Step 1: Installing Libraries and Data Loading

Step 1 involves installing essential libraries and loading data, laying the technical foundation for fine-tuning the stable diffusion model.

```
from google.colab import drive
import os
drive.mount('/content/drive')
```

In the provided code snippet, we first import the necessary libraries and then establish a connection between Google Drive and the Google Colab notebook.

```
os.chdir("/content/drive/My Drive/Colab Notebooks")
print(os.listdir())
```

The previous code changes the current working directory to a specific folder in Google Drive and then prints a list of files and directories in it.

```
!git clone https://github.com/huggingface/diffusers
```

As a next step, we clone the diffusers repository from Hugging Face. Cloning this repository downloads its content to the current working directory of your environment.

```
!pip install accelerate>=0.16.0
!pip install torchvision
!pip install datasets
!pip install ftfy
!pip install tensorboard
!pip install diffusers
!pip install diffusers["torch"] transformers
!pip install transformers -U
!pip install wandb
!pip install export
```

Now we install all the essential libraries in our environment.

- `accelerate` is used to simplify running Python scripts on different hardware configurations (like CPU, GPU, TPU) for machine learning tasks.
- `torchvision` provides datasets, model architectures, and common image transformations for computer vision.
- `ftfy` (Fix Text For You) is used to clean up and correct encoding issues in text.
- `TensorBoard` is a tool for providing the measurements and visualizations needed during the machine learning workflow.
- The `diffusers` library, which contains tools and functionalities for working with diffusion models.

CHAPTER 2   TEXT-TO-IMAGE GENERATION

- `transformers` provides datasets, model architectures, and common image transformations for computer vision.
- `wandb` (Weights & Biases) is a tool for tracking and visualizing machine learning experiments

```
!accelerate config default
```

The previous code is used to set up or reset the accelerate library's configuration to a standard, default state. It's particularly useful for adapting machine learning code to run efficiently on various hardware setups (like CPUs, GPUs, and TPUs).

```
!huggingface-cli login
```

Now, we log in to a Hugging Face account via the command-line interface (CLI). We would have to enter a user access token, which can be found in the Hugging Face user profile section.

```
os.environ['MODEL_NAME'] = "CompVis/stable-diffusion-v1-4"
os.environ['DATASET_NAME'] = "m1guelpf/nouns"
```
[3][4]

As a next step, we set a few environment variables that will be utilized in the later part of the code.

Step 2: Model Training

In step 2, we engage in the model training phase, meticulously fine-tuning the stable diffusion model to enhance its generative capabilities and adapt to specific visual tasks.

```
!accelerate launch --mixed_precision="fp16"  diffusers/examples/text_to_image/train_text_to_image.py \
  --pretrained_model_name_or_path={os.environ['MODEL_NAME']} \
  --dataset_name={os.environ['DATASET_NAME']} \
  --use_ema \
  --resolution=512 --center_crop --random_flip \
  --train_batch_size=1 \
  --gradient_accumulation_steps=4 \
  --gradient_checkpointing \
  --max_train_steps=10000 \
  --learning_rate=1e-05 \
```

CHAPTER 2   TEXT-TO-IMAGE GENERATION

```
--max_grad_norm=1 \
--lr_scheduler="constant" --lr_warmup_steps=500 \
--output_dir="fine-tuned-diffusion-model"
```

In the previous command, we start a training process for a text-to-image model using the Hugging Face `diffusers` library. The following is a breakdown of what each specific parameter means:

- `!accelerate launch` is used to start a Python script with the Accelerate library, which optimizes the script to run on different hardware (like CPUs or GPUs).

- `mixed_precision="fp16"` sets the training to use 16-bit floating-point precision (FP16), which can speed up training and reduce memory usage compared to 32-bit precision.

- `diffusers/examples/text_to_image/train_text_to_image.py` is the path to the Python script that will be run. This script is part of the `diffusers` library's examples and is used for training a text-to-image model.

- `pretrained_model_name_or_path={os.environ['MODEL_NAME']}` specifies the pre-trained model to use for training, with the model name retrieved from an environment variable.

- `dataset_name={os.environ['DATASET_NAME']}` sets the dataset for training, also retrieved from an environment variable.

- `use_ema` enables exponential moving average, which helps stabilize training.

- `resolution=512 --center_crop --random_flip` sets image processing options such as resolution, cropping, and random flipping for data augmentation.

- `train_batch_size=1` and `gradient_accumulation_steps=4` set the batch size and how many steps to accumulate gradients before updating model weights. A large batch size increases memory usage and increases training speed but decreases the generalization capability of the model.

- `gradient_checkpointing` reduces memory usage by selectively saving a subset of intermediate activations and recomputing others as needed during the backward pass. This technique enables the training of large models on hardware with limited memory.

- `max_train_steps=10000` limits the training to 10,000 steps.

- `learning_rate=1e-05` sets the learning rate.

- `max_grad_norm=1` sets the maximum gradient norm for gradient clipping.

- `lr_scheduler="constant"` and `lr_warmup_steps=500` define the learning rate scheduler settings.

- `output_dir="fine-tuned-diffusion-model"` specifies where to save the trained model and any output files.

`train_text_to_image.py` from Hugging Face incorporates a wide array of features, from data loading and preprocessing to model training, validation, and saving. At the outset, the script sets up an environment for training, including handling command-line arguments that allow for extensive customization of the training process. These arguments include model and dataset specifications, training hyperparameters (like learning rate, batch size, and number of epochs), and options for image preprocessing. It supports training with mixed precision for efficiency and utilizes the accelerate library to facilitate easy distributed training across CPUs and GPUs.

The core of the script revolves around preparing the dataset, which can be either a custom dataset provided by the user or one fetched from the Hugging Face Hub. It involves tokenizing text captions and applying transformations to images, ensuring they are in the correct format for model training. The script uses a `DataLoader` for batching and shuffling the prepared dataset, setting the stage for the training loop.

In the training loop, images are first encoded into latents using a Variational AutoEncoder (VAE), then the images are perturbed with noise, and finally the UNet model predicts the noise to be removed at each timestep. This process simulates the reverse diffusion process, guiding the model to generate images from the noisy latents conditioned on the text embeddings produced by the CLIP text encoder. The script supports gradient accumulation and mixed precision training, offering flexibility in managing computational resources. Validation runs at specified intervals, generating images from a set of predefined prompts to visually assess the model's performance. Finally, the trained model, along with its configurations, can be saved locally.

Step 3: Model Inference

Step 3 centers on model inference, applying our fine-tuned stable diffusion model to craft images that are more aligned with our targeted aesthetic and thematic goals.

```
from diffusers import StableDiffusionPipeline
import torch
```

Once the training finishes and the final model is saved then, we import essential libraries, which would now be used to generate images utilizing the newly trained model.

```
model_path = "fine-tuned-diffusion-model"
pipe = StableDiffusionPipeline.from_pretrained(model_path, torch_dtype=torch.float16)
pipe.to("cuda")
```

In these lines of code, the model_path variable is assigned to the path's fine-tuned diffusion model, which is the name of our newly trained model. Next StableDiffusionPipeline is initialized, and then we move the pipeline to a CUDA-enabled GPU for faster processing.

```
image = pipe(prompt="a pixel art character with square dark green glasses, a film strip-shaped head and a yellow-colored body on a warm background").images[0]
image.save("output_from_fine_tuned_diffusion_model.png")
image
```

In this code snippet, an image is generated and saved in the current directory. The generated image from our fine-tuned model, compared with the output from the pre-trained model, is as shown in Figure 2-14.

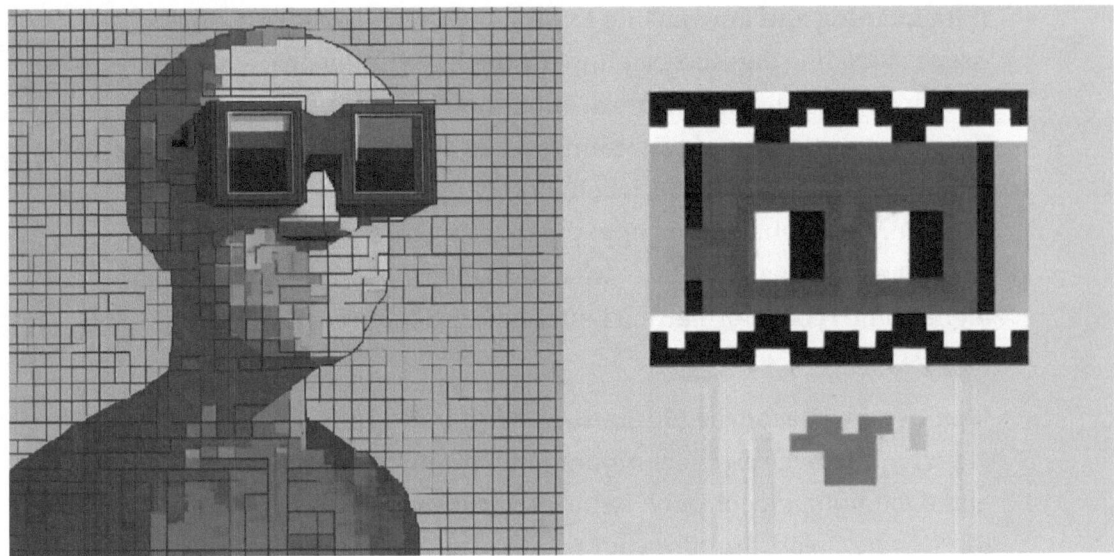

Figure 2-14. *Image generated from pre-trained model (left) and fine-tuned model (right)*

We have successfully fine-tuned a stable diffusion model on a m1guelpf/nouns dataset. This process involved adapting a pre-trained model to better align with our specific set of text-image pairs. By fine-tuning the model to our unique dataset, we have enabled the creation of new images from custom text descriptions, showcasing the potential of Generative AI in producing tailored visual content. This achievement opens up a myriad of possibilities for generating novel images that align closely with user-generated textual inputs, further extending the creative and practical applications of Generative AI in various domains.

Common Challenges and Troubleshooting Tips

Fine-tuning text-to-image models, such as the stable diffusion model, involves complex processes that can present several challenges. This section aims to outline common obstacles you might encounter and provide practical troubleshooting tips to navigate these issues effectively.

- **Data Loading and Formatting Errors:** Incorrect data formatting or issues in loading datasets can impede the fine-tuning process, often resulting in unexpected errors or suboptimal model performance. Verify that the format of your training data aligns with the model's requirements. For instance, when modifying `train_text_to_image.py`, ensure your changes cater to the specific structure and format of your dataset. Utilizing dataset validation tools or scripts can also help identify and rectify formatting issues before beginning the training process.

- **Memory and Resource Limitations:** Fine-tuning advanced models often requires significant computational resources, and running out of memory is a common issue, especially when using platforms like Google Colab. Monitor your resource usage closely and adjust the batch size or use gradient accumulation steps to manage memory usage efficiently. Additionally, consider using mixed precision training (`--mixed_precision="fp16"`) to reduce memory consumption without significantly impacting model performance.

- **Model Performance and Output Quality:** After fine-tuning, the model's output may not meet expected quality standards, possibly due to overfitting, underfitting, or inadequate learning rate adjustments. Regularly evaluate the model's performance on a validation set to monitor progress and adjust hyperparameters accordingly. Experimenting with different learning rates, fine-tuning durations, and regularization techniques can help optimize model output quality.

Custom modifications to scripts like `train_text_to_image.py` can introduce bugs or incompatibilities, particularly if the training data format varies from the original design. Start with minor modifications and test each change thoroughly. It's helpful to maintain a version control system (e.g., Git) to track changes and revert to previous versions if issues arise. Fine-tuning text-to-image models is a nuanced process that requires careful attention to detail and a willingness to experiment and troubleshoot. By anticipating common challenges and applying strategic tips, you can enhance the effectiveness of your fine-tuning efforts and achieve better results.

Conclusion

In this chapter, we traversed the innovative landscape where language meets visual creation, delving deep into the realms of CLIP models and the transformative technology of diffusion models. This journey not only unveiled the sophisticated architecture and functioning of the CLIP model but also guided us through hands-on implementation, providing a practical understanding of bridging textual descriptions with vivid imagery. Furthermore, our foray into diffusion models, from conceptual foundations to fine-tuning pre-trained models using the m1guelpf/nouns dataset, illustrates the chapter's commitment to melding theoretical insights with actionable knowledge.

We have thoroughly explored the stages of data loading, model training, and inference, providing a detailed understanding of the essential steps needed to fine-tune and utilize these sophisticated generative models. This chapter was designed to be a foundational resource for those looking to apply text-to-image generation techniques effectively, covering the critical processes and methodologies in depth.

As we reflect on the insights and learnings garnered throughout this chapter, it's essential to contemplate the future directions of text-to-image generation technology. Advances in artificial intelligence and machine learning are leading to the creation of models with enhanced capabilities to interpret human creativity and emotion from text, promising more precise, realistic, and creatively versatile applications. This progress is set to enrich the blend of art and AI, meeting the demand for visuals born from complex and imaginative prompts. Furthermore, merging text-to-image generation with virtual and augmented reality opens new realms for interactive media and storytelling, where users can craft visuals in real time, and educational materials become more engaging and tailored to improve learning experiences.

The potential applications of these advancements extend across various disciplines, from visualizing scientific data to creating custom content for digital marketing, highlighting the technology's versatility. As we navigate these developments, there's a parallel evolution of ethical standards addressing copyright, representation, and bias, ensuring these innovations contribute positively to society. This ongoing journey in text-to-image generation not only fuels our creativity but also encourages a thoughtful consideration of our impact as innovators in this evolving field, marking an exciting chapter yet to be fully written. As we pivot toward the future, our journey progresses from static imagery to dynamic narratives in the upcoming chapter on text-to-video generation. This leap forward explores more complex and immersive content creation methods, where text not only inspires images but also brings them to life, introducing new storytelling and digital interaction dimensions. Throughout this chapter, we aimed to establish a robust foundation in text-to-image generation, arming you with the necessary knowledge and skills to innovate and experiment within this thrilling field.

CHAPTER 3

From Script to Screen: Unveiling Text-to-Video Generation

Introduction

This chapter continues our journey through Generative AI, advancing from the groundwork established in Chapter 2 with its focus on text-to-image generation. This progression from static imagery to dynamic, moving visuals represents a significant leap forward in the field, highlighting the remarkable capability of AI to not just create images from text but also weave together sequences of images into coherent, engaging videos. Text-to-video generation stands at the forefront of technological innovation, offering a powerful tool that transforms written narratives into visual stories, thereby bridging the gap between the written word and cinematic storytelling. This technology encapsulates a unique blend of natural language processing, computer vision, and machine learning, pushing the boundaries of how we create and consume content in the digital age.

Text-to-video generation is revolutionizing several industries by providing innovative solutions to long-standing challenges. In filmmaking, this technology streamlines the previsualization process, allowing directors and screenwriters to convert scripts into animated storyboards quickly, facilitating more effective communication of vision and intent. The advertising industry benefits from the ability to swiftly produce engaging video content that resonates with target audiences, significantly reducing production times and costs. Educational sectors are witnessing a transformation in content delivery methods, where complex subjects can be taught through captivating visual stories, enhancing understanding and retention. Meanwhile, virtual reality environments gain a richer narrative layer, with text-to-video technologies enabling the creation of immersive experiences that respond dynamically to user inputs. These applications underscore the versatility and transformative potential of text-to-video technologies, opening up new creative horizons and operational efficiencies.

In the business world, text-to-video generation is not just an innovative tool but a strategic asset. Companies leverage this technology for marketing, creating personalized video content that engages consumers more effectively than traditional media. Training and development programs benefit immensely, as customized training videos can be produced on-demand, aligning precisely with the learning objectives and the learner's pace. Moreover, customer service departments utilize automated video responses to address frequently asked questions, providing a more interactive and informative customer experience. The ability to generate video content swiftly and cost-effectively allows businesses to stay agile, respond to market trends promptly, and maintain a competitive edge.

The historical development of text-to-video generation technology is a testament to the rapid evolution and interdisciplinary nature of advancements in artificial intelligence. Initially, the field of generative AI focused on understanding and processing text through natural language processing (NLP) techniques, laying the groundwork for more complex applications. Early efforts in computer vision aimed at creating static images from textual descriptions marked the preliminary steps toward visualizing the text, with significant projects like Google's DeepDream offering insights into how AI could interpret and "dream up" visuals based on learned patterns.

As technology progressed, researchers began exploring the potential of combining sequences of images to create motion, leading to the first endeavors in text-to-image synthesis. Projects like AttnGAN demonstrated the ability to generate detailed and relevant images from textual descriptions, setting the stage for more ambitious goals. The transition from generating single images to compiling them into sequences for video generation required advancements in understanding temporal coherence and narrative flow, a challenge that researchers addressed by developing models that could predict and generate not just isolated frames but sequences that follow a logical progression. Significant milestones in the journey toward effective text-to-video generation include the development of models capable of understanding the nuances of motion and time. Techniques in deep learning, specifically generative adversarial networks (GANs), have been pivotal. GANs, which involve a system of two competing neural networks (a generator and a discriminator), have been adapted to not only generate convincing still images but also to produce sequences of images that mimic the continuity of video content.

One of the landmark projects in this area was VideoBERT, a model designed to understand video content at a high level by learning correspondences between visual sequences and textual descriptions. While not directly generating video from text, VideoBERT and similar projects laid the foundational principles for correlating textual input with visual sequences, an essential step toward generating cohesive video content from textual prompts.

The field continues to evolve, with recent developments focusing on enhancing the quality, realism, and emotional depth of generated videos. Techniques such as improved temporal modeling, better integration of audio with visual content, and advances in unsupervised learning are at the forefront of current research. These efforts aim to refine the process of text-to-video generation, making it more seamless, intuitive, and capable of producing content that is increasingly indistinguishable from videos produced by human creators.

This chapter will guide readers through a comprehensive journey from the foundational aspects of understanding video data to the practical application and customization of pre-trained models. We begin by delving into the complexities of video data, a medium that, unlike static images, incorporates dimensions of time and motion, adding layers of complexity to data processing and interpretation. This section will address the challenges inherent in working with video data, such as the high computational costs, the need for temporal coherence, and the difficulties in capturing and generating nuanced human expressions and natural movements. Further, we will explore the intricate link between video and text data, shedding light on how textual descriptions can be transformed into dynamic visual narratives and how AI models interpret and bridge these two disparate forms of information.

The chapter then transitions to a hands-on demonstration of a pre-trained model, specifically `ali-vilab/modelscope-damo-text-to-video-synthesis`[5], to illustrate the current capabilities of text-to-video generation. This section will provide you with an opportunity to understand the operational mechanics behind the model, showcasing how it processes textual prompts to generate video content. The demonstration will serve as a practical example of the technology in action, offering insights into the model's architecture, the types of inputs it requires, and the quality of video output it can produce.

Building on this practical foundation, the next section of the chapter focuses on fine-tuning the pre-trained model for custom applications. Here, you will learn how to adapt and modify the model to suit specific needs or to improve its performance in generating video content for unique or niche applications. This section will cover the principles of model fine-tuning, including data preparation, parameter adjustment, and techniques for enhancing the model's understanding of specific types of text prompts, thereby extending the model's applicability beyond its initial training scope.

Through this structured flow, the chapter aims to equip you with a deep understanding of the text-to-video generation process, from the theoretical underpinnings to practical applications and customization strategies. By the end of this journey, you should be well-prepared to engage with this transformative technology, applying it creatively and effectively in your own projects or areas of interest.

Understanding Video Data

Understanding video data opens a gateway to the digital storytelling realm, where the intricacies of frames, formats, and compression converge to create motion illusion. This initial foray sets the stage for delving deeper into the technical nuances that define video quality, distribution, and the overall viewing experience, laying the groundwork for the comprehensive exploration that follows.

Videos are essentially a sequence of still images, or *frames*, displayed at a speed that creates the illusion of motion. The continuity of these frames, when played at an appropriate frame rate, results in smooth motion that mimics real-life movements. Let's go through a comprehensive overview of the intricacies involved in video data.

- **Video Formats and Codecs:** The format of a video file determines how data is stored and encoded, with common formats including MP4, AVI, and MOV. Each format has specific characteristics and compatibilities that make it suitable for different applications. *Codecs*, which stand for "compressor-decompressor," are used to reduce the file size of videos through compression algorithms without significantly degrading quality. Examples of widely used codecs include H.264 and H.265 (see Figure 3-1). These codecs play a key role in determining the balance between video quality and file size, affecting both storage requirements and streaming performance. Diving into the impact of codecs on video quality and

CHAPTER 3   FROM SCRIPT TO SCREEN: UNVEILING TEXT-TO-VIDEO GENERATION

compression reveals notable differences. For example, encoding a high-definition video for streaming with the H.264 codec strikes a balance between quality and efficiency, fitting for widely compatible platforms. Switching to the newer H.265 (HEVC) codec, the same video compresses to about half the size with little quality loss, enabling quicker streaming and reduced bandwidth use, an advantage for 4K and higher resolution content.

H.264   H.265

H.264 (AVC) vs. H.265 (HEVC) Comparison

Figure 3-1. Comparing H.264 and H.265 (source: `https://savicontrols.com/h-264-vs-h-265-comparison/`)

- **Frame Rate:** The frame rate of a video significantly affects its appearance and quality. A higher frame rate results in smoother motion (Figure 3-2), making it especially important for videos that capture fast-moving objects or actions. Standard frame rates vary by medium: cinematic productions typically use 24fps for a film look, while television standards often call for 30fps. High-definition gaming and some online video content may use 60fps or higher to achieve an even smoother appearance.

CHAPTER 3   FROM SCRIPT TO SCREEN: UNVEILING TEXT-TO-VIDEO GENERATION

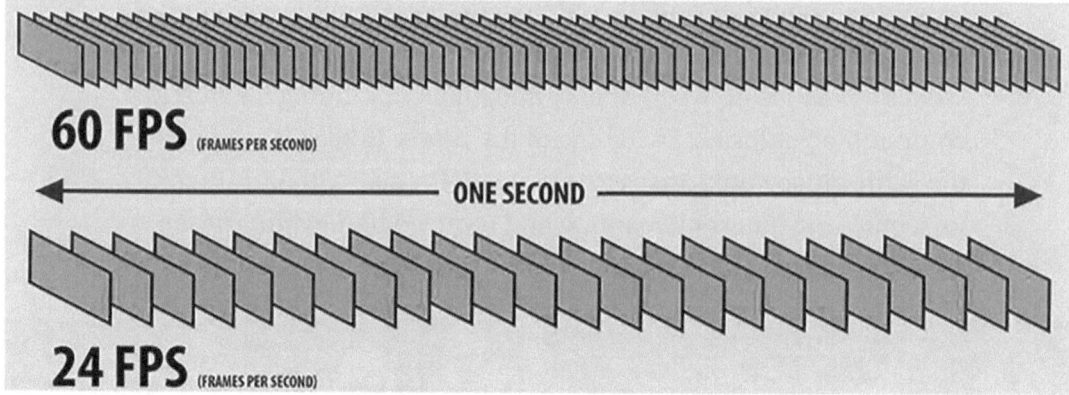

Figure 3-2. Comparing 60 FPS and 24 FPS (source: `https://www.videoproc.com/resource/frame-rate.htm`*)*

- **Resolution and Aspect Ratio:** Video resolution is a critical determinant of its quality, with higher resolutions offering greater clarity and detail. Common resolutions include 1080p (Full HD) and 4K (Ultra HD), with the latter providing a significantly sharper image. The aspect ratio, which is the proportional relationship between a video's width and height, affects how content is framed and displayed. The most common aspect ratio is 16:9, which is considered the standard for widescreen displays.

- **Color Depth and Bitrate:** Color depth, measured in bits per pixel, indicates the number of bits used to represent the color of a single pixel. Higher color depth allows for a broader range of colors and more detailed color gradation, enhancing the visual quality of the video. Bitrate, expressed in bits per second (bps), refers to the amount of data processed over a given time period. A higher bitrate generally means better video quality, as more data can capture more detail, but it also requires more storage space and bandwidth for streaming.

- **Audio Components:** In video files, audio is synchronized with the visual frames to create a cohesive viewing experience. The audio track within a video file can be encoded in various formats, each with its own characteristics regarding quality, compression, and compatibility. Effective integration and synchronization of audio with video frames are crucial for maintaining the viewer's immersion and ensuring that the sound matches the on-screen action.

- **Compression and Storage:** Compression is essential for managing the inherently large file sizes of video data. Through the use of codecs, video files can be compressed into smaller sizes to facilitate easier storage and faster transmission over networks. There are two types of compression: lossy, which reduces file size by permanently eliminating certain information, and lossless, which compresses data without any loss of quality. The choice of compression method affects both the quality of the video and the amount of storage space required.

- **Video Metadata:** Metadata in video files includes a range of information beyond the visible content, such as the video's duration, the codec used, its resolution, and potentially the creator or copyright information. This metadata is crucial for organizing, managing, and playing back video content efficiently. It enables software and services to optimize playback settings, categorize content, and improve searchability, making it easier for users to find and view videos.

Transitioning from the foundational aspects of video data, we now confront the challenges inherent in working with video content, from the technical hurdles of encoding and compression to the creative challenges of storytelling and engagement. This sets the stage for a deeper examination of the link between text and video data, a critical aspect of modern content creation where narratives are woven across visual and textual mediums, enriching the viewer's experience and broadening the horizon of digital storytelling.

Challenges in Working with Video Data

As we venture deeper into the fascinating world of video data, it's crucial for us to acknowledge the hurdles that come with it. Video data, with its rich and dynamic nature, opens a plethora of opportunities for creativity, innovation, and analysis. However, this complexity is a double-edged sword. Working with video data is not without its challenges, some of which can be quite daunting, especially as we push the boundaries of what's possible with modern technology.

In this section, we'll explore these challenges together, understanding not just what makes video data uniquely difficult to work with but also why overcoming these hurdles is worth the effort. From the sheer volume of data each video contains to the intricate details that need to be accurately captured and processed, the obstacles are significant. But fear not! With every challenge discussed, remember that each represents an opportunity for innovation, a chance to solve a puzzle that can lead to breakthroughs in how we create, consume, and comprehend video content.

So, let's roll up our sleeves and dive into the nitty-gritty of working with video data. We'll look at the technical, computational, and creative challenges that we face, understanding them not as roadblocks but as stepping stones toward mastering this complex domain.

- **High Computational and Storage Requirements:** When we talk about video data, we're dealing with files that are significantly larger and more complex than static images or text. Each frame of a video can be seen as an individual image, and when you consider that videos are often comprised of tens of thousands of these frames, the scale of data we're working with becomes clear. This presents a significant challenge in terms of both computational power and storage capacity. Processing video data, especially in high resolutions or over long durations, requires robust hardware capable of handling intensive tasks such as rendering, editing, and encoding. Similarly, storing this data demands vast amounts of disk space, which can escalate costs and logistical considerations. The balance between maintaining high-quality video data and managing these resources efficiently is a continuous challenge for professionals in the field. Netflix tackles storage and bandwidth challenges by dynamically optimizing video encoding, reducing file sizes without sacrificing quality. This approach demonstrates efficient resource management for global video streaming.

- **Temporal Dimension:** Unlike static images, videos capture the dimension of time, adding a layer of complexity to their analysis and generation. This temporal aspect means that understanding a video isn't just about analyzing individual frames but also how these frames relate to each other over time. It introduces challenges in tracking movement, changes in scenery, and the evolution of events within the video. Algorithms must not only be adept at recognizing patterns within a single frame but also capable of interpreting the progression of these patterns through time. This can significantly complicate tasks such as object detection, motion analysis, and behavior prediction, requiring sophisticated models that can effectively process temporal information. Technologies in sports analytics, like those used by STATS Perform, analyze player and ball movements in videos, showcasing how temporal data analysis enhances game strategies and player evaluations.

- **Contextual and Semantic Understanding:** Videos often convey complex narratives or scenarios that rely on an understanding of context and semantics. For a machine, deciphering the subtle cues that contribute to the overall meaning of a video is a formidable challenge. This includes interpreting body language, facial expressions, and the interplay between different elements within a scene. Beyond just recognizing objects or individuals, machines must learn to grasp the narrative or emotional tone of a video, which involves a deep understanding of human culture, language, and social dynamics. Achieving this level of comprehension is critical for applications like content recommendation, automated moderation, and interactive media. YouTube enhances user experience by recommending videos through algorithms that analyze visual, textual, and interaction data, showcasing advanced contextual and semantic understanding.

- **Audio Integration:** Audio is an integral part of most video content, adding a layer of information that is essential for a complete understanding of the content. Integrating audio with visual data presents unique challenges, as it requires algorithms to not only recognize speech and music but also to understand how these auditory elements interact with visual cues. This can be crucial for tasks such as automated captioning, content analysis, and emotion recognition. The synchronization of audio and video data, ensuring that they are processed in harmony to accurately reflect the content's intent and mood, is a complex task that necessitates advanced techniques in signal processing and machine learning. Platforms like Facebook implement automated captioning for videos, using speech recognition to enhance accessibility by accurately synchronizing audio content with visual cues.

- **Ethical and Privacy Concerns:** Working with video data often involves navigating sensitive ethical and privacy issues. Videos can capture personal and private moments, and the misuse of video data can lead to significant breaches of privacy and ethical standards. Ensuring that video data is collected, stored, and used in a manner that respects individual rights and complies with legal regulations is paramount. This includes implementing robust data protection measures, obtaining consent from individuals featured in videos, and carefully considering the implications of surveillance and facial recognition technologies. The responsibility to use video data ethically is a critical challenge for all stakeholders in the field. The European Union's GDPR sets a high standard for video data privacy, mandating strict consent and data protection measures and emphasizing the importance of ethical handling and legal compliance.

- **Data Annotation and Quality of Training Data:** For machine learning models to effectively understand and generate video content, they require large volumes of annotated training data. However, accurately annotating video data is significantly more time-consuming and complex than labeling static images. It involves not only identifying objects or actions within a frame but also tracking these across the video's duration. The quality of training data directly impacts the performance of models, making the task of data annotation both critical and challenging. Ensuring that the data is diverse, representative, and accurately labeled is a continuous struggle, especially as we strive to build models that can understand the vast array of content seen in videos worldwide. Companies like Waymo use manual and semi-automated tools for annotating video data from cameras, ensuring high-quality training data crucial for safe autonomous vehicle operation.

The Synergy of Video and Textual Data

This section explores the dynamic relationship between visual and written content. This section delves into how combining video with textual data—ranging from metadata and annotations to transcripts and semantic tagging—enhances content accessibility, engagement, and comprehension. It highlights the pivotal role of text in making video content searchable, accessible, and interactive, paving the way for innovative applications in content recommendation, machine learning, and beyond. This synergy not only improves user experiences but also drives technological advancements in content analysis and personalization.

- **Descriptive Metadata and Annotations:** Textual metadata and annotations provide essential descriptions and context for video content. This can include titles, descriptions, tags, and captions, which help in categorizing, searching, and understanding the content at a glance.

- **Transcripts and Subtitles:** Transcripts offer a text-based representation of the audio content within a video, including dialogue and relevant nonspeech audio information. Subtitles, on the other hand, not only transcribe spoken dialogue but also include timing information, allowing viewers to follow along with the video. Both enhance accessibility and comprehension for diverse audiences.

- **Semantic Analysis and Tagging:** Using NLP techniques, videos can be semantically analyzed to extract themes, entities, and sentiments, which are then tagged as text data. This process aids in more sophisticated search and discovery experiences, allowing users to find content based on conceptual and thematic queries.

- **Video Summarization:** Text-based summaries of video content provide quick insights into the video's narrative or informational content without requiring full viewership. These summaries can be generated through automated analysis of both the visual and audio components, distilled into concise textual form.

- **Content Accessibility:** Text data linked with video, such as closed captions and descriptive audio transcripts, makes content accessible to individuals with hearing or visual impairments. This not only complies with accessibility standards but also broadens the audience reach.

- **Interactive and Enhanced Experiences:** Integrating text with video supports interactive experiences, such as clickable transcripts that navigate to specific video parts or augmented reality (AR) experiences where text overlays provide additional information.

- **Machine Learning and AI Applications:** The correlation between video and text data enables sophisticated recognition and classification tasks in AI-driven applications like video surveillance and content analysis. For instance, object and action recognition in videos can be enhanced with textual annotations to train more accurate machine learning models.

- **Content Recommendation and Personalization:** Textual analysis of user-generated content, comments, and reviews linked with video data can inform recommendation algorithms, offering personalized content suggestions based on user preferences and interactions.

Tagging Videos with Semantic Metadata

Enhancing the searchability and accessibility of video content through semantic metadata tagging is a crucial step in leveraging the synergy between video and textual data. This practical guide outlines a simple process using YouTube as an example platform, given its widespread use and comprehensive features for content creators:

1. **Select a Video Platform:** For this guide, we'll use YouTube. Ensure you have a video ready for upload or select an existing video in your YouTube Studio.

2. **Upload and Basic Info:** Upload your video to YouTube. In the Details section, fill in the basic information: title, description, and tags. Use descriptive keywords and phrases that accurately reflect your video's content.

3. **Add Descriptive Metadata:**

    a. **Title and Description:** Craft a compelling title and detailed description using relevant keywords. This text serves as primary metadata, aiding search engines and recommendation algorithms in understanding your video's content.

    b. **Tags:** Add relevant tags related to your video's topics, themes, and subjects. Tags function as searchable keywords that improve video discoverability.

4. **Transcripts and Subtitles:**

    a. YouTube can automatically generate subtitles for your video, but for accuracy and inclusivity, consider uploading a custom transcript.

    b. In YouTube Studio, under your video, select the Subtitles option and choose to upload a transcript file or manually enter subtitles.

5. **Review and Publish:** Before publishing, review your metadata to ensure it accurately represents your video content and utilizes keywords effectively. Adjust your privacy settings, and then publish your video.

6. **Monitor and Update:** Monitor your video's performance and update metadata as needed based on viewer feedback and changing trends to maintain relevance and discoverability.

In this chapter, we'll also explore generating captions for our videos using BLIP-2[6], an advanced tool that enhances our semantic metadata efforts by providing precise, descriptive captions tailored to our video content. This process further refines our video's searchability and viewer engagement.

Hands-On: Demonstrating a Pre-Trained Model

In this section, we delve into the practical application of a cutting-edge AI model that exemplifies the incredible advancements in text-to-video synthesis. We will be working with the `modelscope-damo-text-to-video-synthesis` model, hosted on Hugging Face, which represents a significant leap in our ability to generate dynamic, visually compelling videos directly from textual descriptions. The architecture of the `modelscope-damo-text-to-video-synthesis` model involves a multistage process that includes extracting text features, translating text features to video latent space, and converting video latent space into visual space. It uses a U-Net 3D structure for video generation, relying on an iterative denoising process. This approach allows the model to generate videos from textual descriptions by progressively refining the output from Gaussian noise to a coherent video representation.

We will be leveraging the Hugging Face API, and this hands-on section aims to equip you with the knowledge and tools necessary to implement this pre-trained model effectively. By the end of this section, you will have a comprehensive understanding of how to interact with and utilize this model to bridge the gap between textual concepts and their video representations, opening up new possibilities for creative and technical endeavors.

Step 1: Installing Libraries

In step 1, we'll begin by installing the necessary libraries and loading the data required to work with the `modelscope-damo-text-to-video-synthesis` model, setting the foundation for our text-to-video synthesis project.

```
!pip install modelscope==1.4.2
!pip install open_clip_torch
!pip install pytorch-lightning
```

In the previous code snippet, we are installing the `modelscope`, `open_clip_torch`, and `pytorch-lightning` libraries, which are essential for text-to-video generation.

```
from huggingface_hub import snapshot_download
from modelscope.pipelines import pipeline
from modelscope.outputs import OutputKeys
import pathlib
```

Now, we import the necessary modules and functions for working with models hosted on Hugging Face and utilizing the Modelscope library for AI tasks. It facilitates downloading model snapshots from Hugging Face, setting up pipelines for various AI processes with Modelscope, managing output keys for result retrieval, and working with file paths using the `pathlib` module for file system navigation.

Step 2: Model Inference

In step 2, we'll execute model inference to transform textual descriptions into corresponding videos using the pre-trained `modelscope-damo-text-to-video-synthesis` model.

```
model_dir = pathlib.Path('weights')
snapshot_download('damo-vilab/modelscope-damo-text-to-video-synthesis',repo_type='model', local_dir=model_dir)
```

The previous code snippet downloads the model from the Hugging Face repository into a local directory named `weights`, using `pathlib` to handle the directory path and `snapshot_download` for the model acquisition.

```
pipe = pipeline('text-to-video-synthesis', model_dir.as_posix())
test_text = {'text': 'a close up of a pink flower in a vase against a yellow background'}
```

Now we initialize a pipeline for text-to-video synthesis using the Modelscope library, specifying the directory where the model is stored. It then defines a test input with a textual description of a scene involving "a close-up of a pink flower in a vase against a yellow background," which is prepared to be fed into the pipeline for generating a corresponding video.

```
output_video_path = pipe(test_text,)[OutputKeys.OUTPUT_VIDEO]
print('output_video_path:', output_video_path)
```

This code snippet runs the previously defined `pipe` with the `test_text` as input to generate a video, accessing the output through the `OutputKeys.OUTPUT_VIDEO` key to retrieve the path of the generated video. It then prints the path of the output video, providing a way to locate and view the synthesized video.

The output video is hosted on GitHub, and its link is readily available. Figure 3-3 shows a screen from this video to give you a glimpse of the content. Because of computational constraints, the videos have been limited to a duration of two seconds. However, by altering the default settings, it's possible to extend the length of the videos.

CHAPTER 3   FROM SCRIPT TO SCREEN: UNVEILING TEXT-TO-VIDEO GENERATION

Figure 3-3. *Screenshot from video generated by pre-trained model*

In Figure 3-3, it's evident that the video captures elements such as pink flowers and a vase, alongside hints of yellow. However, the background doesn't entirely match our specified criteria, and the detail level doesn't quite achieve the close-up perspective we requested. In the following section, we'll explore methods to fine-tune the model, aiming to enhance the video quality and better align the output with our expectations. During the process of transforming text into video with the `modelscope-damo-text-to-video-synthesis` model, you might encounter issues like erratic video output, model loading errors, or computational limitations. To address these, ensure your textual descriptions are clear and specific, check that all necessary libraries are correctly installed, and verify the model's directory path. For performance issues, consider adjusting the video's resolution or utilizing more powerful computational resources.

Fine-Tuning for Custom Applications

Fine-tuning text-to-video models in the dynamic field of artificial intelligence presents an innovative path for generating content that precisely meets the unique demands of various user groups. This advanced approach significantly enhances the creation of customized videos, directly aligning with specific audience interests or goals across numerous fields, including education, marketing, and entertainment. By customizing pre-trained models to fit unique niches, developers and creators are capable of transcending the limitations of standard video outputs. This customization facilitates the production of distinctive visuals that deeply connect with intended audiences, thereby elevating user engagement and enriching the overall experience.

However, the journey toward harnessing the full potential of fine-tuned text-to-video technology is fraught with obstacles. The primary challenge in fine-tuning lies in the limited availability of high-quality, specialized text-to-video datasets. Many niche applications suffer from a dearth of appropriate data, impeding the model's ability to accurately capture and reproduce the nuances of specialized content. Compounding this issue is the considerable computational expense associated with training complex models on video data. Adjusting the myriad parameters of a text-to-video model requires substantial processing power, often rendering the fine-tuning process impractical for those without access to advanced computing infrastructures.

To tackle the inherent challenges of fine-tuning text-to-video models, we have embarked on an innovative journey utilizing a bespoke dataset of 22 videos, all of which have been personally collected and feature a diverse array of flowers. This unique collection provides a controlled yet highly relevant context for experimenting with fine-tuning techniques, aiming to significantly enhance the model's performance in generating videos that capture specific visual themes accurately.

To augment the value of our dataset, we've turned to Bootstrapped Language Image Pretraining 2 (BLIP-2), a cutting-edge model renowned for its ability to generate precise and descriptive captions. By applying BLIP to our collection of flower videos, we ensure each piece of content is paired with a text description that mirrors its visual elements accurately. This crucial step not only enriches our dataset but also primes it for a more effective fine-tuning process, setting the stage for the creation of videos that are both visually appealing and contextually relevant.

BLIP-2, building on the foundational principles of its predecessor, represents a more advanced convergence of visual and textual data processing, setting new standards in the field of artificial intelligence and machine learning. This enhanced model extends the capabilities of BLIP by incorporating state-of-the-art techniques to more deeply understand both the content of images and the context of their corresponding captions, facilitating an even more nuanced and precise generation of text from visual inputs. The primary aim of BLIP-2 is to elevate the interaction between visual perception and language understanding to unprecedented levels.

Central to BLIP-2's architecture (Figure 3-4) is an evolved integration of visual and linguistic components. It utilizes an advanced vision transformer (ViT) to meticulously analyze visual inputs, extracting even richer feature sets from images. Concurrently, an upgraded language model processes textual data, delving deeper into the semantics, syntax, and contextual nuances of the language. Through enhanced attention mechanisms and neural network layers, BLIP-2 achieves a more sophisticated synthesis of visual and textual information, enabling the generation of captions that are not only accurate but also rich in contextual and emotional depth.

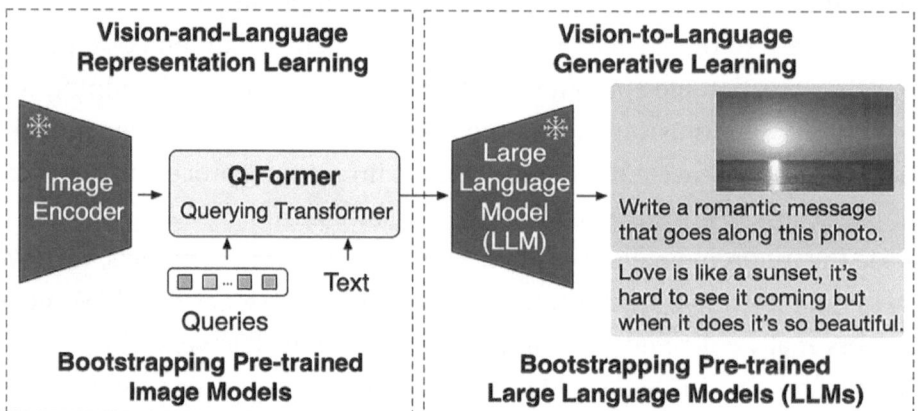

Figure 3-4. Flow of BLIP-2 (source: https://huggingface.co/blog/blip-2)

The training methodology of BLIP-2 incorporates an improved contrastive learning approach, refined to optimize the model's ability to accurately pair images with their correct captions and effectively distinguish them from mismatched pairs. An extensive pre-training regime complements this on a vast and diverse dataset of image-text pairs, which broadens the model's understanding of a wide array of visual concepts and linguistic styles. Such comprehensive training endows BLIP-2 with superior capabilities in text generation from images, making it an invaluable asset for a spectrum of applications, from enhanced automated captioning to creative content generation and beyond.

Step 1: Installing Libraries

To kickstart our fine-tuning journey, the first step involves setting up our environment by installing the necessary libraries. These libraries include those required for handling video and image processing and model training. This foundational step ensures we have all the tools ready for our fine-tuning process.

```
from google.colab import drive
drive.mount('/content/drive')
```

First we start with importing the `drive` module from `google.colab` and mounting the Google Drive to the `/content/drive` directory in a Colab notebook, allowing access to files stored in Google Drive.

```
%cd /content/drive/MyDrive
!git clone https://github.com/ExponentialML/Text-To-Video-Finetuning.git
%cd Text-To-Video-Finetuning/
!pip install -r requirements.txt
!pip install triton
!pip install compel
```

The previous code snippet performs the following actions in a sequence:

1. Changes the current working directory to `/content/drive/MyDrive` in a Google Colab notebook environment.

2. Clones the repository `Text-To-Video-Finetuning` from GitHub into the current working directory.

3. Changes the current working directory to the newly cloned `Text-To-Video-Finetuning` directory.

4. Installs the Python dependencies listed in the `requirements.txt` file found within the `Text-To-Video-Finetuning` directory.

5. Installs the `triton` & `compel` library. Triton is designed to help write highly efficient GPU code.

```
!git clone https://github.com/ExponentialML/Video-BLIP2-Preprocessor.git
%cd Video-BLIP2-Preprocessor/
!pip install -r requirements.txt
```

CHAPTER 3   FROM SCRIPT TO SCREEN: UNVEILING TEXT-TO-VIDEO GENERATION

In the previous code we first clone the repository named Video-BLIP2-Preprocessor from GitHub to the local machine. This operation creates a copy of the repository's content in a directory named after the repository itself. Following the successful cloning, the command %cd Video-BLIP2-Preprocessor/ changes the current working directory to the newly cloned Video-BLIP2-Preprocessor directory, ensuring that subsequent commands operate within the context of this directory. Finally, the command !pip install -r requirements.txt installs the Python dependencies listed in the requirements.txt file located in the directory.

Step 2: Data Loading and Preprocessing

In step 2, we focus on importing our collection of 22 videos and undertaking essential preprocessing tasks. Our primary goal is to standardize the format of all videos to a resolution of 384x256 pixels, ensuring uniformity and compatibility with our text-to-video synthesis model. Following this, we will proceed to generate captions for each video, enriching our dataset with descriptive textual information that accurately reflects the content of the videos. This step is crucial for preparing our data for the fine-tuning process, laying a solid foundation for enhanced model performance.

```
from moviepy.editor import VideoFileClip
import os
```

In the previous code we import the VideoFileClip class from the moviepy.editor module, which is used for reading video files and performing operations on them, such as editing or processing video content. Then we import the os module, which provides functions for interacting with the operating system.

```
input_path = '/content/drive/MyDrive/sampe_videos_for_fine_tuning'
output_path = '/content/drive/MyDrive/sampe_videos_for_fine_tuning_384_256'

os.makedirs(output_path, exist_ok=True)

for filename in os.listdir(input_path):
    if not filename.endswith('.mp4'):
        continue
    video_path = os.path.join(input_path, filename)
    output_video_path = os.path.join(output_path, f"processed_{filename}")
    clip = VideoFileClip(video_path)
```

```
    processed_clip = clip.resize(newsize=(384, 256)).without_audio()
    processed_clip.write_videofile(output_video_path, codec='libx264',
    audio_codec=None)
```

```
print("Video processing completed.")
```

By utilizing the previous code, we are importing a collection of videos, specifically adjusting their resolution to 384x256 pixels, a step crucial for ensuring consistency across our dataset. First, we define two paths: input_path points to the location where our original sample videos are stored, and output_path designates where we'll save the processed videos. The os.makedirs function is employed to create the output directory, with exist_ok=True ensuring that no error is thrown if the directory already exists.

We then iterate over every file in the input_path directory using a for loop. With a simple conditional check, we ensure that we're only working with MP4 files, skipping any file that does not meet this criterion. For each video file, we construct its full path and designate an output path that includes processed_ prefixed to the original filename, indicating that this file has been processed.

Utilizing the VideoFileClip class from the moviepy.editor module, we load each video file. We apply the resize method to alter the video's resolution to our desired dimensions of 384x256 pixels and remove the audio track with without_audio(), as our focus is purely on visual content. The processed clip is then saved to the specified output directory using write_videofile, employing the libx264 codec for video compression to ensure high-quality output without audio. Upon completing the loop, we signal the end of the process with a simple print statement: "Video processing completed."

```
%cd /content/drive/MyDrive/Text-To-Video-Finetuning/Video-BLIP2-
Preprocessor
```

```
!python preprocess.py \
--video_directory /content/drive/MyDrive/sampe_videos_for_fine_
tuning_384_256 \
--config_name "captions_for_videos" \
--config_save_name "captions_for_videos" \
--prompt_amount 8 \
--max_prompt_length 100
```

CHAPTER 3   FROM SCRIPT TO SCREEN: UNVEILING TEXT-TO-VIDEO GENERATION

The previous code snippet is generating captions for our videos using BLIP-2. We begin by navigating to the specific directory within our Google Drive where our preprocessing script resides, using `%cd /content/drive/MyDrive/Text-To-Video-Finetuning/Video-BLIP2-Preprocessor`. Next, we execute a Python script named `preprocess.py` with several parameters tailored to our project's needs. The `--video_directory` parameter points to the location where our sample videos are stored. These videos, crucial for our fine-tuning process, are short clips, typically ranging from four to six seconds. We specify a configuration through `--config_name` and `--config_save_name`, both set to `captions_for_videos`, which indicates the naming convention for both our input configuration and the output JSON file where our processed data will be stored.

The `--prompt_amount` parameter, set to 8, reflects our decision to generate a total of eight prompts for each video. This choice is informed by the brief duration of our videos; however, we note that this number can be adjusted based on the length of the videos being processed. Additionally, we limit the maximum length of each prompt to 100 characters via `--max_prompt_length`, ensuring conciseness and relevance. Ultimately, this process compiles our video data and associated captions into a JSON file named `captions_for_videos.json`.

The `preprocess.py` script is a comprehensive tool designed for preprocessing video data to generate prompts or captions using the BLIP2 model. At the core of this script is the `PreProcessVideos` class that is initialized with a variety of parameters that allow customization of the preprocessing workflow, such as the directory containing the videos, parameters related to the generation of prompts (like the number of prompts per video, the minimum and maximum length of prompts), and settings for handling video frames (such as selecting random frames or clipping frame data). The script operates by first setting up the BLIP2 model and preparing the environment for video processing. It then iterates through each video in the specified directory, applying a series of transformations and utilizing the BLIP2 model to generate captions for selected frames. These frames can be chosen either randomly or deterministically, based on the script's configuration. For each video, it creates a structured representation in the form of a JSON file, which includes the path to the video, the number of frames processed, and the generated prompts for each frame. Optionally, if enabled, it can also save the selected frames as video clips alongside the JSON data. This structured output facilitates easy integration with downstream tasks.

Step 3: Model Training (Fine-Tuning)

In step 3, we embark on the pivotal phase of model training, specifically focusing on fine-tuning. This stage is crucial for adapting a pre-trained model to our specific task, enhancing its ability to understand and generate content that aligns closely with our dataset.

```
!git lfs install
!git clone https://huggingface.co/damo-vilab/text-to-video-ms-1.7b
```

In the previous code first we install the Git Large File Storage (LFS) extension for Git, a version control system, designed to improve handling of large files by storing references to these files in the repository, while the actual file contents are stored on a remote server. This command sets up Git LFS on the user's system, ensuring that any subsequent cloning of repositories with large files will handle these files appropriately, downloading the actual contents rather than just their references. After setting up Git LFS, the second command, !git clone https://huggingface.co/damo-vilab/text-to-video-ms-1.7b, is used to download the pre-trained model from Hugging Face's model hub.

```
%%writefile /content/drive/MyDrive/Text-To-Video-Finetuning/configs/my_config_hq.yaml
pretrained_model_path: "/content/drive/MyDrive/Text-To-Video-Finetuning/text-to-video-ms-1.7b"
output_dir: "./outputs"
dataset_types:
  - 'json'
use_offset_noise: False
rescale_schedule: False
extend_dataset: False
cache_latents: True
cached_latent_dir: null #/path/to/cached_latents
train_text_encoder: False
lora_version: "cloneofsimo"
use_unet_lora: True
use_text_lora: True
lora_unet_dropout: 0.1
```

```yaml
lora_text_dropout: 0.1
save_lora_for_webui: True
only_lora_for_webui: False
save_pretrained_model: True
unet_lora_modules:
  - "UNet3DConditionModel"
text_encoder_lora_modules:
  - "CLIPEncoderLayer"
lora_rank: 16
train_data:
  width: 384
  height: 256
  use_bucketing: True
  sample_start_idx: 1
  fps: 24
  frame_step: 1
  n_sample_frames: 8
  single_video_path: ""
  single_video_prompt: ""
  fallback_prompt: ''
  json_path: "/content/drive/MyDrive/Text-To-Video-Finetuning/Video-BLIP2-Preprocessor/train_data/captions_for_videos.json"

validation_data:
  prompt: ""
  sample_preview: True
  num_frames: 16
  width: 384
  height: 256
  num_inference_steps: 25
  guidance_scale: 9

learning_rate: 1e-5
adam_weight_decay: 1e-2
extra_unet_params: null
extra_text_encoder_params: null
```

```yaml
train_batch_size: 1
max_train_steps: 7500
checkpointing_steps: 2500
validation_steps: 100
trainable_modules:
  - "all"
trainable_text_modules:
  - "all"
seed: 64
mixed_precision: "fp16"
use_8bit_adam: False
gradient_checkpointing: True
text_encoder_gradient_checkpointing: False
enable_xformers_memory_efficient_attention: False

enable_torch_2_attn: True
```

By utilizing the previous code, we are meticulously crafting a YAML configuration file to guide the fine-tuning process of a text-to-video model. At the heart of this configuration is the specification of the `pretrained_model_path`, which points to the location of the pre-trained model assets necessary for fine-tuning. The `output_dir` denotes where the fine-tuned model's outputs will be stored. This configuration underscores the importance of JSON as the dataset format, indicated under `dataset_types`.

Significant to this configuration are various settings that tailor the fine-tuning process to specific requirements. For example, `use_offset_noise`, `rescale_schedule`, and `extend_dataset` are Boolean flags that control aspects of data handling and training dynamics, whereas `cache_latents` enables caching of intermediate representations to speed up training. The `train_text_encoder`, `use_unet_lora`, and `use_text_lora` flags are pivotal for adjusting the training focus, specifically whether to train the text encoder and apply Low-Rank Adaptation (LoRA) to the UNet and text encoder for enhanced model adaptability.

The configuration also outlines parameters for the LoRA adaptation, such as `lora_version`, dropout rates (`lora_unet_dropout`, `lora_text_dropout`), and module names indicating where LoRA is to be applied (`unet_lora_modules`, `text_encoder_lora_modules`). The `lora_rank` setting specifies the rank for the LoRA adaptation, impacting how model parameters are adjusted.

Training data specifics are delineated, including dimensions (`width`, `height`), frame handling (`fps`, `frame_step`, `n_sample_frames`), and path to the training dataset (`json_path`). Similarly, validation settings are provided to ensure the model's performance is regularly assessed against a set of criteria during training.

Lastly, the file specifies hyperparameters like `learning_rate` and `adam_weight_decay`, training configurations such as batch size (`train_batch_size`), training steps (`max_train_steps`, `checkpointing_steps`, `validation_steps`), and modules eligible for training. Advanced settings like `mixed_precision` for computational efficiency, `gradient_checkpointing` for memory management, and `enable_torch_2_attn` for leveraging specific attention mechanisms underscore the comprehensive nature of this configuration, ensuring a tailored and efficient fine-tuning process.

Selecting optimal hyperparameters is crucial for fine-tuning text-to-video models effectively. Best practices suggest starting with hyperparameters from similar successful projects or pre-trained models and then iteratively adjusting them based on your dataset's specific characteristics and performance metrics. Key hyperparameters to focus on include learning rate, which should be lowered from the pre-trained settings to avoid overfitting; batch size, adjusted based on computational resources and memory constraints; and the number of training steps, determined by the size and complexity of the dataset. Additionally, employing techniques like learning rate schedulers can help in adapting the learning rate based on training progress. Regular evaluation against a validation set not only informs the effectiveness of these parameters but also guides further refinement, ensuring the model learns to generate high-quality video content that aligns closely with textual descriptions.

```
%cd /content/drive/MyDrive/Text-To-Video-Finetuning
!python train.py --config ./configs/my_config_hq.yaml
```

In the previous code we are setting up and starting the training process for a text-to-video fine-tuning task. First, the command `%cd /content/drive/MyDrive/Text-To-Video-Finetuning` changes the working directory to the specified path within Google Drive, ensuring that operations are performed in the correct project folder. Then, `!python train.py --config ./configs/my_config_hq.yaml` executes a Python script named `train.py`, which initiates the model training process using the settings and parameters defined in `my_config_hq.yaml`. Fine-tuning text-to-video models is a highly resource-intensive task, as evidenced by the fact that processing merely 22 videos, each ranging from 4 to 6 seconds, requires approximately 5 hours on an A100 GPU via Google Colab. This duration underscores the significant computational demands associated with these models.

The `train.py` script orchestrates the fine-tuning process for a text-to-video model, utilizing advanced deep-learning libraries such as Transformers and Diffusers. It begins by setting up the environment with necessary configurations and initializing the Accelerator for efficient distributed training. This setup includes loading pre-trained models (like UNet, AutoencoderKL, and CLIPTextModel), defining optimization parameters, and managing datasets for training. The script supports various dataset formats (e.g., JSON, single videos) and implements functionality to extend datasets for balanced training. It leverages the LoRA technique to fine-tune specific parts of the model, enhancing its ability to generate relevant video content from textual prompts.

During the training loop, the script processes batches of data, applies transformations, and computes losses to update model weights. It utilizes gradient checkpointing and mixed precision training for memory efficiency and speed. The training process includes generating video frames from textual prompts and evaluating the model's performance at specified intervals. The script dynamically adjusts learning rates, employs caching techniques for latents to expedite training, and integrates with Hugging Face's Accelerate library for streamlined distributed training. Finally, the trained model and configurations are saved, allowing for the generation of videos from new textual prompts. This script represents a comprehensive approach to fine-tuning text-to-video models, emphasizing efficiency, scalability, and adaptability to different training datasets and model configurations.

Step 4: Model Inference

In step 4, we transition from the meticulous process of training and fine-tuning our text-to-video model to leveraging its capabilities to generate video content from text prompts. This phase is where the fruits of our labor become evident, as we input textual descriptions and observe the model's ability to interpret and visualize these descriptions as coherent video sequences.

```
import random
import numpy as np
import torch
```

CHAPTER 3   FROM SCRIPT TO SCREEN: UNVEILING TEXT-TO-VIDEO GENERATION

```python
from diffusers import DPMSolverMultistepScheduler, DDPMScheduler, TextToVideoSDPipeline
from compel import Compel
import imageio
```

The previous code snippet imports the essential libraries and modules random and numpy for numerical operations, torch for deep learning, diffusers for utilizing diffusion models (including schedulers and pipelines for text-to-video synthesis), compel for enhancing or comparing multimedia content, and imageio for reading and writing images and videos.

```python
def save_frames_to_video(file_path: str, frames_array: np.ndarray, frames_per_second: int) -> str:
    video_writer = imageio.get_writer(file_path, format='FFMPEG', fps=frames_per_second)
    for frame_index in range(frames_array.shape[0]):
        current_frame = frames_array[frame_index]
        if current_frame.dtype != np.uint8:
            current_frame = (current_frame * 255).astype(np.uint8)
        video_writer.append_data(current_frame)
    video_writer.close()
    return file_path
```

In the previous code we define the function save_frames_to_video, which takes a file path, an array of video frames, and a frame rate (frames per second) as input and saves these frames as a video file at the specified location. The function initializes a video writer object from the imageio library, specifying the output file path, the video format as 'FFMPEG', and the frame rate. It then iterates over each frame in the input array, converting the frame to the appropriate data type if necessary, and appends it to the video file. After all frames are processed and appended, the video writer is closed, finalizing the video file creation. The function then returns the file path of the saved video, indicating the location where the video has been stored.

```
%cd /content/drive/MyDrive/Text-To-Video-Finetuning
```

```python
trained_model_directory = "/content/drive/MyDrive/Text-To-Video-Finetuning/outputs/train_2024-02-18T05-32-38/checkpoint-7500"
text_to_video_pipeline = TextToVideoSDPipeline.from_pretrained(trained_model_directory, torch_dtype=torch.float16, variant="fp16", low_cpu_mem_usage=False)
```

CHAPTER 3   FROM SCRIPT TO SCREEN: UNVEILING TEXT-TO-VIDEO GENERATION

```
content_enhancer = Compel(tokenizer=text_to_video_pipeline.tokenizer, text_
encoder=text_to_video_pipeline.text_encoder)
text_to_video_pipeline.scheduler = DPMSolverMultistepScheduler.from_
config(text_to_video_pipeline.scheduler.config)
text_to_video_pipeline.enable_model_cpu_offload()
```

In the previous code, we set up the process for generating video content from text using a pre-trained model. Initially, the working directory is changed to /content/drive/MyDrive/Text-To-Video-Finetuning, ensuring that all operations are executed in the context of the project's directory on Google Drive. A path to the trained model directory is defined, pointing to a specific checkpoint. This path is used to initialize a TextToVideoSDPipeline, specifying the use of half-precision floating-point numbers (FP16) for tensor operations to reduce memory consumption and potentially speed up computation. Additionally, a Compel content enhancer is instantiated with the pipeline's tokenizer and text encoder, aiming to refine the generated content further. The pipeline's scheduler is set to a DPMSolverMultistepScheduler based on its configuration, optimizing the diffusion process for video generation. Lastly, the pipeline is configured to offload model parameters to CPU memory when not in use, reducing GPU memory usage and facilitating the handling of larger models or datasets.

```
random_seed = random.randint(0, 1000000)
random_generator = torch.Generator().manual_seed(random_seed)

video_prompt = "a close up of a pink flower in a vase against a yellow
background"
avoidance_prompt = "text, blurry, ugly, username, url, low resolution, low
quality"
positive_prompt_embeddings = content_enhancer.build_conditioning_
tensor(video_prompt)
negative_prompt_embeddings = content_enhancer.build_conditioning_
tensor(avoidance_prompt)
generated_frames = text_to_video_pipeline(prompt_embeds=positive_
prompt_embeddings, negative_prompt_embeds=negative_prompt_embeddings,
num_frames=100, width=384, height=256, num_inference_steps=50, guidance_
scale=9, generator=random_generator).frames
output_filename = f"/content/{video_prompt.replace(' ', '-')}-seed{random_
seed}.mp4"
```

```
final_video_frames = generated_frames[0]
generated_video_path = save_frames_to_video(output_filename, final_video_
frames, 24)
print(f"Generated video saved to {generated_video_path}")
```

The previous code demonstrates the process of generating a video from a text description using a pre-trained text-to-video model. It begins by setting a random seed to ensure reproducibility, creating a deterministic environment for video generation. The script defines two prompts: a `video_prompt` for the desired scene ("a close up of a pink flower in a vase against a yellow background") and an `avoidance_prompt` listing elements to exclude from the video (e.g., text, blurry images). These prompts are then converted into embeddings using a content enhancement tool, which prepares them for the model.

Using the text-to-video pipeline, the script generates frames based on these embeddings, specifying the number of frames, resolution, number of inference steps, and a guidance scale to control the influence of the positive and negative prompts. The output is a sequence of frames, which are saved as a video file with a filename reflecting the original prompt and the seed used for generation. The video is saved to a specified path, and the script concludes by printing the location of the generated video. This process illustrates the integration of content specification and avoidance in creative video generation, leveraging advanced machine-learning techniques to produce customized visual content.

Figure 3-5 offers a visual confirmation that the adjustments made have successfully resulted in a more detailed close-up, aligning perfectly with the prompt's specifications. However, it's evident that there's still a pathway open for enhancing the overall quality of the videos produced. Because of the constraints on computational resources, the training of the model was limited to 7,500 steps. For achieving even better results, it's advisable to extend the training over more steps and utilize a broader dataset.

Figure 3-5. *Screenshot from video of fine-tuned model*

Before we conclude our journey through fine-tuning, it's essential to discuss the valuation metrics critical for assessing the performance of text-to-video models. Common metrics include the Structural Similarity Index (SSIM) and Peak Signal-to-Noise Ratio (PSNR) for video quality, along with content relevance scores to measure how well the generated videos align with textual prompts. Comparing these metrics against established benchmarks allows us to gauge the model's improvement and its alignment with industry standards or specific project goals. This evaluation phase not only validates the efficacy of our fine-tuning efforts but also identifies potential areas for further refinement, ensuring our model achieves both high-quality and contextually accurate video outputs.

This brings us to the end of our fine-tuning section, through which we've not just acquired the skills to generate captions for videos but also ventured into the intricacies of fine-tuning algorithms within the realm of text-to-video machine learning. This journey underscores the potential for improvement and the importance of resource availability in pushing the boundaries of what these sophisticated models can achieve.

Conclusion

In this chapter, we went on a comprehensive journey through the intricate process of transforming textual data into vivid video content. Together, we explored the foundational aspects of video data, delved into the unique challenges it presents, and celebrated the potent synergy between video and textual data. Through hands-on demonstrations and step-by-step guides, we not only witnessed the capabilities of pre-trained models but also ventured into the realm of fine-tuning these models to cater to our custom applications.

Our exploration has equipped you with the knowledge and skills necessary to navigate the complexities of video data processing and the art of text-to-video generation. From the initial steps of installing essential libraries to the critical phases of data loading, preprocessing, and model training, we've covered substantial ground. The journey through fine-tuning has illuminated the path toward tailoring pre-trained models to our specific needs, allowing us to push the boundaries of creativity and innovation in video content creation.

As we close this chapter on text-to-video generation, it's pivotal to look forward to the possibilities that lie ahead. The field of text-to-video technology is on the cusp of remarkable advancements, driven by relentless innovation and research. Emerging areas such as hyper-realistic video synthesis, AI-driven storytelling, and interactive video content creation are set to redefine the boundaries of digital media. Anticipated breakthroughs in deep learning algorithms and computational power promise to further enhance the fidelity and efficiency of video generation from text, opening up unprecedented opportunities for creators and technologists alike. The integration of augmented reality and virtual reality with text-to-video synthesis also looms on the horizon, offering immersive experiences that blur the line between the virtual and the real. As we venture into these new realms, the potential for groundbreaking applications in entertainment, education, and beyond is boundless. Embracing these advancements, we stand on the brink of a new era in digital storytelling, where the written word seamlessly transforms into visual narratives that captivate and inspire.

As we prepare to transition to the next chapter, where we'll explore the fascinating world of text-to-audio and audio-to-text generation, let's carry forward the insights and experiences gained from this chapter. The skills we've developed in manipulating video and textual data serve as a foundation upon which we'll build as we continue to explore the broader landscape of media generation and transformation. Our adventure into the realms of audio and text promises to be equally enlightening, offering new perspectives and opportunities for growth. Together, we'll continue to unlock the potential of AI in transforming the way we create, interpret, and interact with various forms of media.

CHAPTER 4

Bridging Text and Audio in Generative AI

Brief History

We live in an era where our spoken words are seamlessly converted into text, and language barriers are easily surmounted with the aid of technology. This is no longer a scenario confined to the realm of science fiction but a reality made possible by the remarkable advancements in text-to-audio and audio-to-text generation.

Let's take a closer look at some of the everyday magic these technologies bring into our lives. Have you ever enjoyed an audiobook? That's text-to-audio generation at work, transforming written words into engaging narratives. Or think about the last time you asked a question to a voice assistant on your phone or smart speaker. This seamless interaction is powered by audio-to-text technology, interpreting your spoken words into digital text.

For those with visual impairments, text-to-audio conversion has opened up a world of information that was previously inaccessible. Similarly, audio-to-text technologies assist in transcribing lectures for students, making meetings accessible through real-time captioning and even breaking down language barriers with instant translation services.

The evolution of audio and text processing technologies is marked by significant milestones. From the early days of mechanical speech synthesis in the 1930s to the first computer-based speech synthesis in the 1960s, each step brought us closer to replicating the nuances of human speech. The advent of digital signal processing in the 1970s and 1980s further advanced speech synthesis, allowing for more natural-sounding voices.

In parallel, the development of speech recognition technologies made significant strides. The shift from simple command recognition to more complex natural language processing allowed computers not just to hear but to understand and respond to human speech. As we delve further into this chapter, our next step will be to understand the complexities of audio data. This exploration is crucial as it forms the backbone of today's sophisticated text-to-audio and audio-to-text systems. From there, we will navigate the fascinating world of machine learning models that make text-to-audio and audio-to-text translation possible, focusing mainly on transformer-based models. Additionally, we'll explore the art of fine-tuning these models to achieve precision and efficiency. Prepare to immerse yourself in a technological adventure where we decode how machines understand and replicate human language, transforming how we interact with the world around us, one word at a time. See Figure 4-1.

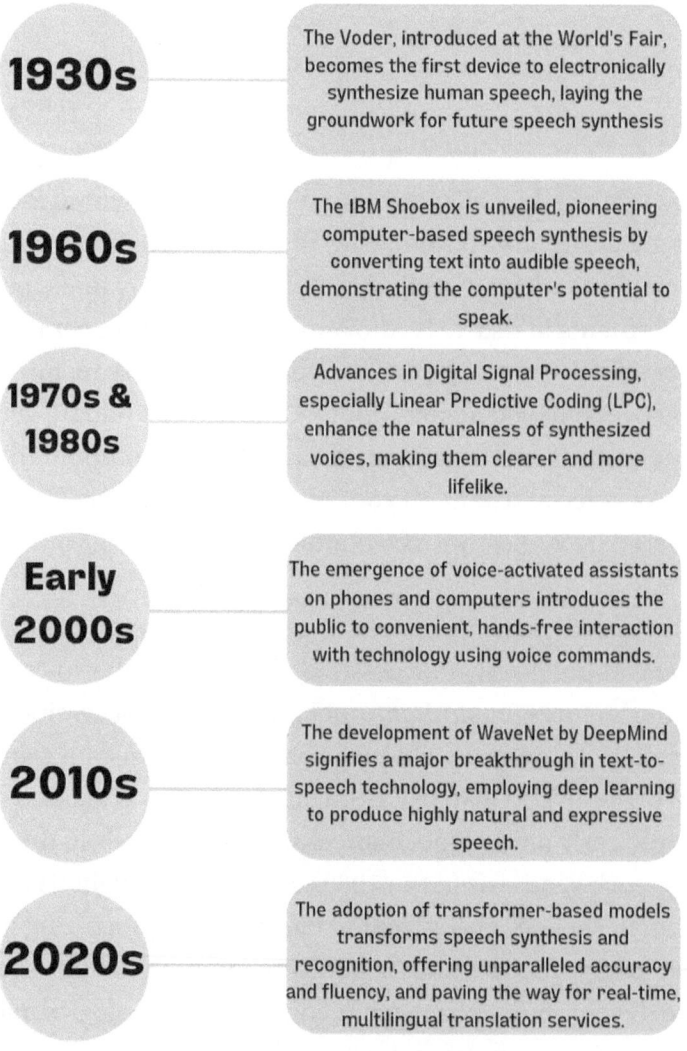

Figure 4-1. *Evolution of audio generation technologies*

Fundamentals and Challenges

In this section we'll explore the key aspects of audio signals, from their digital representation to the hurdles faced in effectively capturing, storing, and extracting meaningful information from diverse audio environments. This exploration is essential for anyone looking to navigate the nuanced landscape of audio data and leverage its potential in various applications.

Understanding Audio Data

Sound waves exist as a continuous and uninterrupted flow, constantly carrying a vast array of signal values at any moment. This inherent continuity presents a challenge for digital devices, which are inherently designed to process and manage finite, quantifiable datasets. Sound waves must transform to bridge this gap between the analog and digital realms. They must be converted from their natural, continuous state into a distinct sequence of values. This process results in what we know as the digital representation of sound, making it suitable for digital processing, storage, and transmission.

You will often encounter digital files encapsulating various sound segments in audio datasets. These can range from spoken text to musical compositions and are stored in different file formats such as `.wav` (Waveform Audio File), `.flac` (Free Lossless Audio Codec), or `.mp3` (MPEG-1 Audio Layer 3). Each format has its unique way of compressing the digital audio signal. The journey from an analog sound to a digital file begins with its capture via a microphone, which transforms sound waves into an electrical signal. Following this, an analog-to-digital converter takes over, digitizing the electrical signal through a process called *sampling*.

Sampling (Figure 4-2) meticulously breaks down the continuous sound wave into a series of discrete, digital instances, effectively creating audio snapshots in a format that digital devices can understand and manipulate. The sampling rate, measured in Hz, determines how often these measurements occur. In Generative AI models tasked with synthesizing speech, a higher sampling rate allows for the creation of audio that captures the nuanced intonations of human speech more accurately. This is crucial in applications like virtual assistants, where clarity and naturalness are key.

CHAPTER 4   BRIDGING TEXT AND AUDIO IN GENERATIVE AI

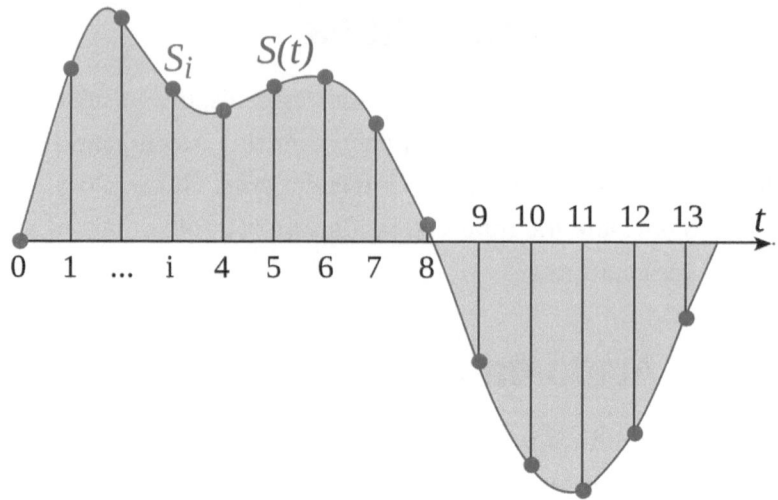

Figure 4-2. *Signal sampling representation (source: https://en.wikipedia.org/wiki/Sampling_(signal_processing))*

The sampling rate indicates the frequency of taking samples, but it's the amplitude in each sample that conveys the sound's volume and intensity. Amplitude, measured in decibels (dB), reflects sound pressure levels, with higher dB indicating louder sounds. In digital audio, the amplitude at each sample point is captured, and the bit depth determines the precision of this measurement. Higher bit depths like 16-bit or 24-bit offer finer gradations of sound intensity, leading to a more accurate digital representation. In Generative AI, using a higher bit depth can dramatically improve the realism of generated sounds, such as the richness of musical notes or the clarity of synthesized speech, by capturing subtle differences in sound intensity more precisely.

Amplitude, which reflects the loudness of sound at any moment, is visually represented in a waveform. A waveform (Figure 4-3) can show the structure of the sound, like the timing of notes or speech. By examining a waveform, one can identify specific characteristics and anomalies in the audio, such as loudness, duration, and noise. Software like librosa in Python can generate waveforms, helping to analyze and understand audio data visually. Waveforms are essential for audio processing and analysis, providing a clear visual understanding of the sound's properties. In the context of AI-generated music, analyzing the waveform of existing tracks helps AI understand and replicate the dynamic range and emotional intensity of music, leading to creations that resonate more deeply with listeners.

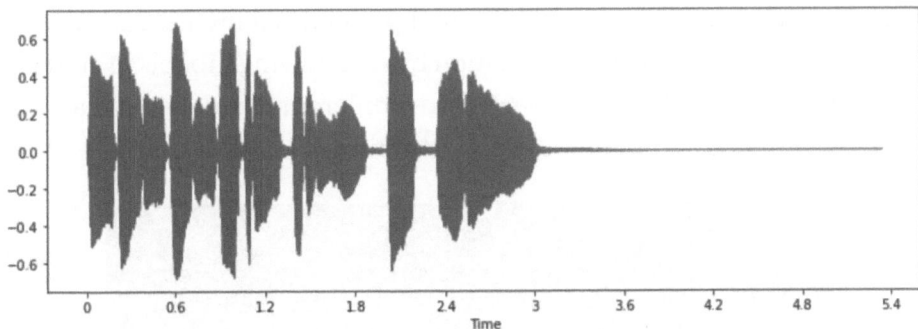

Figure 4-3. *Plot of a waveform*

Moving beyond waveforms, the frequency spectrum offers another perspective on audio data. Using the discrete Fourier transform (DFT), it breaks down an audio signal into its constituent frequencies and their strengths. This frequency domain representation is especially useful for analyzing the composition of sounds, such as identifying the harmonics in musical notes. Generative AI models, when trained on the frequency spectrum of various musical genres, can discern and generate distinct musical styles by recognizing and applying unique frequency patterns, enhancing the diversity of AI-generated music. See Figure 4-4.

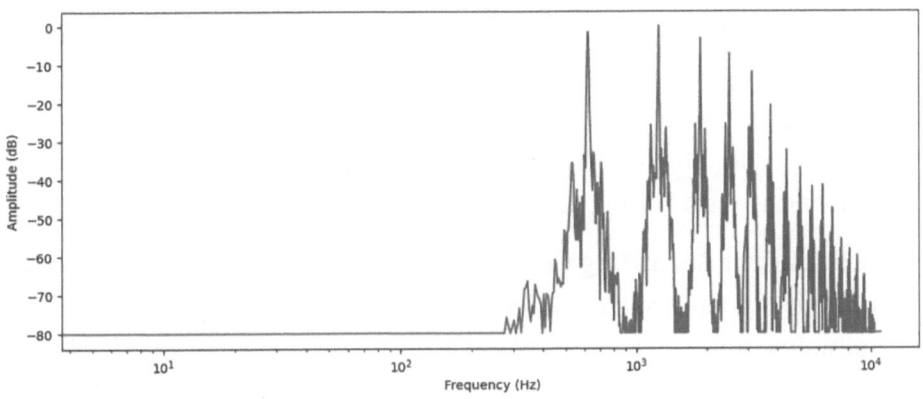

Figure 4-4. *Plot of frequency spectrum*

Then there's the spectrogram (Figure 4-5), a more dynamic tool that illustrates how frequencies in an audio signal change over time. It is created by applying multiple DFTs to successive time segments of the audio, resulting in a visual representation that combines time, frequency, and amplitude. Spectrograms are invaluable in various applications, from music analysis, which reveals the interplay of instruments, to speech processing, where they help distinguish different phonetic elements. Spectrograms play a vital role in speech synthesis and recognition by AI, allowing models to capture and reproduce the intricacies of human speech across different languages and accents by visualizing how phonetic components evolve over time.

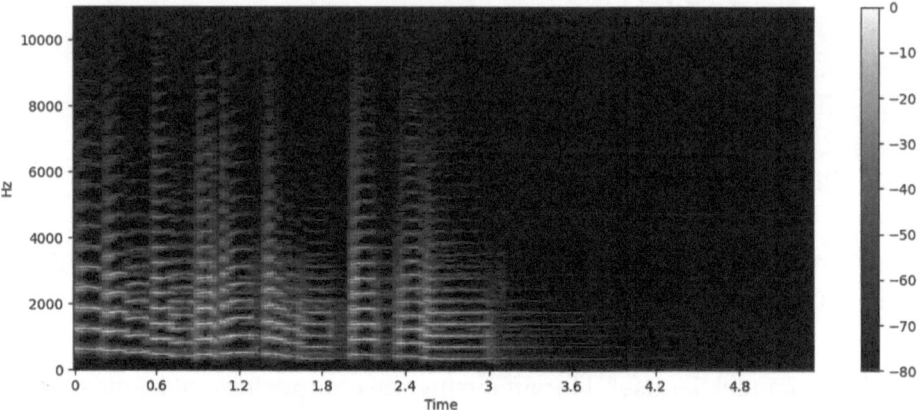

Figure 4-5. Plot of spectrogram

Challenges in Working with Audio Data

Transitioning from the technical foundations of audio processing, it's crucial to address the inherent challenges of working with audio data. While tools like spectrograms offer profound insights into the structure and characteristics of audio signals, translating these insights into actionable information is fraught with obstacles. The following are challenges of working with audio data:

- **Variability in Recording Conditions:** Different microphones, spaces, and settings can result in varied audio qualities, making it hard to maintain a uniform standard for data.

- **Background Noise and Interference:** Audio often contains unwanted sounds, such as traffic noise or people talking, which can mask the primary signal.

- **High Dimensionality:** Audio data often has high dimensionality, meaning it contains a large amount of information per unit of time, leading to significant computational demands.

- **Temporal Dependencies:** Audio signals have strong temporal dependencies, making it challenging to effectively capture and analyze the context and sequence of sounds.

- **Speech Variabilities:** In speech data, variations in accents, dialects, speaking styles, and speed add complexity to tasks like speech recognition and natural language processing.

- **Nonlinearities and Echoes:** Echoes and other nonlinear audio characteristics can further complicate signal processing and recognition tasks.

- **Data Labeling and Annotation:** Manually labeling audio data for training machine learning models can be time-consuming and requires expertise.

Mitigating Challenges in Audio Data Processing

Overcoming the inherent challenges of working with audio data requires a combination of advanced technologies, innovative methodologies, and practical strategies. Here's how some of these challenges can be addressed:

- **Adapting to Variability in Recording Conditions:** Employing audio enhancement techniques like normalization and equalization can help standardize audio levels and qualities across different recordings.

- **Reducing Background Noise and Interference:** Techniques such as noise reduction algorithms and machine learning models trained on identifying and isolating the primary audio signal can significantly minimize unwanted sounds.

- **Managing High Dimensionality:** Dimensionality reduction techniques, like Principal Component Analysis (PCA) or autoencoders in deep learning, can effectively reduce computational demands without losing critical information.

- **Handling Temporal Dependencies:** Recurrent neural networks (RNNs), especially long short-term memory (LSTM) networks, are designed to capture temporal dependencies and sequences in audio data, making them ideal for tasks like speech recognition.

- **Accommodating Speech Variabilities:** Utilizing a diverse training dataset that includes a wide range of accents, dialects, and speaking styles can enhance the robustness of speech recognition systems.

- **Addressing nonlinearities and echoes:** Advanced signal processing techniques, such as adaptive filters, can be employed to mitigate the effects of echoes and non-linear audio characteristics.

- **Efficient Data Labeling and Annotation:** Semi-supervised learning approaches and crowdsourcing can both expedite the labeling process while ensuring the quality of data annotation.

By adopting these solutions, researchers and developers can effectively navigate the complexities of audio data processing, enabling the development of more accurate, efficient, and robust Generative AI systems for audio applications.

Bridging Text and Audio: The CLAP Model Implementation

In this section of the chapter, we venture into the cutting-edge realm of connecting the dots between text and audio data, leveraging the capabilities of the Contrastive Language-Audio Pretraining (CLAP) model. This breakthrough approach stands at the forefront of audio-text alignment technologies, enabling an unprecedented correlation between spoken words and their textual counterparts.

The CLAP model represents a significant advancement in machine learning, employing contrastive learning techniques to effectively "learn" the relationship between audio samples and related text descriptions. Doing so opens up many possibilities, from enhancing automated transcription services to improving voice-activated search functionalities and beyond. This model's capacity to understand and interpret the nuances of spoken language in relation to its textual form is a game-changer for developers, researchers, and end users alike.

CHAPTER 4   BRIDGING TEXT AND AUDIO IN GENERATIVE AI

Let's understand the infrastructure for the CLAP model (Figure 4-6) before we delve deeper into its implementation. CLAP is a sophisticated neural network model that has been meticulously trained across a diverse array of (audio and text) pairs. This model is uniquely designed to proficiently predict the most appropriate text snippet corresponding to a given audio input, achieving this without being directly fine-tuned for any specific task. At the heart of the CLAP model lies a SWIN Transformer, tasked with extracting audio features from log-Mel spectrogram inputs, while a RoBERTa model is employed to derive text features. These extracted features from both modalities are subsequently projected into a shared latent space of the same dimensionality, facilitating the alignment of audio and text representations. The similarity between an audio input and a text snippet is quantified through the dot product of their projected features in this latent space, measuring their relevance to each other.

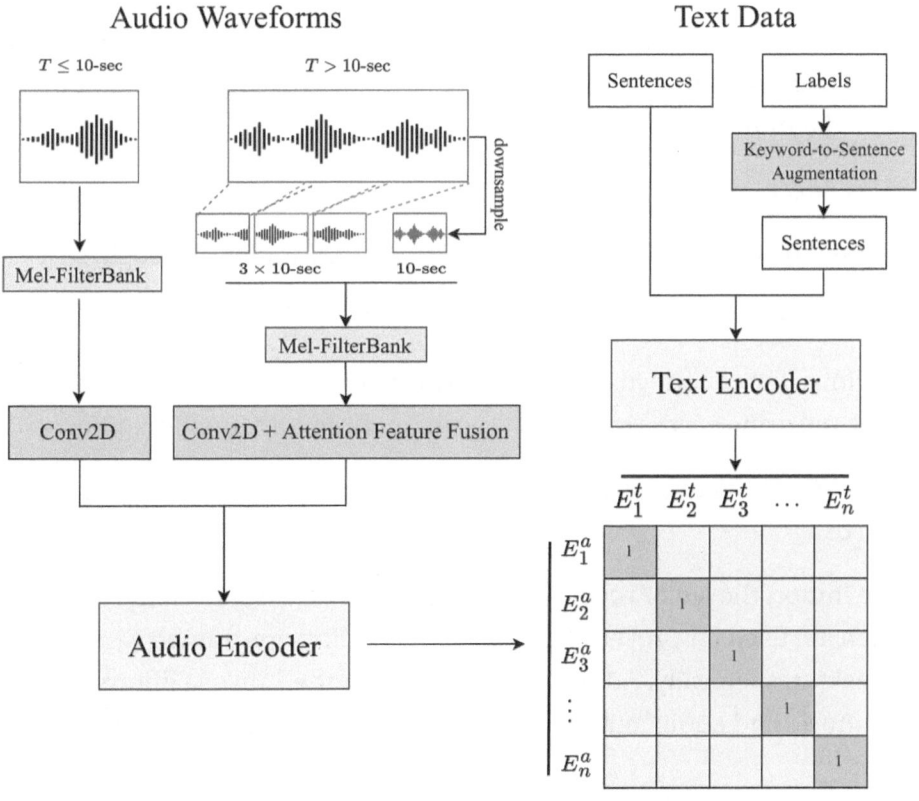

Figure 4-6. CLAP architecture[7] (source: `https://github.com/LAION-AI/CLAP?tab=readme-ov-file`)

Let's implement the CLAP model via the HuggingFace API in a Google Colab Notebook, illustrating the model's capacity to forge meaningful connections between audio and textual data.

Step 1: Installing Libraries and Data Loading

We'll begin by loading our dataset, focusing on an audio recording of a cow mooing. To ensure compatibility with the CLAP model, we'll adjust the audio's sampling rate through preprocessing.

```
from google.colab import drive
import os
drive.mount('/content/drive')
```

In the previous code block, we are mounting Google Drive to the Colab environment. It imports the necessary library to interact with Google Drive and then mounts the drive at the specified mount point (/content/drive).

```
os.chdir("/content/drive/My Drive/Colab Notebooks")
```

Then we set the working directory to the appropriate folder in Google Drive.

```
!pip install transformers
```

The previous line of code installs the latest version of the `transformers` library in the environment, ensuring access to its most recent features and functionalities.

```
from transformers import AutoProcessor, ClapModel
import librosa
```

Then we import the `AutoProcessor` and `ClapModel` classes from the `transformers` library, which are used for processing inputs and loading pre-trained CLAP models for audio-text tasks, respectively. Additionally, we import the `librosa` library, a Python package for music and audio analysis, enabling the loading, analysis, and manipulation of audio files.

```
wav_file_path = 'cow_moo.wav'
loaded_audio, _ = librosa.load(wav_file_path, sr=48000)
```

CHAPTER 4 BRIDGING TEXT AND AUDIO IN GENERATIVE AI

We imported our cow mooing file, and this will be our audio sample for this task. The previous line of code uses the `librosa.load` function to read an audio file named `cow_moo.wav` from Google Drive into memory, converting it to a digital audio signal. The file is resampled at a sampling rate of 48,000 Hz, the sampling rate at which CLAP is trained. The `librosa.load` function returns two values: the audio data as a numpy array (`loaded_audio`) and the sampling rate used for the audio file.

Step 2: Model Inference

In this part, we'll import a pre-trained model, prepare our input according to the model's requirements, and then delve into interpreting the model's output.

```
clap_model =ClapModel.from_pretrained("laion/clap-htsat-unfused")
clap_processor = AutoProcessor.from_pretrained("laion/clap-htsat-unfused")
```

`clap_model` loads a pre-trained CLAP model named `laion/clap-htsat-unfused` from Hugging Face's model hub. `clap_processor` loads the processor for the CLAP model, also named `laion/clap-htsat-unfused`. The processor is responsible for preparing both the audio and text data in the format expected by the model. This includes tasks such as tokenizing text, normalizing audio data, and converting both types of data into tensors that can be fed into the model for training or inference.

```
audio_descriptions = ["Sound of a cow", "Sound of a human", "Sound of a cat"]
```

Now, we create a list named `audio_descriptions` containing string elements. Each string describes a distinct sound: a cow mooing, a human speaking or making noise and a cat meowing. We will be comparing these inputs with our audio sample using CLAP.

```
processed_inputs = clap_processor(text=audio_descriptions, audios=loaded_audio, return_tensors="pt", padding=True, sampling_rate=48000)
```

In the previous line of code, we prepare the input data for the CLAP model by processing both the audio and text data. The text parameter is given a list of audio descriptions, and the `audios` parameter receives the loaded audio data. It specifies the return of the processed data as PyTorch tensors (`return_tensors="pt"`), applies padding to ensure uniform input size, and sets the audio sampling rate to 48,000 Hz. The result `processed_inputs` contains the processed audio and text data ready for model input.

CHAPTER 4   BRIDGING TEXT AND AUDIO IN GENERATIVE AI

```
model_predictions = clap_model(**processed_inputs)
similarity_scores = model_predictions.logits_per_audio
probability_scores = similarity_scores.softmax(dim=-1)
```

`model_predictions = clap_model(**processed_inputs)` sends the processed input data (which includes both the audio file and the text descriptions) to the CLAP model for prediction. The `**processed_inputs` syntax unpacks the dictionary of processed inputs (audio and text) so they can be accepted by the model as arguments.

`similarity_scores = model_predictions.logits_per_audio` extracts the logits associated with each audio-text pair from the model's predictions. These logits represent the model's raw scores before any normalization like `softmax` is applied, indicating the similarity between the provided audio and each text description. `probability_scores = similarity_scores.softmax(dim=-1)` applies the `softmax` function to the similarity scores. This normalization step converts the raw scores into probabilities, making it easier to interpret how likely each text description is to match the audio input based on the model's assessment. The `dim=-1` parameter ensures that the `softmax` is applied across the dimension representing the different text descriptions, resulting in a probability distribution for each audio-text pair comparison.

```
probabilities = probability_scores.detach().numpy()[0]
for audio_description, probability in zip(audio_descriptions,
probabilities):
    print(f"{audio_description} - {probability*100:.0f}%")
```

In the previous block of code, we convert the model's probability scores from PyTorch tensors to a NumPy array and retrieve the first set of probabilities. We then iterate over each audio description and its corresponding probability, printing them in a human-readable format where each label is paired with its probability percentage. The `detach()` method is used to remove the probabilities from the computation graph, allowing them to be converted into a NumPy array without keeping track of the operations that led to their calculation.

Our output is as follows:

```
Sound of a cow - 99%
Sound of a human - 1%
Sound of a cat - 0%
```

The output clearly indicates that the CLAP model has accurately matched the audio of a cow mooing to the "Sound of a cow," validating its efficacy in correlating audio with relevant text descriptions. This success underscores our learning on bridging audio and textual data, showcasing the model's potential in precise audio identification. As we progress, we'll delve into the realm of transformer-based models, focusing on converting text to audio and vice versa, further expanding our toolkit for multimodal data interaction.

Understanding AI-Driven Text and Audio Conversion Models

In this section, we explore the transformative technologies that underpin audio-to-text and text-to-audio conversions, starting with an insightful overview of Connectionist Temporal Classification (CTC) models. These models are pivotal for translating audio signals into textual representations, enabling the transcription of spoken language into written form. Following this, we'll delve into the intricacies of sequence-to-sequence (seq2seq) architectures, examining their pivotal role in audio-to-text conversion and the nuanced process of generating spoken language from text. As part of our exploration, we will implement pre-trained models from the HuggingFace API, showcasing practical applications for text-to-audio and audio-to-text transformations. Our journey through these implementations will provide a concrete foundation for understanding multimodal communication, offering insights into the practical integration of advanced machine-learning techniques in real-world scenarios.

Understanding CTC Architectures

Connectionist Temporal Classification (CTC) is a method utilized in automatic speech recognition by encoder-only transformer models like Wav2Vec2, HuBERT, and M-CTC-T. These models, focusing solely on the encoder part of the transformer architecture (Figure 4-7), process the audio waveform input into a series of hidden states or output embeddings. By employing an additional linear transformation on these hidden states, CTC models predict class labels, which correspond to the alphabet's characters. This approach allows for predicting any word in the target language with a compact classification layer, requiring only a minimal set of characters and special tokens for the vocabulary.

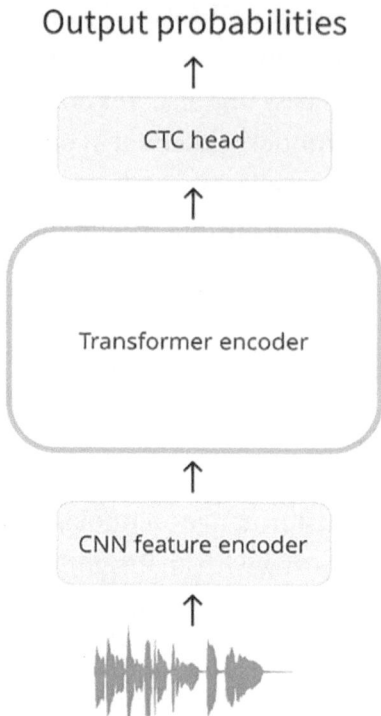

Figure 4-7. Flow for CTC architectures (https://huggingface.co/learn/audio-course/chapter3/ctc)

Unlike NLP models, which use a vast vocabulary comprising tokens that represent parts of words or entire words, CTC models operate with a significantly smaller vocabulary, typically less than 50 characters. This limited vocabulary, which ignores letter casing and includes a few special tokens like a space for word separation and a padding token for silence, is optimized for speech recognition's unique challenges, particularly the unknown alignment between audio inputs and text outputs. Integrating CTC into a transformer encoder model involves a linear layer that maps the encoder's output to a vocabulary, refined with CTC loss for accurate prediction. However, CTC's focus on characters over words can lead to phonetically correct but misspelled outputs, a drawback mitigated by supplementing it with an external language model for spellchecking.

A key feature of CTC models is their elegant solution to the alignment problem in speech recognition, a challenge stemming from the lack of direct correspondence between segments of audio (audio frames) and the characters or words they represent. Let's go through the following example:

Input Phrase: "Hello AI"
CTC Model Processing Example: H_E_LL_OO_||A_I_I
Detailed Breakdown of the CTC Process:

1. **Audio Encoding:** The audio input, such as Hello AI, is encoded by a transformer encoder into a series of hidden states, capturing the nuances of the spoken words.

2. **CTC's Strategic Decoding:** A subsequent linear transformation applied to these hidden states assigns probabilities to each character in the model's vocabulary, including a unique blank token (_) that plays a pivotal role in sequence alignment.

3. **Alignment and Deduplication:**

    a. The blank token (_) is ingeniously utilized to separate and manage character occurrences, effectively dealing with variations in speech rate and pronunciation.

    b. Through CTC's logic, the initial complex sequence H_E_LL_OO_||A_I_I is streamlined, removing duplicates and blanks, to yield a coherent and accurate textual representation: Hello AI.

4. **Final Output Generation:** The refined process culminates in the model outputting "Hello AI" and demonstrating CTC's capability to transform raw audio transcriptions into clear, intelligible text.

This illustration not only showcases CTC's proficiency in handling the inherent alignment problem but also underscores its broader application in enhancing speech recognition systems' clarity and accuracy. By bridging the divide between audio inputs and textual outputs, CTC models play an indispensable role in advancing the field of speech recognition, making them a cornerstone of modern audio processing technologies.

Models like Wav2Vec2, HuBERT, and M-CTC-T all utilize a transformer encoder with a CTC head, varying mainly in input processing and training goals. Wav2Vec2 processes raw audio, whereas M-CTC-T uses mel spectrograms and includes a diverse vocabulary for multilingual recognition. Both Wav2Vec2 and HuBERT share architecture but differ in training approaches, showcasing the adaptability of CTC models to various speech recognition tasks.

CHAPTER 4 BRIDGING TEXT AND AUDIO IN GENERATIVE AI

Understanding Seq2Seq Architectures

We explored how CTC architecture-based models leverage only the encoder component of the transformer architecture, making a prediction for each input segment, which results in input and output sequences of identical length. For example, in models like Wav2Vec2, the audio input is downsampled, yet there remains a direct prediction for every 20 ms sound segment. However, when we integrate both the encoder and decoder to form what's known as a sequence-to-sequence (seq2seq) model, we enable the transformation of sequences from one form to another without the constraint of matching sequence lengths. This flexibility makes seq2seq models particularly powerful for a wide range of tasks in natural language processing (NLP), such as summarizing texts or translating between languages, and extends their utility to audio processing tasks like speech recognition.

Delving into the seq2seq models for audio-to-text conversion reveals a sophisticated two-part process. Initially, these models employ a transformer encoder that intricately processes log-mel spectrograms, extracting vital features embedded in spoken language into a series of detailed hidden states. This meticulous encoding not only captures the essence of the audio's content but also facilitates the direct interpretation of audio waveforms, laying a robust foundation for accurate transcription.

Following the encoding stage, the journey continues with the transformer decoder. This component employs cross-attention mechanisms to sequentially produce text tokens, carefully ensuring that predictions for each token are influenced solely by preceding elements, without any foresight into future tokens. This characteristic ability of seq2seq models to generate contextually nuanced transcriptions markedly surpasses CTC models, which may struggle with contextual depth. By incorporating a dynamic language model within the decoder, seq2seq architectures achieve a harmonious balance between precision and contextual relevance, creating a streamlined path from spoken language to written text.

In contrast to the encoder-focused approach of CTC architectures, which excel in generating real-time transcriptions by mapping each audio segment to a character, seq2seq models thrive in scenarios demanding a higher level of comprehension and contextual integration, such as translating speech or generating audio from text. While CTC models offer speed and efficiency in applications where immediate output is crucial, they may falter in capturing the full context or nuances of speech. Seq2seq models, with their encoder-decoder setup, provide a more comprehensive understanding, making them better suited for tasks requiring detailed interpretation

CHAPTER 4 BRIDGING TEXT AND AUDIO IN GENERATIVE AI

and synthesis of language, albeit at the cost of increased computational complexity. This delineation between the models highlights the importance of choosing the appropriate architecture based on the specific requirements of the task at hand, whether it be the immediacy of real-time transcription or the depth of audio-to-text conversion. You might find it unsurprising that the mechanism for text-to-speech (TTS) in seq2seq models mirrors the process outlined for audio-to-text, albeit with input and output roles inverted. Here, a transformer encoder interprets a series of text tokens, creating hidden states that encapsulate the textual information. Then, the transformer decoder, leveraging cross-attention with the encoder's outputs, generates a spectrogram.

A spectrogram, essentially, is constructed by amalgamating the frequency spectrums of consecutive audio waveform slices, translating to a sequence of frequency spectra represented in log-mel for each time step. Unlike the automatic speech recognition (ASR) model that initiates decoding with a special "start" token, text-to-speech (TTS) begins with a blank spectrogram, effectively serving as the start signal. This initial step allows the decoder to progressively extend the spectrogram, one slice at a time, by interpreting the encoder's hidden states. See Figure 4-8.

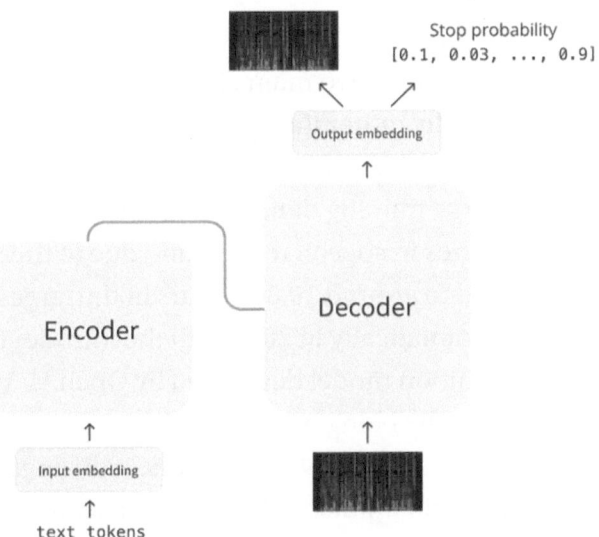

Figure 4-8. *Flow of Seq2Seq architectures (https://huggingface.co/learn/audio-course/chapter3/seq2seq)*

CHAPTER 4 BRIDGING TEXT AND AUDIO IN GENERATIVE AI

As we've gained insights into the foundational architectures and the intricacies of handling audio data, we're now poised to explore the practical aspects of AI-driven text and audio conversion models. Moving forward, we'll delve into the application of pre-trained models, as well as the fine-tuning processes. This will illustrate how you can customize these models to fit your unique requirements.

Implementation AI-Driven Text and Audio Conversion Models

In the forthcoming section, we embark on a practical journey to implement pre-trained and fine-tune cutting-edge AI-driven text and audio conversion technologies, namely, Whisper and SpeechT5. Through hands-on examples, we will guide you through the process of integrating these models into your projects, optimizing their performance, and tailoring their capabilities to meet specific needs.

Speech to Text

Seq2Seq models, while powerful, face two main challenges: their slower decoding speed, as they decode sequentially rather than in parallel, and their substantial appetite for data, necessitating vast amounts of training data for effective learning. These issues, particularly the demand for large training datasets, have historically impeded the progress of Seq2Seq architectures in speech recognition due to the scarcity of labeled speech data, which was limited to around 10,000 hours in the largest datasets available. This landscape transformed dramatically in 2022 with the introduction of Whisper, a groundbreaking speech recognition model developed by OpenAI. Whisper marked a departure from previous models by being pre-trained on an unprecedented scale of labeled audio-transcription data, totaling 680,000 hours. It distinguished itself by its adeptness at processing long audio inputs, resilience against background noise, and capability to generate cased and punctuated transcriptions, positioning it as an ideal solution for practical speech recognition applications.

Whisper is available in five different configurations catering to various requirements and data scopes, with the four smallest models trained on exclusively English or multilingual datasets and the largest being strictly multilingual.

Using Pre-trained Model for STT

Now, we will commence a hands-on exploration, utilizing the advanced features of the pre-trained OpenAI Whisper model, particularly the `whisper-small`[8] variant, for converting speech to text. Our effort will concentrate on applying this sophisticated speech-to-text model to voice data from the Common Voice 13 dataset, particularly targeting the Hill Mari language. Subsequently, we will employ a stringent validation metric to evaluate the precision and dependability of the converted text outputs.

Step 1: Installing Libraries and Data Loading

Let's proceed by importing Hill Mari audio files from the Common Voice 13 dataset[9]. For the purposes of this section, we will specifically focus on retrieving files located within the test folder.

```
!pip install jiwer
!pip install datasets
!pip install transformers
```

In the previous lines of code, we are going to install the necessary libraries for working with speech-to-text models, evaluating their performance, handling datasets, and importing pre-trained models.

```
from jiwer import wer
import torch
from transformers import pipeline
from datasets import load_dataset, DatasetDict
import re
```

Now we will import specific functions and modules from different Python libraries, each serving a unique purpose in data processing and machine learning tasks. The `from jiwer import wer` line imports the word error rate (WER) calculation function from the `jiwer` library, which is used to evaluate the accuracy of speech recognition systems. `import torch` brings in PyTorch, a deep learning framework that enables building and training neural networks. `from transformers import pipeline` imports the `pipeline` function from the `transformers` library, a tool for easily deploying pre-trained models for natural language processing tasks. Lastly, `from datasets import load_dataset,`

DatasetDict imports functions from the datasets library for loading and managing datasets, and import re imports Python's built-in module for regular expressions, useful for text processing and manipulation.

```
from huggingface_hub import notebook_login
notebook_login()
```

The code enables the user to log in to the Hugging Face Hub from a notebook by prompting for an API token through notebook_login, allowing access to private resources and the ability to upload content.

```
common_voice = DatasetDict()
common_voice_test_subset = load_dataset("mozilla-foundation/common_voice_13_0", "mrj", split="test")
common_voice["test"] = common_voice_test_subset
print(common_voice)
```

Let's go through the previous code step-by-step.

- common_voice = DatasetDict(): This line initializes an empty DatasetDict, a Hugging Face datasets library class designed to store and manage multiple subsets of a dataset (like training, validation, and test splits) in a dictionary-like structure.

- common_voice_test_subset = load_dataset("mozilla-foundation/common_voice_13_0", "mrj", split="test"): Here, the load_dataset function from the datasets library is called to load the test split of the Common Voice version 13.0 dataset for the Hill Mari language (denoted by "mrj"). This function retrieves the specified subset of the dataset from the Hugging Face datasets repository.

- common_voice["test"] = common_voice_test_subset: This line assigns the loaded test subset to the common_voice DatasetDict under the key "test", effectively storing this subset within the common_voice dictionary for easy access.

- print(common_voice): Finally, printing common_voice displays the contents of the DatasetDict, including the test subset we've just loaded.

The loaded dataset features a test split encompassing 4,428 audio files.

Step 2: Model Inference

In this section, we will initially preprocess our data, refining the audio files for optimal input quality. Following this, we will input the processed data into the pre-trained Whisper small model for conversion into text. Finally, we'll assess the outcomes using the word error rate (WER) metric.

```
common_voice = common_voice.select_columns(["audio", "sentence"])
```

This line of code filters the common_voice dataset to keep only the audio and sentence columns, discarding any other data it may contain.

```
def normalize_text(text):
    text = text.lower()
    text = re.sub(r'[^\w\s]', '', text)
    return text
```

The normalize_text function is designed to preprocess text data by converting it to lowercase and removing any characters that are not letters, numbers, or whitespace. Here's a breakdown:

- text = text.lower(): This line converts all characters in the input text to lowercase, ensuring uniformity and reducing complexity for text analysis.

- text = re.sub(r'[^\w\s]', ", text): This line uses a regular expression to remove any characters that are not word characters (letters, digits, or underscores) or whitespace. The pattern [^\w\s] matches any character that is not a word character or whitespace, and re.sub replaces these matched characters with an empty string, effectively removing them.

This ensures the data is uniformly formatted, making it more suitable for comparison with predictions.

```
device = "cuda:0" if torch.cuda.is_available() else "cpu"
pipe = pipeline("automatic-speech-recognition", model="openai/whisper-small", device=device)
```

The code snippet dynamically assigns the computation device based on the availability of CUDA-enabled GPU support, defaulting to using the GPU ("cuda:0") if available or falling back to the CPU otherwise. It then initializes an `automatic-speech-recognition` pipeline from the Hugging Face transformers library, specifying the use of the `openai/whisper-small` model for speech-to-text tasks. This pipeline is set to run on the selected computation device, optimizing performance by leveraging GPU acceleration if possible, thus enhancing the efficiency of processing and converting speech to text.

Step 3: Model Evaluation

Before we advance, it's essential to familiarize ourselves with the word error rate, a crucial metric that we'll employ to validate the effectiveness of our speech-to-text conversion task. The WER stands as the de facto standard for assessing the performance of speech recognition systems. It quantifies the accuracy of transcribed text by calculating the number of substitutions, insertions, and deletions at the word level, effectively annotating errors on a word-by-word basis. This approach provides a clear and detailed measure of how well a speech recognition model can replicate exact spoken content in written form, making it an invaluable tool for our evaluation process. Here's an example:

Reference	the	dog	is	running	very	fast
Prediction	the	dog.	a	run		fast
Label	✓	✓	S	S	D	✓

If we compare this reference and prediction, then we have this:

> 2 substitution (S) ("is" and "running")
>
> 0 insertions (I)
>
> 1 deletion (D) ("very")

WER is calculated by summing all the different errors and dividing it by the total words in the sentence. We have a total of three errors this example.

> WER=(S+I+D)/N
>
> WER=(2+1+1)/6
>
> WER=0.5

Since WER calculates errors in its numerator, a lower WER indicates better performance. Now, let's return to our implementation to evaluate how the Whisper small model fares with the Hill Mari language.

```
total_wer = 0
for sample in common_voice["test"]:
    reference = sample["sentence"]
    reference = normalize_text(reference)
    hypothesis = pipe(sample["audio"].copy(), generate_kwargs={"task": "transcribe"})
    hypothesis = normalize_text(hypothesis['text'])
    total_wer += wer(reference, hypothesis)
```

The previous lines of code are executing a process to evaluate the WER of transcriptions produced by a speech-to-text model against reference transcriptions in the `common_voice["test"]` dataset. Here's a step-by-step explanation:

- `total_wer = 0`: Initializes a variable to accumulate the total word error rate across all samples in the test dataset.

- `for sample in common_voice["test"]`: Iterates over each sample in the test portion of the Common Voice dataset.

- `reference = sample["sentence"]`: Retrieves the reference transcription for the current audio sample.

- `reference = normalize_text(reference)`: Normalizes the reference transcription text by converting it to lowercase and removing nonalphanumeric characters, ensuring consistency for comparison.

- `hypothesis = pipe(sample["audio"].copy(), generate_kwargs={"task": "transcribe"})`: Passes the audio data through the speech-to-text pipeline (referred to as pipe) to generate a transcription. The `copy()` method ensures the original audio data is not modified, and `generate_kwargs={"task": "transcribe"}` specifies the task for the pipeline.

- `hypothesis = normalize_text(hypothesis['text'])`: Normalizes the generated transcription in the same way as the reference text.

CHAPTER 4   BRIDGING TEXT AND AUDIO IN GENERATIVE AI

- `total_wer += wer(reference, hypothesis)`: Calculates the word error rate for the current sample by comparing the normalized hypothesis (generated transcription) against the normalized reference transcription. This value is then added to the `total_wer` variable, accumulating the error rate across all samples.

This procedure effectively evaluates the performance of the speech-to-text model by quantifying the discrepancy between its transcriptions and the actual correct transcriptions.

```
total_length = len(common_voice['test'])
average_wer = total_wer / total_length
print(f"Average WER: {average_wer}")
```

In the previous lines of code calculate and display the average WER for the speech-to-text transcriptions performed by the model on the test set of the Common Voice dataset for the Hill Mari language. Here's a breakdown of what each line does:

- `total_length = len(common_voice['test'])`: This line calculates the total number of samples in the test subset of the Common Voice dataset and stores it in the variable `total_length`.

- `average_wer = total_wer / total_length`: Here, the total accumulated word error rate (`total_wer`) is divided by the total number of test samples (`total_length`) to calculate the average word error rate. This provides a measure of the model's performance across the entire test dataset.

- `print(f"Average WER: {average_wer}")`: This line prints out the average WER.

The explanation reveals that the average WER obtained from this calculation is 1.29. This value indicates the model's average performance in terms of the accuracy of its speech-to-text transcriptions for the Hill Mari language. In the next section, we will attempt to improve the model's understanding and handling of the Hill Mari language by fine-tuning it on similar sentences, potentially lowering the WER by making the model more attuned to the nuances of the Hill Mari language.

Fine-Tuning a Speech-To-Text Model for a Specific Language

In this section, we delve into the nuanced process of fine-tuning the Whisper small model for the Hill Mari language, a linguistic venture aimed at enhancing speech recognition capabilities for a language with limited representation in mainstream models. Fine-tuning a model like Whisper, originally pre-trained on vast and diverse datasets, to cater to a specific language such as Hill Mari, involves several compelling reasons. First, it significantly improves the model's understanding and transcription accuracy for the language, bridging the gap in performance disparities observed with underrepresented languages. Second, fine-tuning allows the model to capture unique linguistic nuances, accents, and dialectical variations inherent to Hill Mari, ensuring a more inclusive and equitable technology. Lastly, by tailoring the model to a specific linguistic context, we enable a wider range of applications, from voice-activated technologies to digital archiving, thereby preserving and promoting linguistic diversity. Throughout this section, we will explore the methodology, challenges, and outcomes of fine-tuning the Whisper small model for the Hill Mari language and compare its performance against the pre-trained version.

Step 1: Installing Libraries and Data Loading

In this section, we focus on acquiring the Common Voice dataset for the Hill Mari language, downloading both training and testing datasets.

```
!pip install datasets
!pip install jiwer
!pip install evaluate
!pip install transformers[torch]
!pip install accelerate -U
```

First, we start with installing essential Python libraries for NLP and speech recognition projects. `datasets` provide access to a wide range of datasets, `jiwer` measures speech recognition accuracy, `evaluate` offers metrics for model performance evaluation, `transformers` enables the use of pre-trained NLP models with PyTorch support, and `accelerate` optimizes model training across different hardware.

```
from huggingface_hub import notebook_login
notebook_login()
```

CHAPTER 4   BRIDGING TEXT AND AUDIO IN GENERATIVE AI

By utilizing the previous code, we are authenticating to the Hugging Face Hub within a notebook environment.

```
from dataclasses import dataclass
from typing import Any, List, Dict, Union
import re
import evaluate
import torch
from datasets import load_dataset, Audio, metric, DatasetDict
from transformers import WhisperProcessor, WhisperForConditionalGeneration,Seq2SeqTrainingArguments, Seq2SeqTrainer, WhisperFeatureExtractor, WhisperTokenizer
from dataclasses import dataclass
from typing import Any, Dict, List, Union
```

The previous block code is being used to import the necessary tools and libraries for speech recognition tasks using the Whisper model from Hugging Face's Transformers library. It includes utilities for data processing, model training, and evaluation, as well as PyTorch for deep learning operations. Specifically, it sets up the environment to load audio datasets, preprocess them with Whisper's components, and evaluate model performance. It facilitates tasks from dataset manipulation to model fine-tuning and evaluation in a machine-learning pipeline.

```
common_voice = DatasetDict()
common_voice["train"] = load_dataset("mozilla-foundation/common_voice_13_0", "mrj", split="train")
common_voice["test"] = load_dataset("mozilla-foundation/common_voice_13_0", "mrj", split="test")
```

In the previous code, we initialize an empty `DatasetDict`, a structure for managing multiple datasets, and then load the training and testing splits of the Common Voice version 13.0 dataset for the Hill Mari language (denoted by `mrj`) from the Mozilla Foundation's collection hosted on Hugging Face's dataset repository. The `load_dataset` function is used to fetch the specified language dataset splits, which are then assigned to the `train` and `test` keys of the `DatasetDict`.

Step 2: Data Preprocessing

Now we will prepare our training data, ensuring it is in the optimal format for input into the Whisper model for effective training.

```
common_voice = common_voice.select_columns(["audio", "sentence"])
common_voice = common_voice.cast_column("audio", Audio(sampling_rate=16000))
```

Now we refine the `common_voice` dataset by narrowing it down to essential `audio` and `sentence` columns for speech recognition tasks and standardize the audio data by setting its sampling rate to 16,000 Hz. This process ensures the dataset is optimally formatted for processing by the Whisper model.

```
feature_extractor = WhisperFeatureExtractor.from_pretrained("openai/whisper-small")
tokenizer = WhisperTokenizer.from_pretrained("openai/whisper-small", language="Russian", task="transcribe")
```

In previous code first we create a `WhisperFeatureExtractor` object by loading pretrained settings from the `openai/whisper-small` model available on Hugging Face's model repository. This feature extractor is responsible for preparing the audio input by extracting relevant features that the Whisper model can understand and process efficiently. In the second line we initialize a `WhisperTokenizer`, also from the `openai/whisper-small` model, configuring it specifically for transcribing text in Russian. Russian was selected for importation as it shares linguistic similarities with Hill Mari, aiding in the model's understanding and processing of the language. The tokenizer plays a vital role in transforming the model's output into legible text, conforming to the linguistic characteristics and textual conventions of Russian.

```
def data_preprocessing(batch):
 audio = batch["audio"]
 batch["input_features"] =feature_extractor(audio["array"],sampling_rate=audio["sampling_rate"]).input_features[0]
  batch["labels"] =tokenizer(batch["sentence"]).input_ids
    return batch
```

The `data_preprocessing` function processes a batch of data by extracting audio features and converting sentences into tokenized forms suitable for model training. It takes the audio signal from the batch, applies the `feature_extractor` with the audio's sampling rate to obtain input features, and assigns these features back to the batch under the key `input_features`. Simultaneously, it tokenizes the sentence associated with the audio, using the `tokenizer` to convert the text into a sequence of input IDs (numeric tokens representing the text), and stores these tokens under the `labels` key in the batch. This process prepares each batch of data for the Whisper model by aligning audio inputs with their corresponding textual representations.

```
common_voice = common_voice.map(data_preprocessing, remove_columns=common_voice.column_names["train"], num_proc=4)
```

Now we apply the `data_preprocessing` function across the `common_voice` dataset, transforming audio and sentence data into model-ready formats, while removing all original columns except for the processed ones. We execute this transformation in parallel using four processes to expedite the preprocessing.

Step 3: Model Training (Fine-Tuning)

Now we will embark on training the Whisper small model using the preprocessed data from earlier steps.

```
@dataclass
class SpeechToTextDataCollator:
    speech_processor: Any

    def __call__(self, samples: List[Dict[str, Union[List[int], torch.Tensor]]]) -> Dict[str, torch.Tensor]:
        audio_input_features = [{"input_features": sample["input_features"]} for sample in samples]
        processed_batch = self.speech_processor.feature_extractor.pad(audio_input_features, return_tensors="pt")
        text_label_features = [{"input_ids": sample["labels"]} for sample in samples]
        processed_labels_batch = self.speech_processor.tokenizer.pad(text_label_features, return_tensors="pt")
```

```
processed_labels = processed_labels_batch["input_ids"].masked_
fill(processed_labels_batch.attention_mask.ne(1), -100)
if (processed_labels[:, 0] == self.speech_processor.tokenizer.bos_
token_id).all().cpu().item():
    processed_labels = processed_labels[:, 1:]
processed_batch["labels"] = processed_labels
return processed_batch
```

In the previous code block we define a `SpeechToTextDataCollator` class for preparing batches of data for a speech-to-text model, specifically doing the following:

- **Data Class Initialization:** It's a data class (@dataclass) named SpeechToTextDataCollator that takes a speech_processor as an input, which can be any speech processing model (in this context, likely a Whisper model processor).

- **Callable Method:** Implements a __call__ method that allows instances of this class to be called like a function. This method takes a list of samples, where each sample is a dictionary containing audio input features and text labels.

- **Audio Feature Processing:**
  - Extracts input_features from each sample in the provided list and creates a new list of dictionaries with these features.
  - Uses the speech_processor's feature extractor to pad these audio input features into a uniform format, preparing them as a batch (processed_batch) for the model.

- **Text Label Processing:**
  - Extracts labels (textual representations) from each sample, creating a list of dictionaries containing these labels.
  - Uses the speech_processor's tokenizer to pad these labels into a uniform format, resulting in a processed_labels_batch.

- **Label Adjustment:**
  - Masks the padded labels (processed_labels) to ignore padding during model training by replacing padding indices with -100.

- Checks if the first token of every label in the batch is the beginning-of-sentence token (bos_token_id). If so, it removes this token, adjusting the labels for proper processing.
- **Return Value:** Updates the processed_batch dictionary to include the adjusted labels and returns this dictionary as the output, making it ready for input into a speech-to-text model for training.

```
processor = WhisperProcessor.from_pretrained("openai/whisper-small",
language="Russian", task="transcribe")
```

Now we initialize a WhisperProcessor by loading a pre-trained whisper-small model from OpenAI, configured specifically for transcribing Russian language audio. The processor combines feature extraction and tokenization, streamlining the preparation of audio data for transcription tasks.

```
speech_to_text_data_collator = SpeechToTextDataCollator(speech_
processor=processor)
```

Now we create an instance of the SpeechToTextDataCollator class by the name speech_to_text_data_collator, passing in a processor object as the argument for its speech_processor parameter. speech_to_text_data_collator prepares batches of data for training or inference with a speech-to-text model by performing tasks such as padding and tokenization.

```
def normalize_text(text):
    text = text.lower()
    text = re.sub(r'[^\w\s]', '', text)
    return text
```

Now we define the normalize_text function to preprocess text data similar to the way that was done during the evaluation of the pre-trained model.

```
wer_metric = evaluate.load("wer")
def compute_wer_metrics(prediction):
    predicted_ids = prediction.predictions
    true_label_ids = prediction.label_ids
    true_label_ids[true_label_ids == -100] = tokenizer.pad_token_id
```

```
    predicted_texts = tokenizer.batch_decode(predicted_ids, skip_special_
    tokens=True)
    true_texts = tokenizer.batch_decode(true_label_ids, skip_special_
    tokens=True)
    predicted_texts = [normalize_text(text) for text in predicted_texts]
    true_texts = [normalize_text(text) for text in true_texts]
    word_error_rate = 100 * wer_metric.compute(predictions=predicted_texts,
    references=true_texts)
    return {"wer": word_error_rate}
```

Now we define the function to calculate WER our validation metrics. First, we start by loading the WER metric from the `evaluate` library. Then we define a function called `compute_wer_metrics` to calculate the WER of the predictions made by a speech-to-text model. The function receives a prediction object containing both the model's predicted output IDs and the true label IDs. It replaces any ignored token IDs (marked as -100) in the true labels with the tokenizer's pad token ID for proper comparison and then decodes both predicted and true label IDs into their textual representations using the tokenizer. These textual representations are passed to the WER metric computation, which calculates the percentage of error between the predicted texts and the true texts. The function returns the WER as a percentage, providing insight into the model's performance by quantifying the rate at which it incorrectly transcribes speech to text.

```
whisper_model = WhisperForConditionalGeneration.from_pretrained("openai/
whisper-small")
whisper_model.config.forced_decoder_ids = None
whisper_model.config.suppress_tokens = []
```

The previous lines of code perform the following actions related to initializing and configuring a Whisper model for speech-to-text tasks:

- **Model Initialization:** `whisper_model = WhisperForConditionalGeneration.from_pretrained("openai/whisper-small")` loads a pre-trained Whisper model named whisper-small from OpenAI's collection. This model is designed for conditional generation tasks, such as speech-to-text, allowing it to generate text based on audio input.

- **Configuration Adjustment 1:** whisper_model.config.forced_decoder_ids = None sets the model's configuration to not force any specific decoder token IDs during the generation process. This means the model will not be biased toward producing any predetermined tokens, allowing for more natural and contextually appropriate text generation.

- **Configuration Adjustment 2:** whisper_model.config.suppress_tokens = [] specifies that no tokens should be suppressed in the model's output. In other words, it ensures that the model does not intentionally omit any tokens from the generated text, based on this configuration setting.

Together, these lines of code prepare a Whisper model for generating text from audio inputs, with configurations ensuring unbiased and unrestricted text generation.

```
whisper_hill_mari_finetuning_args  = Seq2SeqTrainingArguments(
    output_dir="dkhublani/whisper_small_model_fine_tuned ",
    hub_model_id="dkhublani/whisper_small_model_fine_tuned",
    per_device_train_batch_size=16,
    gradient_accumulation_steps=1, decrease in batch size
    learning_rate=1e-5,
    lr_scheduler_type="constant_with_warmup",
    warmup_steps=20,
    max_steps=500,  # increase to 4000 if you have your own GPU or a Colab paid plan
    gradient_checkpointing=True,
    fp16=True,
    fp16_full_eval=True,
    evaluation_strategy="steps",
    per_device_eval_batch_size=16,
    predict_with_generate=True,
    generation_max_length=225,
    save_steps=250,
    eval_steps=250,
    logging_steps=25,
    report_to=["tensorboard"],
```

```
    load_best_model_at_end=True,
    metric_for_best_model="wer",
    greater_is_better=False,
    push_to_hub=True,
)
```

In the previous block of code, we define the training arguments for fine-tuning the Whisper model on the Hill Mari language dataset using the `Seq2SeqTrainingArguments` class. These arguments configure various aspects of the training process:

- `output_dir`: Specifies the directory where training outputs like model checkpoints will be saved.
- `per_device_train_batch_size`: Sets the batch size for training per device to 16.
- `gradient_accumulation_steps`: Indicates that gradients should be accumulated over one step. This can be adjusted to accumulate over more steps if the batch size is reduced, effectively simulating a larger batch size.
- `learning_rate`: Sets the learning rate to 1e-5.
- `lr_scheduler_type`: Uses a constant learning rate with warmup at the beginning of training.
- `warmup_steps`: Specifies 20 warmup steps where the learning rate gradually increases to the maximum set.
- `max_steps`: Limits the training to 500 steps, with a suggestion to increase this for more extensive resources.
- `gradient_checkpointing`: Enables gradient checkpointing to reduce memory usage at the cost of slower training.
- `fp16`: Enables training with mixed precision (FP16), reducing memory usage and potentially speeding up training with compatible hardware.
- `fp16_full_eval`: Ensures full evaluation is also conducted in FP16.
- `evaluation_strategy`: Evaluates the model every specified number of steps.

- **per_device_eval_batch_size**: Sets the batch size for evaluation per device.

- **predict_with_generate**: Enables generation of predictions for evaluation.

- **generation_max_length**: Sets the maximum length of the generated sequences.

- **save_steps**: Determines after how many steps a model checkpoint should be saved.

- **eval_steps**: Specifies after how many steps the model should be evaluated.

- **logging_steps**: Configures the frequency of logging for training progress.

- **report_to**: Designates `tensorboard` as the destination for logging metrics.

- **load_best_model_at_end**: Loads the best model based on the metric specified in `metric_for_best_model` at the end of training.

- **metric_for_best_model**: Uses the WER metric to determine the best model.

- **greater_is_better**: Indicates that a lower WER is better (false).

- **push_to_hub**: Disables automatic pushing of the model to the Hugging Face Model Hub.

These settings collectively aim to optimize the fine-tuning process for effective learning and evaluation.

```
processor.save_pretrained(whisper_hill_mari_finetuning_args.output_dir)
whisper_hill_mari_trainer  = Seq2SeqTrainer(
    args=whisper_hill_mari_finetuning_args ,
    model=whisper_model,
    train_dataset=common_voice["train"],
    eval_dataset=common_voice["test"],
    data_collator=speech_to_text_data_collator,
    compute_metrics=compute_wer_metrics,
```

```
        tokenizer=processor.feature_extractor,
)
whisper_hill_mari_trainer.train()
```

In the previous code we first initialize a Seq2SeqTrainer object named `whisper_hill_mari_trainer` with a specific configuration for fine-tuning a Whisper model on the Hill Mari language dataset and then start the training process:

- Initialization:
    - `args`: Sets training arguments using the `whisper_hill_mari_finetuning_args` object, which includes various training parameters such as batch size, learning rate, and evaluation strategy.
    - `model`: Specifies the `whisper_model` to be fine-tuned.
    - `train_dataset`: Uses the `train` split of the `common_voice` dataset for training.
    - `eval_dataset`: Uses the `test` split of the `common_voice` dataset for evaluation.
    - `data_collator`: Assigns `speech_to_text_data_collator` to format batches of data correctly for training.
    - `compute_metrics`: Utilizes `compute_wer_metrics` to calculate the WER for evaluating model performance.
    - `tokenizer`: Sets the feature extractor from the `processor` as the tokenizer, which is a bit unconventional as typically, a tokenizer meant for text processing is used, but in the context of Whisper, the feature extractor plays a similar role in preparing data.
- Training: The `.train()` method is called on the `whisper_hill_mari_trainer` to start the fine-tuning process based on the specified parameters and datasets.

The training process was conducted for 500 steps and completed in approximately 1 hour on an A100 GPU using Google Colab. The subsequent sections will focus on evaluating the performance of the fine-tuned model, comparing it against the pre-trained Whisper model, and conducting inference tasks to assess improvements in speech recognition accuracy.

Step 4: Model Evaluation and Inference

In step 4, we turn our focus toward the critical phase of model evaluation and inference, where we assess the performance of our fine-tuned Whisper model against its pre-trained counterpart. This evaluation is not merely about benchmarking accuracy but also understanding how fine-tuning impacts the model's ability to interpret and transcribe audio data under various conditions. Through meticulous analysis and comparison, we aim to uncover insights into the enhancements achieved and the potential areas for further improvement. Let's begin by comparing quantitative metrics between fine-tuned and pre-trained models.

Step	WER
250	0.411
500	0.308

The tables (Tables 4-1, 4-2, 4-3) clearly show that our fine-tuned model has markedly surpassed the performance of the pre-trained model. After 500 training steps, the WER of the fine-tuned model stands at 0.308, in contrast to the pre-trained model's WER of 1.29. To further assess the improvements, let's also conduct a qualitative evaluation by comparing the performance of both models on a few sentences from the test dataset.

Original statement: колжы миде сеткӓшкӹ

Table 4-1. *Comparing Output from Pre-trained and Fine-Tuned Models*

Group type	Sentence	WER
Pre trained	koža midė sėt keška	1.33
Fine Tuned	колжы миде сет кӓшкӹ	0.66

Original statement: йынгы йыла стихотворени отважный морякреволюционер макаров лӹмеш сирӹмӹ

Table 4-2. *Comparing Output from Pre-trained and Fine-Tuned Models*

Group type	Sentence	WER
Pre trained	янга ела стихотворение отважный моряк революционер макаров лемеш сирами	0.88
Fine Tuned	йынгы йыла стихотворени утважный моряк революционер макаров лӹмеш сир	0.38

Original statement: кырыквлä äнгӹрвлäм йоктарат

Table 4-3. Comparing Output from Pre-trained and Fine-Tuned Models

Group type	Sentence	WER
Pre trained		1.33
Fine Tuned	карыквлä äнгӹрвлäм йыктарат	0.66

From the examples provided, it's evident that the fine-tuned model has significantly outperformed the pre-trained model. Through this process, we have gained valuable insights into customizing a speech-to-text model to better meet our specific needs. With this knowledge in hand, let's transition to exploring the realm of text-to-speech conversion and learn how to develop a tailored text-to-speech model that aligns with our objectives.

Text to Speech

In the preceding section, we delved into utilizing Whisper for converting speech to text. Additionally, we trained it on a language that was previously not included in its training dataset, thereby improving its performance. Now, we're turning our attention to text-to-speech (TTS), a process that converts written text into human-like audio. This technology has a variety of applications, including helping visually impaired individuals access digital content, creating audiobooks, powering virtual assistants like Siri and Google Assistant, and enhancing entertainment, gaming, and language learning by providing voice content.

However, the capabilities of TTS technology also raise ethical concerns. It could be misused to create convincing fraudulent audio recordings with someone's voice without consent. It's important to use TTS technology ethically and responsibly.

We are set to employ the SpeechT5 model, crafted by Junyi Ao and his team at Microsoft, for our text-to-speech (TTS) applications. At its core, SpeechT5 utilizes a conventional Transformer encoder-decoder framework, adept at capturing the subtleties of sequence transformations by formulating hidden representations. This foundational approach is a staple across the diverse range of tasks the model is engineered to perform. Enhancing this core, SpeechT5 integrates six specialized pre-nets and post-nets, each designed to process either speech or text inputs. These networks are pivotal in preparing the input for the Transformer, ensuring that it is in an optimal state for processing, which in turn facilitates the production of the final output in the required format by the post-net.

To effectively conclude the speech generation process, SpeechT5 incorporates a unique predictive mechanism within its decoder, tasked with determining whether a given time step could signify the end of output generation. This process halts when it reaches a predefined confidence level, ensuring a seamless conclusion to speech synthesis. Following this, a convolutional post-net comes into play to refine the output spectrogram further, ensuring the highest quality of the generated speech. The training regime of the TTS model is centered around minimizing the variance between the predicted and the actual spectrograms, employing either L1 or MSE loss metrics for this purpose. The culmination of this process sees the transformation of the refined spectrogram back into an audio waveform, a task executed by a vocoder that operates independently from the model's seq2seq architecture, ensuring a crisp and natural auditory output. See Figure 4-9.

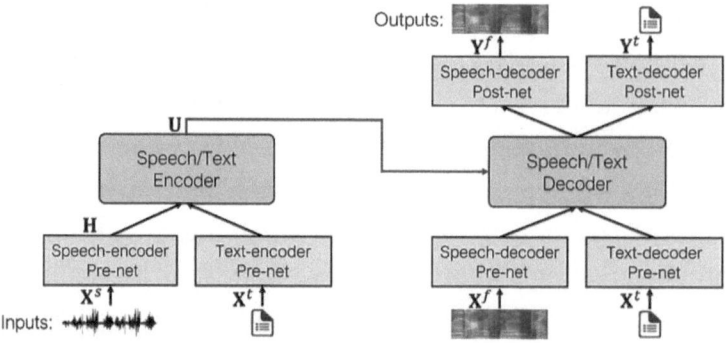

Figure 4-9. *Architecture of SpeechT5 (source: arXiv:2110.07205)*[10]

Initially, SpeechT5 undergoes pre-training on a vast corpus of unlabeled speech and text data to learn a universal representation of these modalities, utilizing all pre-nets and post-nets. Subsequent fine-tuning tailors the model to specific tasks by activating only the relevant nets for the task at hand, such as text-to-speech, which uses specific pre-nets for text input and post-nets for generating speech output. See Figure 4-10.

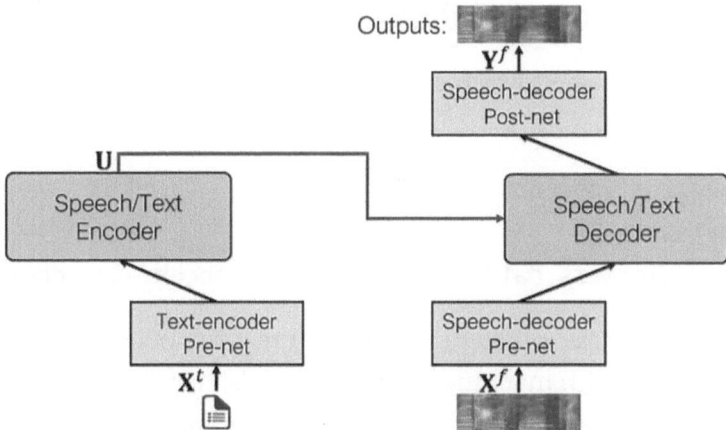

Figure 4-10. *Architecture of SpeechT5 during fine-tuning (source: arXiv:2110.07205)*

This approach results in a collection of models that are precisely calibrated for a range of speech-related tasks, all benefiting from an initial broad pre-training phase. While these models start from the same pre-trained foundation, they evolve distinctly as they are fine-tuned, highlighting the flexibility of the system to adapt to specific tasks without the models being interchangeable among different functions. This adaptability is a testament to the design's versatility, allowing for specialized optimization based on the unique requirements of each speech task.

The challenge inherent in text-to-speech (TTS) systems stems from their one-to-many relationship, where a single piece of text can be translated into various audio outcomes, making the evaluation of such models complex. Consequently, the assessment of TTS models primarily relies on human judgment, utilizing the mean opinion score (MOS) to gauge performance instead of solely depending on conventional loss metrics. This method acknowledges the subjective nature of speech perception and ensures that the models meet a high standard of auditory quality. Having now established a solid understanding of how SpeechT5 functions, we are prepared to proceed with the practical application of the pre-trained SpeechT5 model.

Using a Pre-trained Model for TTS

In this section, we explore the use of a pre-trained SpeechT5 model to transform Welsh text into speech, mimicking the voice of a Welsh speaker. We initiate this process by generating speaker embeddings to capture the unique vocal characteristics, coupled with Welsh sentences as input to the SpeechT5 model. Utilizing the Common Voice 13 dataset, rich in Welsh language data.

CHAPTER 4   BRIDGING TEXT AND AUDIO IN GENERATIVE AI

Step 1: Installing Libraries and Data Loading

Let's generate a speaker embedding for a specific sample from the Common Voice dataset in Welsh.

```
!pip install huggingface-hub
!pip install datasets
!pip install transformers datasets soundfile speechbrain accelerate
```

We first import all the essential libraries for our text-to-speech task. These packages are essential for working with machine learning models, handling datasets, audio processing, and accelerating AI model training and inference tasks.

```
from transformers import SpeechT5Processor, SpeechT5ForTextToSpeech, SpeechT5HifiGan
from datasets import load_dataset, Audio, DatasetDict, Audio
from collections import defaultdict
from transformers import
import os
import torch
from speechbrain.pretrained import EncoderClassifier
```

These lines of code import various libraries and modules necessary for processing and converting text to speech using AI models. Specifically, they import `SpeechT5Processor, SpeechT5ForTextToSpeech,` and `SpeechT5HifiGan` from the `transformers` library for handling text-to-speech tasks with the SpeechT5 model, along with `load_dataset, Audio,` and `DatasetDict` from `datasets` for dataset management and audio handling. Additionally, `defaultdict` from the collections module is imported for dictionary creation with default values. The `torch` for working with tensors and performing computations, and `EncoderClassifier` from `speechbrain.pretrained` for audio classification tasks.

```
from huggingface_hub import notebook_login
notebook_login()
```

The previous code prompts for Hugging Face Hub authentication within a Google Collab notebook, allowing users to access their Hugging Face account and private models or datasets directly from the notebook environment.

```
welsh_voice_train_dataset = load_dataset("mozilla-foundation/common_
voice_13_0", "cy", split="train")
welsh_voice_train_dataset = welsh_voice_train_dataset.cast_column("audio",
Audio(sampling_rate=16000))
```

In the previous code we load a training dataset from the mozilla-foundation/common_voice_13_0 collection, specifically targeting the Welsh language version ("cy"), from the Hugging Face datasets library. After successfully loading the dataset, we then transform the audio column of the dataset to ensure that all audio files it contains are processed with a sampling rate of 16000 Hz, using the Audio class from the datasets library. This transformation is crucial for standardizing the audio data, making it consistent in terms of sampling rate, which is important for subsequent audio processing.

Step 2: Data Processing

In step 2, we focus on the data preprocessing required to optimize the performance of the pre-trained SpeechT5 model for our specific use case.

```
client_id_to_filter = 'ccdf75e8e5f1a3c84918a8616f520e605825c284dc124df
adeb3f1d36e26f4a86a893075dc03589ff4996b449f273627a9fb0c39f2b76a20227567
faee343016'
welsh_voice_train_filtered_dataset = welsh_voice_train_dataset.
filter(lambda example: example['client_id'] == client_id_to_filter)
```

Now we filter the Welsh voice training dataset to include only the audio samples from a specific speaker, identified by the client_id. By applying this filter, the resultant dataset, welsh_voice_train_filtered_dataset, contains entries solely from the chosen speaker. We are undertaking this step because, during our fine-tuning process, it's crucial to maintain consistency in the speaker's voice. This will allow us to make a fair comparison between the pre-trained model and the fine-tuned model.

```
speaker_recognition_model_name = "speechbrain/spkrec-xvect-voxceleb"
computation_device = "cuda" if torch.cuda.is_available() else "cpu"
speaker_recognition_model = EncoderClassifier.from_hparams(
    source=speaker_recognition_model_name,
    run_opts={"device": computation_device, "timeout": 30},
    savedir=os.path.join("/tmp", speaker_recognition_model_name),
)
```

CHAPTER 4 BRIDGING TEXT AND AUDIO IN GENERATIVE AI

```
def generate_speaker_embedding(audio_waveform):
    with torch.no_grad():
        embeddings = speaker_recognition_model.encode_batch(torch.
        tensor(audio_waveform))
        normalized_embeddings = torch.nn.functional.
        normalize(embeddings, dim=2)
        flattened_embeddings = normalized_embeddings.squeeze().
        cpu().numpy()
    return flattened_embeddings
```

Now, we are setting up and utilizing a speaker recognition model from `SpeechBrain`, specifically the `spkrec-xvect-voxceleb`[11] model, to generate speaker embeddings. First, we determine the computational device to use, preferring CUDA for GPU acceleration if available, otherwise defaulting to CPU. This choice ensures our process is as efficient as possible. Next, we initialize the speaker recognition model with its name, specifying our device of choice and a timeout for operations. We also set a directory in the temporary folder to save any required files related to the model. This setup is crucial for preparing the model to process audio data.

Within the `generate_speaker_embedding` function, we process an audio waveform to produce speaker embeddings. We do this by first converting the waveform into a tensor, which is then fed into the model to generate embeddings. These embeddings are then normalized and reshaped (flattened) to ensure they are in a consistent format suitable for further processing or analysis. The function concludes by returning these processed embeddings. This step is vital for capturing the unique characteristics of the speaker's voice from the audio waveform.

```
def process_audio_data(sample):
    audio_data = sample["audio"]
    processed_sample = tts_processor(
        text=sample["sentence"],
        audio_target=audio_data["array"],
        sampling_rate=audio_data["sampling_rate"],
        return_attention_mask=False,
    )
```

```
processed_sample["labels"] = processed_sample["labels"][0]
processed_sample["speaker_embeddings"] = generate_speaker_
embedding(audio_data["array"])
return processed_sample
```

In the previous code snippet, we're taking a step-by-step approach to prepare audio data for text-to-speech (TTS) processing. Starting with a given audio sample, we first extract the audio information it contains. We then feed this audio data, along with the sample's text (a sentence), into our TTS processing function. This function, tts_processor, is designed to handle converting text to speech. It takes the text of the sentence, the audio array (the actual audio data), and the audio's sampling rate as inputs and processes these without generating an attention mask. Once we've processed the sample through the TTS processor, we adjust the structure of the processed sample by removing the batch dimension from the labels. This is necessary because the processing might introduce an extra dimension meant for batch processing, which we don't need for individual samples.

Furthermore, we enhance the processed sample by generating speaker embeddings from the audio array. The function generate_speaker_embedding is used here to extract features from the audio data that represent the unique characteristics of the speaker's voice. By the end of this process, each audio sample is enriched with labels and speaker embeddings, making it ready for further TTS tasks or analysis.

```
tts_processor = SpeechT5Processor.from_pretrained("microsoft/speecht5_tts")
```

We initialize a TTS processor by loading a pre-trained SpeechT5[12] model from Microsoft, designed to convert text into speech, preparing it for use on our dataset.

```
welsh_voice_train_filtered_dataset = welsh_voice_train_filtered_dataset.
map(process_audio_data, remove_columns=welsh_voice_train_filtered_dataset.
column_names)
```

We then apply a function to our Welsh voice training dataset to process each sample for TTS, removing all original columns to focus solely on the enriched data essential for speech synthesis.

```
sample_data = welsh_voice_train_filtered_dataset[1]
sample_speaker_embeddings = torch.tensor(sample_data["speaker_
embeddings"]).unsqueeze(0)
```

CHAPTER 4   BRIDGING TEXT AND AUDIO IN GENERATIVE AI

In the previous code, we first select a sample from our dataset. Next, we take the speaker embeddings from this selected sample and convert them into a PyTorch tensor. By using `torch.tensor()`, we convert the embeddings into a format that our model can work with. The `unsqueeze(0)` function is then applied to add an extra dimsension to the tensor, essentially converting it from a 1D array into a 2D array with one row.

Step 3: Model Inference

Let's load our pre-trained model and evaluate the audio quality it produces.

```
tts_input_parameters = tts_processor(text="Doedd hi ddim wedi arfer gyda'r math yma o beth chwaith.", return_tensors="pt")
speech_synthesis_model = SpeechT5ForTextToSpeech.from_pretrained("microsoft/speecht5_tts")
```

In the previous lines of code, we're preparing to synthesize speech from text. First, we process a Welsh sentence, through our text-to-speech (TTS) processor. This step converts the text into a format that our model can understand, specifying that the output should be PyTorch tensors ("pt"). Next, we load a pre-trained SpeechT5 model from Microsoft, designed for converting text into speech, and assign it to `speech_synthesis_model`. This model is now ready to generate speech from the processed text input.

```
audio_generator = SpeechT5HifiGan.from_pretrained("microsoft/speecht5_hifigan")
generated_audio = speech_synthesis_model.generate_speech(tts_input_parameters["input_ids"], sample_speaker_embeddings, vocoder=audio_generator)
```

Now we're finalizing the process of generating speech from text. First, we initialize a high-quality audio generator, SpeechT5HifiGan[13], which we load with a pre-trained model from Microsoft. This model specializes in converting the raw, synthesized speech signal into more natural and higher-quality audio. Next, we use our previously loaded speech synthesis model to transform the processed text and speaker embeddings into speech. We pass the text's input IDs and the speaker embeddings to the model, specifying the `audio_generator` as the vocoder. The vocoder is a critical component that enhances the output audio's realism, ensuring the synthesized speech sounds more like a human voice. The result of this operation is `generated_audio`, which contains the audio representation of our input text.

```
from IPython.display import Audio
Audio(generated_audio, rate=16000)
```

Now we utilize Python's IPython library to play the audio that we've generated from text. By importing the Audio function and then calling it with our `generated_audio` and a sample rate of 16000 Hz, we enable the playback of the synthesized speech directly within our notebook environment. The audio output mentioned will be accessible in our GitHub repository. The audio quality is not optimal, with some instances of voice breaking and overall clarity not being very clear. Now, let's proceed to fine-tune the SpeechT5 model to see if we can enhance the audio quality.

Fine-Tuning Text-To-Speech Model

Fine-tuning the SpeechT5 model represents a tailored approach to enhancing text-to-speech (TTS) translation, particularly for languages with less representation in pre-trained models, such as Welsh. By adapting the SpeechT5 model, which is already proficient in general speech synthesis tasks, to focus specifically on the Welsh language, we aim to refine its understanding and output quality for this linguistic context. Fine-tuning offers several benefits, including improved accuracy in voice and intonation, greater adaptability to language-specific nuances, and, potentially, a more natural-sounding speech output. This process leverages the same dataset to train the SpeechT5 model on Welsh, providing an opportunity to directly compare performance enhancements. With the anticipation of improved TTS translations, let's embark on the journey of data loading and set the stage for fine-tuning our SpeechT5 model to better cater to the Welsh language.

Step 1: Installing Libraries and Data Loading

Before we load the dataset, let's start with installing essential libraries.

```
!pip install transformers datasets soundfile speechbrain accelerate
```

Now we have the libraries installed, let's import libraries that we will be using during our fine-tuning process.

```
from datasets import load_dataset, Audio
from transformers import SpeechT5Processor,SpeechT5ForTextToSpeech,
Seq2SeqTrainingArguments, SpeechT5HifiGan, Seq2SeqTrainer
```

CHAPTER 4   BRIDGING TEXT AND AUDIO IN GENERATIVE AI

```
import os
import torch
from speechbrain.pretrained import EncoderClassifier
from dataclasses import dataclass
from typing import Any, Dict, List, Union
from functools import partial
from collections import defaultdict
```

Now, let's sign into the Hugging Face Hub from Google Colab using the following code snippet:

```
from huggingface_hub import notebook_login
notebook_login()
```

Now we're set to load our dataset.

```
welsh_voice_train_dataset = load_dataset("mozilla-foundation/common_voice_13_0", "cy", split="train")
```

By utilizing the previous code, we have loaded the training split of the Welsh language dataset from Mozilla's Common Voice version 13.0 into our environment.

Step 2: Data Preprocessing

Let's prepare our dataset to ensure it's ready for model training.

```
welsh_voice_train_dataset = welsh_voice_train_dataset.cast_column("audio", Audio(sampling_rate=16000))
```

First we will start with converting the `audio` column of our Welsh voice training dataset to have a uniform sampling rate of 16000 Hz, making it consistent for model training.

```
model_checkpoint = "microsoft/speecht5_tts"
tts_processor = SpeechT5Processor.from_pretrained(model_checkpoint)
tts_tokenizer = tts_processor.tokenizer
```

In the previous code snippet, we're setting up the groundwork for our text-to-speech (TTS) system. First, we specify the model we'll be using, the SpeechT5 model from Microsoft, by assigning its identifier to `model_checkpoint`. Then, we initialize the tts_processor, which is responsible for preparing our input data (text) for the model, by loading this processor using the model's checkpoint. This processor encapsulates both the model's tokenizer and feature extraction capabilities. Finally, we explicitly access the tokenizer component of our processor through tts_processor.tokenizer, which is essential for converting our text input into a format that the SpeechT5 model can understand and process into speech.

```
def compile_dataset_vocabulary(batch):
    concatenated_sentences = " ".join(batch["sentence"])
    unique_vocab = list(set(concatenated_sentences))
    return {"unique_vocab": [unique_vocab], "concatenated_sentences":
    [concatenated_sentences]}

extracted_vocab = dataset.map(
    compile_dataset_vocabulary,
    batched=True,
    batch_size=-1,
    keep_in_memory=True,
    remove_columns=dataset.column_names,
)
```

In previous lines of code, we're performing a comprehensive analysis of our dataset to extract and compile its unique vocabulary. First, through the compile_dataset_vocabulary function, we concatenate all sentences from our dataset into a single string and then identify and list every unique character (or vocab element) present. This process results in two key pieces of information: a list of unique vocabulary items and the combined text of all sentences, both stored for each batch processed.

Following this, we apply the compile_dataset_vocabulary function to our entire dataset using the .map method. This operation is executed in batch mode, processing the entire dataset in one go (batch_size=-1), ensuring efficiency and keeping the processed data in memory for immediate access. We also remove the original columns of the dataset since our focus is solely on the extracted vocabulary and concatenated text data. This procedure allows us to systematically understand the diversity of characters or vocab elements in our dataset.

```
complete_dataset_vocab = set(extracted_vocab["unique_vocab"][0])
tts_processor_vocab = {k for k, _ in tts_processor.tokenizer.get_vocab().
items()}
missing_vocab_in_processor = complete_dataset_vocab - tts_processor_vocab
missing_vocab_in_processor
```

SpeechT5 is trained on English, and we are using Welsh dataset so the previous code would help us find those Welsh characters that don't exist in English. Initially, we compile a unique set of characters (or vocabulary) from all the sentences in our dataset, creating `complete_dataset_vocab`. We then extract the vocabulary present in the TTS processor's tokenizer, forming `tts_processor_vocab`. By subtracting the tokenizer's vocabulary from the dataset's vocabulary, we pinpoint `missing_vocab_in_processor`, which are the characters present in our dataset but not recognized by the TTS processor's tokenizer.

```
welsh_to_english_replacements = {
    '¬': '',
    'Â': 'A',
    'Ô': 'O',
    'à': 'a',
    'á': 'a',
    'â': 'a',
    'ä': 'a',
    'ë': 'e',
    'î': 'i',
    'ï': 'i',
    'ò': 'o',
    'ô': 'o',
    'ö': 'o',
    'û': 'u',
    'Ŵ': 'W',
    'ŵ': 'w',
    'ŷ': 'y',
    '–': '-',
    ''': "'",
    '"': '"',
```

```
    '"': '"'
}
missing_vocab_replacements = [(src, dst) for src, dst in welsh_to_english_
replacements.items()]
```

Now we would replace Welsh characters that are unique to Welsh with something similar in English. We've created a dictionary named `welsh_to_english_replacements` that maps to specific Welsh characters. Following the creation of this dictionary, we then generate a list called `missing_vocab_replacements` from it. This list is composed of tuples, where each tuple contains a source character from our dataset that the TTS processor's tokenizer may not recognize (`src`) and the character or string it should be replaced with (`dst`).

```
def normalize_sentence_characters(sentence_mapping):
    for original_char, replacement_char in missing_vocab_replacements:
        sentence_mapping["sentence"] = sentence_mapping["sentence"].
        replace(original_char, replacement_char)
    return sentence_mapping

welsh_voice_train_dataset = welsh_voice_train_dataset.map(normalize_
sentence_characters)
```

In the previous code block, we're applying a text normalization function, `normalize_sentence_characters`, to our Welsh voice training dataset. This function iterates through a list of character replacements, `missing_vocab_replacements`, and updates each sentence in the dataset by replacing specific Welsh characters with their designated substitutes. The aim is to standardize the text, making it more compatible with the text-to-speech processor by ensuring that all characters are recognized and properly processed.

```
speaker_frequency = defaultdict(int)
for client_identifier in welsh_voice_train_dataset["client_id"]:
    speaker_frequency[client_identifier] += 1

def is_speaker_within_range(client_id):
    return 100 <= speaker_frequency[client_id] <= 400

dataset_with_selected_speakers = welsh_voice_train_dataset.filter(is_
speaker_within_range, input_columns=["client_id"])
```

CHAPTER 4   BRIDGING TEXT AND AUDIO IN GENERATIVE AI

In the previous code snippet, we're analyzing our Welsh voice training dataset to identify speakers with a specific range of contributions. First, we tally the number of recordings each speaker (identified by `client_id`) has contributed to the dataset, storing this information in `speaker_frequency`. This step ensures we understand the distribution of contributions across different speakers. Next, we define a function, `is_speaker_within_range`, to determine whether a given speaker has contributed between 100 and 400 recordings. This range is selected to focus on speakers with a significant, but not overwhelming, number of samples.

Finally, we filter the entire Welsh voice training dataset, retaining only the recordings from speakers who meet our defined criteria. The result, `dataset_with_selected_speakers`, is a curated subset of the original dataset that includes only those speakers with an ideal number of contributions, ready for further processing or model training. This approach helps in creating a more balanced dataset for training purposes, improving the performance of speech synthesis models by focusing on a well-represented set of speakers.

```python
speaker_recognition_model_name = "speechbrain/spkrec-xvect-voxceleb"
computation_device = "cuda" if torch.cuda.is_available() else "cpu"
speaker_recognition_model = EncoderClassifier.from_hparams(
    source=speaker_recognition_model_name,
    run_opts={"device": computation_device, "timeout": 30},
    savedir=os.path.join("/tmp", speaker_recognition_model_name),
)

def generate_speaker_embedding(audio_waveform):
    with torch.no_grad():
        embeddings = speaker_recognition_model.encode_batch(torch.
        tensor(audio_waveform))
        normalized_embeddings = torch.nn.functional.
        normalize(embeddings, dim=2)
        flattened_embeddings = normalized_embeddings.squeeze().
        cpu().numpy()
    return flattened_embeddings

def process_audio_data(sample):
    audio_data = sample["audio"]
    processed_sample = tts_processor(
```

```
        text=sample["sentence"],
        audio_target=audio_data["array"],
        sampling_rate=audio_data["sampling_rate"],
        return_attention_mask=False,
    )
    processed_sample["labels"] = processed_sample["labels"][0]
    processed_sample["speaker_embeddings"] = generate_speaker_
    embedding(audio_data["array"])
    return processed_sample
```

We use the same code block from the pre-trained section to update our training data with labels and speaker embeddings, making it ready for training.

```
dataset_with_selected_speakers = dataset_with_selected_speakers.
map(process_audio_data, remove_columns=dataset_with_selected_speakers.
column_names)
```

We're applying the `process_audio_data` function to our dataset, which processes each selected speaker's audio sample for text-to-speech tasks. This operation removes all original columns, ensuring the dataset only contains the processed data necessary for model training.

Step 3: Model Training (Fine-Tuning)

Our dataset is prepared, and we're now set to begin the model training process.

```
dataset_with_selected_speakers = dataset_with_selected_speakers.train_test_
split(test_size=0.2)
```

First, we start with splitting our dataset into training and testing sets, with 20% of the data reserved for testing. This ensures we have separate datasets to train our model and evaluate its performance.

```
class TTSDataCollatorWithSpeakerEmbedding:
    tts_processor: Any

    def __call__(
        self, samples: List[Dict[str, Union[List[int], torch.Tensor]]]
    ) -> Dict[str, torch.Tensor]:
```

```
text_input_ids = [{"input_ids": sample["input_ids"]} for sample in 
samples]
audio_labels = [{"input_values": sample["labels"]} for sample in 
samples]
speaker_embeddings_list = [sample["speaker_embeddings"] for sample 
in samples]

batched_data = tts_processor.pad(
    input_ids=text_input_ids, labels=audio_labels, return_
    tensors="pt"
)

batched_data["labels"] = batched_data["labels"].masked_fill(
    batched_data.decoder_attention_mask.unsqueeze(-1).ne(1), -100
)

del batched_data["decoder_attention_mask"]

if model.config.reduction_factor > 1:
    target_lengths = torch.tensor(
        [len(sample["input_values"]) for sample in audio_labels]
    )
    adjusted_lengths = target_lengths.new(
        [
            length - length % model.config.reduction_factor
            for length in target_lengths
        ]
    )
    max_length = max(adjusted_lengths)
    batched_data["labels"] = batched_data["labels"][:, :max_length]

batched_data["speaker_embeddings"] = torch.tensor(speaker_
embeddings_list)
return batched_data
```

In the previous code, we're defining a custom data collator class, `TTSDataCollatorWithSpeakerEmbedding`, tailored for preparing batches of data for text-to-speech (TTS) training that includes speaker embeddings. This collator takes a list of samples, where each sample is a dictionary containing text input IDs, audio labels, and speaker embeddings.

- First, we extract and organize these components from the input samples into separate collections for processing.
- We then use the TTS processor's `pad` method to combine the text inputs and audio labels into a single batch, ensuring consistent tensor shapes and converting them into PyTorch tensors.
- To handle sequences of variable lengths, we mask padding in the labels with a value of -100, which is standard practice to ignore these areas during loss calculation.
- We remove the decoder attention mask from the batch as it's not needed for fine-tuning.
- If the model has a reduction factor (used in some TTS models to reduce the temporal resolution of the output for efficiency), we adjust the lengths of our target audio labels to be multiples of this factor, ensuring compatibility with the model's architecture.
- Finally, we add the speaker embeddings to the batch, converting them into a PyTorch tensor for model processing.

This setup ensures that each batch fed into the TTS model during training is properly formatted, including the crucial speaker embeddings that enable the model to learn speaker-specific characteristics.

```
TTSdata_collator = TTSDataCollatorWithSpeakerEmbedding()
tts_model = SpeechT5ForTextToSpeech.from_pretrained(model_checkpoint)
tts_model.config.use_cache = False
tts_model.generate = partial(tts_model.generate, use_cache=True)
```

CHAPTER 4 BRIDGING TEXT AND AUDIO IN GENERATIVE AI

In the previous lines, we're setting up our text-to-speech (TTS) system with a custom collator and configuring our model for generation.

- First, we instantiate TTSDataCollatorWithSpeakerEmbedding, a custom data collator class designed to properly format batches of data, including speaker embeddings, for training our TTS model.

- We then load our TTS model, SpeechT5ForTextToSpeech, from a predefined checkpoint, ensuring we have the latest and most suitable pre-trained model for our text-to-speech tasks.

- We explicitly set use_cache to False in the model's configuration. This usually means we're instructing the model not to use cached states during training, which can affect performance and memory usage.

- Finally, we modify the model's generate method using Python's partial function to ensure use_cache is True during generation. This adjustment enables the model to utilize cached information when generating speech, which can improve generation efficiency and speed by avoiding unnecessary recomputations.

```
tts_training_args = Seq2SeqTrainingArguments(
    output_dir="dkhublani/test_speecht5",
    hub_model_id="dkhublani/test_speecht5",
    hub_strategy="every_save",
    per_device_train_batch_size=4,
    gradient_accumulation_steps=8,
    learning_rate=1e-5,
    warmup_steps=500,
    max_steps=4000,
    gradient_checkpointing=True,
    fp16=False,
    evaluation_strategy="steps",
    per_device_eval_batch_size=2,
    save_steps=1000,
    eval_steps=1000,
    logging_steps=25,
    report_to=["tensorboard"],
```

```
    load_best_model_at_end=True,
    greater_is_better=False,
    label_names=["labels"],
    push_to_hub=True,
    save_total_limit=2,
    save_strategy="steps",
)
```

In this block of code, we are configuring the training parameters for our text-to-speech (TTS) model using the Seq2SeqTrainingArguments from the Hugging Face Transformers library.

- We specify output_dir and hub_model_id to determine where the trained model and its checkpoints will be saved, both locally and on the Hugging Face Hub under the project name dkhublani/test_speecht5.

- The hub_strategy is set to save the model to the Hugging Face Hub at every checkpoint, facilitating version control and sharing.

- We define our training batch size, learning rate, warmup steps, and the maximum number of training steps to ensure efficient training with a gradual learning rate increase and a controlled training duration.

- gradient_checkpointing is enabled to reduce memory usage at the cost of slower backward passes, and fp16 is set to False, indicating we are not using mixed-precision training.

- The model evaluation is configured to occur at specified intervals (evaluation_strategy, save_steps, eval_steps), with additional settings for logging and reporting to track the training progress.

- load_best_model_at_end is set to True with greater_is_better set to False, ensuring that the model with the best performance according to the evaluation metric is loaded at the end of training.

- We also specify label_names for correct loss calculation, enable push_to_hub for automatic model uploading, and set save_total_limit to manage the number of saved checkpoints, optimizing storage.

This comprehensive setup prepares our TTS model for an efficient and effective training process, with an emphasis on model management, performance tracking, and resource optimization.

```
tts_trainer = Seq2SeqTrainer(
    args=tts_training_args,
    model=tts_model,
    train_dataset=dataset_with_selected_speakers["train"],
    eval_dataset=dataset_with_selected_speakers["test"],
    data_collator=TTSdata_collator,
    tokenizer=tts_processor,
)
```

We now initialize a `Seq2SeqTrainer` object to manage the training and evaluation of our text-to-speech (TTS) model. We pass in several key components.

- `args`: The training arguments we previously defined, which include details on how the training should be conducted, such as batch sizes, learning rate, and the strategy for saving and evaluating the model.
- `model`: Our TTS model that we intend to train.
- `train_dataset` and `eval_dataset`: The portions of our dataset designated for training and evaluation, respectively, ensuring that our model learns from one set of data and is tested on another to gauge its performance accurately.
- `data_collator`: The `TTSDataCollatorWithSpeakerEmbedding` instance, which prepares batches of data in the format required by our model, including handling of speaker embeddings.
- `tokenizer`: The processor associated with our model, which is essential for preparing text inputs for the model.

By setting up the `Seq2SeqTrainer` with these components, we are ready to start the training process.

```
tts_trainer.train()
trainer.push_to_hub()
```

With the previous code, we initiate the training process for our text-to-speech model using the configured trainer. After completing the training, we will upload the model to the Hugging Face Hub. Training on Google Colab's A100 GPU took approximately 1.5 hours. Here are the outcomes of the training session:

Step	Training Loss	Validation Loss
1000	0.603200	0.558628
2000	0.565400	0.534424
3000	0.561100	0.529091
4000	0.543700	0.527492

Evaluating TTS models using standard metrics can be challenging. To assess improvements in audio quality, let's proceed with some inference tests.

Step 4: Model Inference

Let's retrieve our model from Hugging Face, use the same speaker embedding and input text as with the pre-trained model, and evaluate its performance.

```
fine_tuned_model = SpeechT5ForTextToSpeech.from_pretrained("dkhublani/test_speecht5")
```

We now load our fine-tuned version of the SpeechT5ForTextToSpeech model from the Hugging Face Hub.

```
sample_speaker = dataset_with_selected_speakers["test"][304]
speaker_embeddings = torch.tensor(sample_speaker["speaker_embeddings"]).unsqueeze(0)
```

In the provided code, we chose the same speaker that was used in our implementation with the pre-trained model. Following that, we retrieve the speaker embeddings for this individual and transform them into a PyTorch tensor.

```
tts_input_parameters = tts_processor(text="Doedd hi ddim wedi arfer gyda'r math yma o beth chwaith.", return_tensors="pt")
audio_generator = SpeechT5HifiGan.from_pretrained("microsoft/speecht5_hifigan")
generated_audio = fine_tuned_model.generate_speech(inputs["input_ids"], speaker_embeddings, vocoder=audio_generator)
```

Just like in the section where we used the pre-trained model, we employ the TTS model here to transform text input into audio output.

```
from IPython.display import Audio
Audio(speech, rate=16000)
```

The audio quality is notably enhanced, offering clearer and more seamless playback compared to the pre-trained model. The audio will be accessible in our GitHub repository. Congratulations, we've mastered the art of fine-tuning models for both speech-to-text and text-to-speech functionalities.

Troubleshooting Common Issues

During model training or inference, you may encounter issues. Here are some common ones:

- **Poor Audio Quality:** This could indicate insufficient training data, inappropriate loss function, or hyperparameter tuning problems.

- **Incorrect Intonation or Pronunciation:** Fine-tuning with a dataset containing Welsh speech data can address these issues.

- **Inaccurate Speaker Embeddings:** Ensure speaker embeddings accurately capture the target voice characteristics.

Conclusion

This chapter navigated the intricate interplay between linguistic expressions and their auditory counterparts, uncovering the foundations and challenges inherent in working with audio data. Our exploration led us from the basics of understanding audio data to the practical implementation of cutting-edge models like CLAP for bridging text and audio and the fine-tuning of models such as Whisper and SpeechT5 for speech-to-text and text-to-speech tasks, respectively. This chapter not only provided a comprehensive overview of the technical challenges and solutions in audio data processing but also offered hands-on experience in fine-tuning pre-trained models for specific languages, showcasing the remarkable potential of AI-driven text and audio conversion models in understanding and generating human-like speech.

The real-life applications of these advancements are vast and varied, ranging from enhancing accessibility with more accurate speech recognition systems for assistive technologies to creating more natural and engaging user interfaces in virtual assistants and entertainment. In the realm of education, these technologies promise to break down language barriers, offering personalized learning experiences through real-time translation and audio content generation. Moreover, the healthcare sector stands to benefit significantly, with applications in patient care through voice-enabled documentation and interactive systems for diagnostics and therapy.

Peering into the future, the horizon of text and audio conversion technologies brims with potential, steering toward a future where the distinction between human and artificial speech becomes increasingly indistinct. We anticipate pioneering developments that enhance the ability of models to grasp and replicate the subtle intricacies of emotion, tone, and the situational context in speech, thereby fostering more authentic and nuanced interactions. Furthermore, the pursuit of more advanced training methodologies, which thrive on minimal data while conserving computational efforts, stands at the forefront of this evolution. The advent of sophisticated multimodal systems, capable of concurrently processing and generating text, audio, and visual content, heralds a transformative shift in our engagement with digital environments, promising a future where technology seamlessly integrates with the fabric of human experience, enriching communication and creativity.

As we prepare to delve into the next chapter, our focus will shift toward understanding and implementing large language models (LLMs). This upcoming exploration promises to expand our comprehension of AI's capabilities further, venturing into how LLMs are shaping the future of natural language understanding and generation and their application across various domains.

CHAPTER 5

Large Language Models

Introduction

Large language models (LLMs) are highly developed and potent neural network models trained on enormous volumes of text data to comprehend and produce language similar to humans. These models have made tremendous strides in recent years thanks to their outstanding performance in various natural language processing applications.

Large language models are intended to capture and understand natural language's intricate patterns, syntax, and semantics, such as OpenAI's Generative Pre-trained Transformer (GPT) series. They process and create text using transformer architectures by paying attention to the connections between words in a sequence.

These models are often trained via unsupervised learning on sizable datasets such as online content or books to get a general knowledge of the language. The models develop their ability to anticipate the next word in a phrase using the context of the previous words supplied.

The success of large language models depends on their capacity to produce logical and contextually appropriate content. They have many uses, including text completion, sentiment analysis, question answering, language translation, and creative writing. Large language models have shown outstanding language creation skills, frequently delivering results identical to paper created by humans.

In recent years, LLMs have achieved extraordinary feats, such as OpenAI's GPT-3 and 4's ability to generate creative fiction and educational content that rivals human quality or Google's BERT, which has significantly improved the relevance of search engine results.

Multimodal Integration: LLMs have extended beyond text-only understanding. Multimodal LLMs now incorporate images, audio, and other modalities, enabling tasks such as image captioning, speech recognition, and cross-modal reasoning. For instance, OpenAI's CLIP combines vision and language to perform zero-shot image classification, as explained in Chapter 2.

Few-Shot Learning: LLMs excel at learning from limited examples. GPT-3, for instance, can perform tasks with minimal training data, making it valuable for personalized applications and niche domains.

However, alongside these advancements, it is crucial to acknowledge the limitations and ethical concerns associated with LLMs.

Bias: LLMs inherit biases from training data, perpetuating stereotypes and discriminatory language.

Data Privacy: Fine-tuning LLMs on sensitive data raises privacy concerns.

Environmental Impact: Training large LLMs consumes significant energy.

Misinformation: LLMs can inadvertently generate false or harmful content.

Explainability: Understanding LLM decisions remains challenging.

In summary, while LLMs offer immense potential, we must navigate their ethical implications and address their limitations to ensure responsible deployment in real-world applications.

The size of the large language models continues to increase. Figure 5-1 illustrates the current state of large language models as of September 2023.

CHAPTER 5 LARGE LANGUAGE MODELS

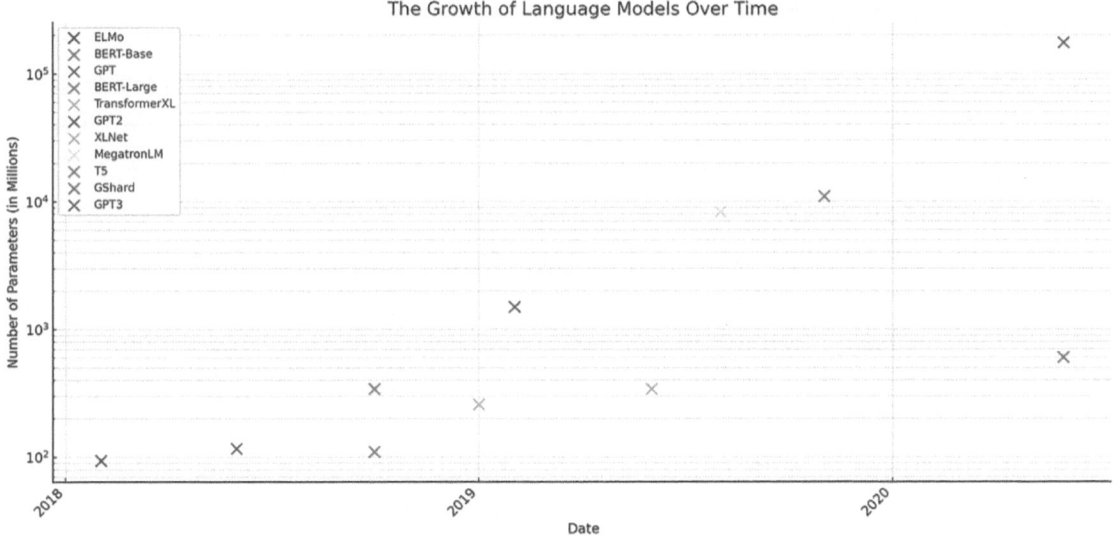

Figure 5-1. *Growth of LLMs over time*

The size of LLMs is increasing because the larger the model, the larger and more expensive the compute requirements. Therefore, the adoption and use of LLMs depend on their size.

Phases of Training and Adoption of Large Language Models

There are a considerable number of steps required, from building a large language model to deploying it in production. Figure 5-2 shows a high-level overview of the steps.

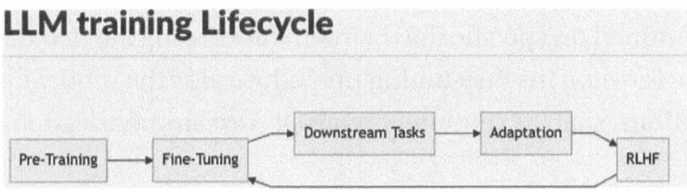

Figure 5-2. *LLM training and adoption life cycle*

Pre-training of LLMs:

LLMs undergo pre-training, the first stage of training when they are exposed to a large quantity of unlabeled text data to understand the fundamental patterns and structures of language. This pre-training phase is crucial for the subsequent fine-tuning and adaptation of the models to downstream tasks.

LLMs are taught unsupervised or self-supervised during pre-training when they learn to anticipate missing words or word sequences, given the context the surrounding text provides. Typically, methods like masked language modeling or next-sentence prediction are used to train the models. In the upcoming sections, we will explore the details of masked language modeling and next-sentence prediction.

In masked language modeling, a particular portion of the input tokens is randomly masked, and the model forecasts their original values. Thanks to this method, the model has a greater comprehension of the language and begins to learn contextual links between words.

Predicting the next sentence includes teaching the model to recognize whether two provided sentences follow one another in the source text. This activity aids the model's ability to represent text coherence and semantic connections.

Large language models must be pre-trained, which consumes much computing power and training time. Usually, parallel processing and distributed training methods are used to train the models on high-performance computer clusters. Numerous model parameters are optimized throughout the training phase using stochastic gradient descent or its derivatives.

Fine-Tuning and Adaptation of LLMs

The LLMs are fine-tuned on specific downstream tasks using labeled data when the pre-training phase is over. This fine-tuning procedure aids the models' adaptation to particular applications, such as sentiment analysis, text summarization, or machine translation.

Pre-training helps LLMs build a solid basis for language understanding, which is then fine-tuned. However, pre-training an LLM is computationally expensive. It requires a huge corpus, whereas, on the other hand, fine-tuning can be achieved using a small, labeled corpus and much less compute power, as shown in Figure 5-3.

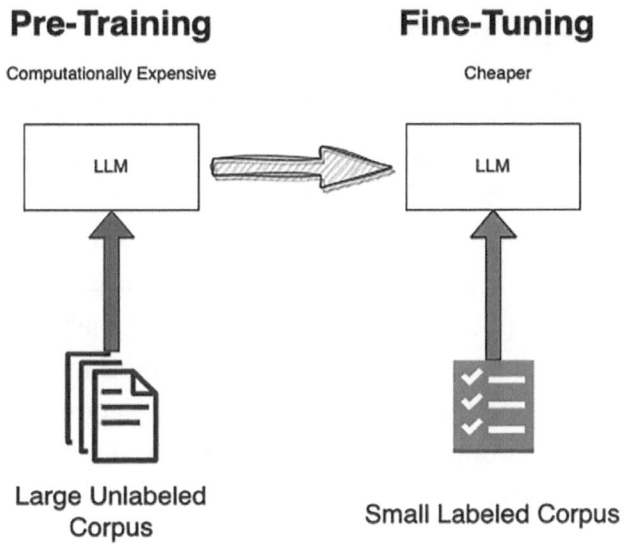

Figure 5-3. Pre-training versus fine-tuning LLM

This two-step process of pre-training and fine-tuning the language models has developed modern language models to produce coherent and contextually appropriate text.

To enhance our grasp of the life cycle of LLMs, let's delve into a tangible example—ChatGPT, developed by OpenAI. Envision the extensive pre-training phase as laying down the groundwork for understanding language intricacies. This phase is akin to the broad education you receive before choosing a specialization. OpenAI's GPT models were introduced to a diverse range of texts, helping it learn the fundamental building blocks of language, including linguistic structures and patterns.

Then, in the fine-tuning phase, ChatGPT, much like a graduate honing in on a chosen field, was meticulously refined for distinct tasks such as conversing, composing essays, or even generating poetry. This process, which utilizes a smaller, focused dataset, sharpens the model's capabilities, adapting it to specific contexts and nuances. Additionally, it also underwent a stage of reinforcement learning from human feedback (RLHF) to generate more accurate responses. Human annotators provided feedback to the responses generated by ChatGPT. The feedback data was used to further fine-tune and align the output of the ChatGPT model.

CHAPTER 5   LARGE LANGUAGE MODELS

Why Large Language Models?

LLMs have gained immense popularity and significance for several reasons:

Language Understanding: LLMs are excellent at deciphering and producing human-like language. They can produce coherent and meaningful material to its context by recognizing intricate patterns, syntax, and semantic links between words. This makes them useful for various NLP applications, including sentiment analysis, text generation, question answering, and machine translation. All of these tasks can be performed using prompt completion. The LLM can complete the prompt (shown in Figure 5-4) sent by a human/user, which can be used for several purposes.

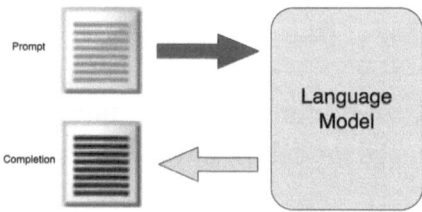

Figure 5-4. Prompt completion

Pre-training and Transfer Learning: LLMs are pre-trained on a lot of text material to build a solid foundation in language understanding. This pre-training makes transfer learning possible, enabling the pre-trained models to be refined on specific downstream tasks with less modelling data. Therefore, pre-training enables quicker construction and deployment of task-specific models than training models from scratch. It also saves a substantial amount of time and money.

Generalization: LLMs are good at generalizing to many areas and activities. Their pre-trained representations include a variety of language ideas and patterns from various data sources. As long as the job is related to language understanding, this generalization ability enables LLMs to do well on activities they were not expressly taught.

Creative Text Generation: LLMs can produce imaginative content that is also acceptable for the situation. So, LLMs can be used for applications that create coherent and exciting text based on specified prompts or conditions, such as creative writing, storytelling, dialogue creation, and others.

Research and Innovation: The field of natural language processing has seen a tremendous increase in both. LLMs may be made better by constructing more effective structures, adding outside information, correcting biases, and creating strategies for greater control and interpretability, among other things.

Benchmarks for Language Modeling: LLMs have established new standards for language modelling and related downstream NLP activities. They have attained cutting-edge results on several benchmarks, proving their usefulness and expanding what is conceivable regarding language understanding and production.

Types of Language Transformers Models

All the language models we have discussed and will continue to build upon in the following sections are built using a transformer architecture. The transformer architecture, first introduced in June 2017, has two main blocks, as shown in Figure 5-5.

CHAPTER 5   LARGE LANGUAGE MODELS

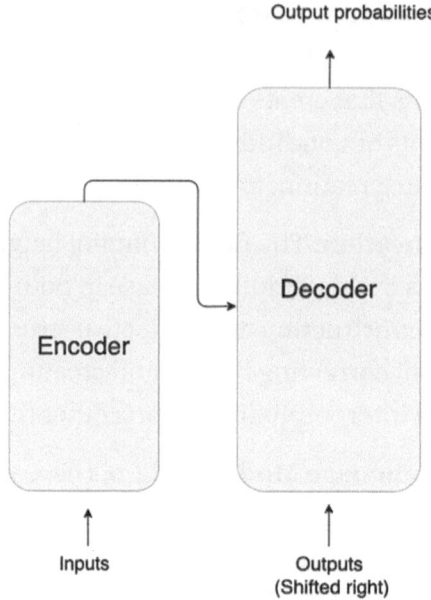

Figure 5-5. Basic language transformer architecture

Encoder (left): The encoder processes incoming data to construct an internal representation, capturing its essential aspects. This is designed to allow the model to comprehend the input effectively.

Decoder (right): Utilizing the encoded features from the encoder alongside additional inputs, the decoder produces the desired sequence. This is tailored to enable the model to create outputs efficiently.

Transformers are distinguished by their utilization of unique components known as *attention layers*. The foundational paper of transformer technology was aptly titled "Attention Is All You Need," highlighting the central role of these layers. They enable the model to focus on specific parts of the input sentence while processing each word, prioritizing certain words over others.

To illustrate, let's look at language translation, for example, from English to French. When translating the phrase "You like this book," the model must consider the word "You" to accurately translate "like" due to the variations in verb conjugation in French based on the subject. Other parts of the sentence are less relevant for translating "like."

Similarly, the word "this" requires consideration of the following noun "book" to determine the appropriate translation, as the gender of the noun in French affects the translation of "this." The translation of "book" is unaffected by the rest of the sentence. As sentences become more complex, the model must strategically focus on particular words that may significantly impact the translation.

This principle of context affecting meaning is fundamental to all natural language tasks; a word's significance is intricately tied to the surrounding words.

Initially crafted for translation, the original transformer model (Figure 5-6) employs a unique training process. The encoder ingests input sentences in one language while the decoder feeds corresponding sentences in another. Within the encoder, attention layers can consider every word in the input sentence, which is essential for accurate translation where context matters. In contrast, the decoder operates stepwise, considering only the words it has already translated. For instance, if the first three words of a sentence have been translated, the decoder will leverage this information, along with the encoder's input, to predict the fourth word.

CHAPTER 5 LARGE LANGUAGE MODELS

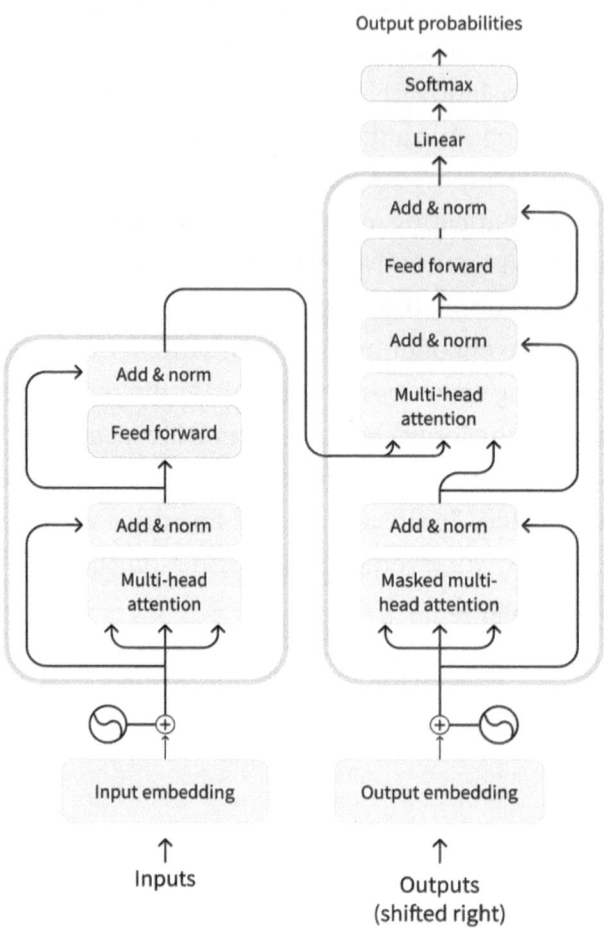

Figure 5-6. *Original transformer architecture[14] (source: https://arxiv.org/pdf/1706.03762.pdf)*

During training, to enhance efficiency, the decoder is provided with the entire target sentence but is restricted from using future words to aid in prediction. This ensures that it can consider only the preceding words when predicting a word. The archetypal transformer setup features an encoder on one side and a decoder on the other to transform input text to a target language.

The initial attention layer scrutinizes all previous decoder inputs in a decoder block. In contrast, the subsequent layer integrates the encoder's output, encompassing the entire input sentence to enhance the prediction of the current word. This feature is crucial for handling the variances in word order and grammatical structures across different languages, where context provided later can influence the translation of earlier words.

Additionally, attention masks are employed within the encoder/decoder framework to selectively focus the model's attention, mainly to disregard certain words like padding tokens that equalize sentence lengths for batch processing.

There are three types of transformer variations, as shown in Table 5-1. We will dive into the details of each type of transformer variation in the upcoming sections. Table 5-1 outlines the key differences, advantages, and use cases of these transformer variations.

Table 5-1. Comparison of Transformer Types

Transformer Type	Key Differences	Advantages	Typical Use Cases
Encoder-only	Encoder block only	Captures relationships within a sequence	Text classification, named entity recognition, etc.
Decoder-only	Decoder block only	Generates sequences based on a provided starting point	Text generation, code generation, etc.
Encoder-decoder	Separate encoder and decoder blocks	Well-suited for tasks requiring an understanding of both input and output sequences	Machine translation, text summarization, question answering, etc.

Encoder Models

Now that we understand the basics of a transformer model, we will explore different types of transformer models, starting with the ones that contain only the encoder component of the transformers. Encoder architectures utilize solely the encoder component of the transformer model, where attention layers concurrently process all elements of the input sentence. This concurrent processing capability has led to these models being described as having "bi-directional" attention, and they are frequently referred to as *auto-encoding models*.

The foundational training of these models typically involves the alteration of input sentences, such as by concealing random words, with the goal for the model to deduce or restore the original sentence.

Such encoder models are particularly adept at tasks that demand a comprehensive understanding of entire sentences. This includes applications like classifying sentences, identifying named entities, and, more broadly, classifying words within sentences and answering questions based on information extracted directly from texts.

BERT

BERT[15] and its variants are the most popular encoder models that utilize a bidirectional approach, unlike traditional language models that process text in a left-to-right or right-to-left manner. It takes into account the entire context of a word by considering both the preceding and succeeding words (as shown in Figure 5-7), enabling a deeper understanding of word meaning.

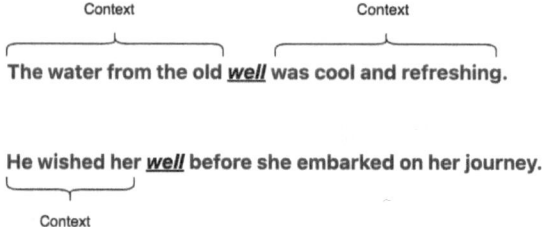

Figure 5-7. *BERT's bidirectional context approach*

This is useful in understanding the meaning, as both the left and right contexts are important. The pre-training objective of BERT is masked language modeling and next-sentence prediction. Let's understand the pre-training objectives one by one.

> **Masked language model (MLM):** BERT leverages a masked language model objective to randomly mask out a certain percentage (k%) of words in the input sequence and trains the model to predict those masked words based on the remaining context (Figure 5-8). This objective encourages BERT to learn contextual representations that accurately fill missing words. The value of k for masking in BERT is 15% uniformly sampled. If you mask too few words, it is computationally expensive, whereas if you mask too many words, there is not enough context in the text to learn from.

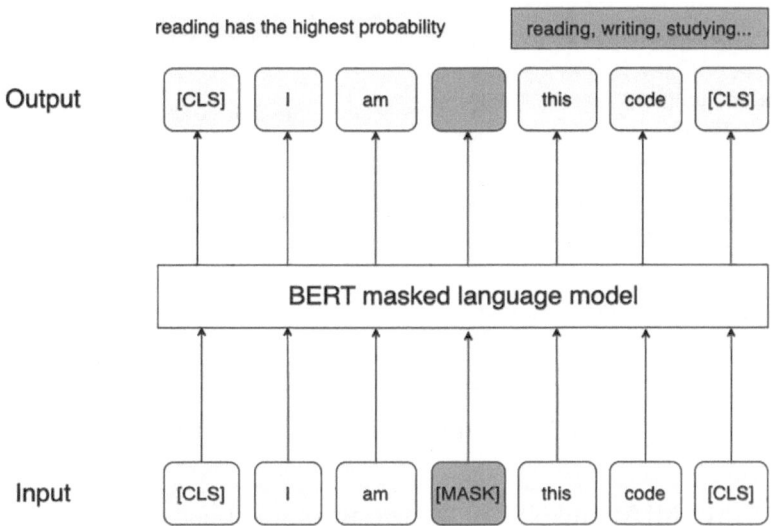

Figure 5-8. Masked language modeling

The process of masking for 15% of the words is distributed as follows:

1. 80% of these are replaced with [MASK] tokens.
   I am reading this book -> I am [MASK] this book

2. 10% of these are replaced with random words in the vocabulary.
   I am reading this book -> I am writing this book

3. 10% of these are kept unchanged.
   I am reading this book -> I am reading this book

The second pre-training objective is next-sentence prediction.

> **Next-sentence prediction (NSP):** BERT also employs an NSP objective during pre-training. It predicts whether two sentences appear consecutively or are randomly sampled (Figure 5-9). This objective helps BERT learn to capture the relationships between sentences, which is useful for downstream tasks that require understanding sentence-level semantics.

CHAPTER 5 LARGE LANGUAGE MODELS

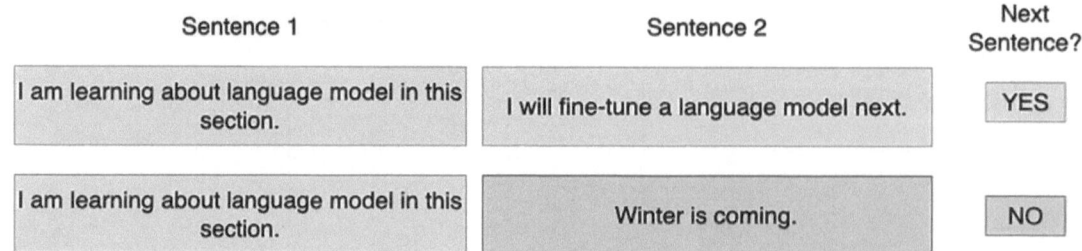

Figure 5-9. Next-sentence prediction

NSP is the base of many downstream tasks such as paraphrase detection, question-answering (QA), etc., which require understanding the relationship between two sentences.

Thus, BERT uses MLM and NSP in the pre-training phase to learn the representations, as shown in Figure 5-10.

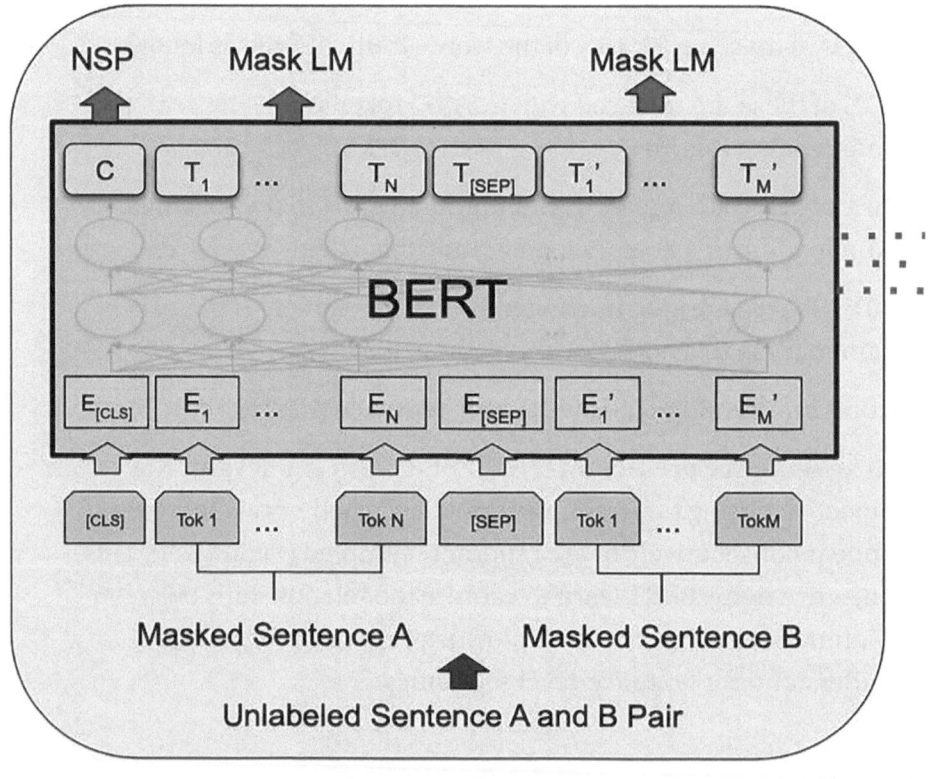

Figure 5-10. BERT pre-training[16] (https://arxiv.org/abs/1810.04805)

To illustrate BERT's practical utility, let's consider its application in sentiment analysis, a common NLP task. Businesses often employ sentiment analysis to gauge customer opinions on their products or services based on reviews.

Here's a simple example showcasing how you can leverage BERT for this purpose:

```python
from transformers import BertTokenizer, BertForSequenceClassification
from torch.nn.functional import softmax
import torch

# Load the pre-trained BERT model and tokenizer
model = BertForSequenceClassification.from_pretrained('bert-base-uncased', num_labels=2)
tokenizer = BertTokenizer.from_pretrained('bert-base-uncased')

# Sample text
text = "The new coffee flavor is amazing!"

# Encode text
encoded_input = tokenizer(text, return_tensors='pt', truncation=True, max_length=512)

# Model prediction
with torch.no_grad():
    outputs = model(**encoded_input)

# Process prediction
predictions = softmax(outputs.logits, dim=-1)

# Output sentiment analysis
sentiments = ['Negative', 'Positive']
sentiment_score = predictions.numpy().flatten()
sentiment_result = sentiments[sentiment_score.argmax()]

print(f"Sentiment: {sentiment_result} (Confidence: {sentiment_score.max():.2f})")
```

In this snippet, we first load the required libraries, a pre-trained BERT model (BertForSequenceClassification is specifically fine-tuned for the binary classification task of sentiment analysis), and its tokenizer. Then, we pass an example text for sentiment analysis and encode it using the BERT tokenizer. Finally, we pass the encoded text to the BERT model and process the prediction to convert the probability into a sentiment class. So, BERT efficiently classifies the sentiment of a given text, discerning it as positive or negative.

CHAPTER 5 LARGE LANGUAGE MODELS

Fine-Tuning BERT

In the previous section, we learned that the pre-trained language models should be fine-tuned to be adapted for a specific task or domain (Figure 5-11). The BERT family models can be fine-tuned for sentence-level tasks such as sentence pair classification (Figure 5-11 (a)), token-level tasks such as extractive question answering (Figure 5-11 (c)), named entity recognition (NER) (Figure 5-11 (d)), and other NLP tasks.

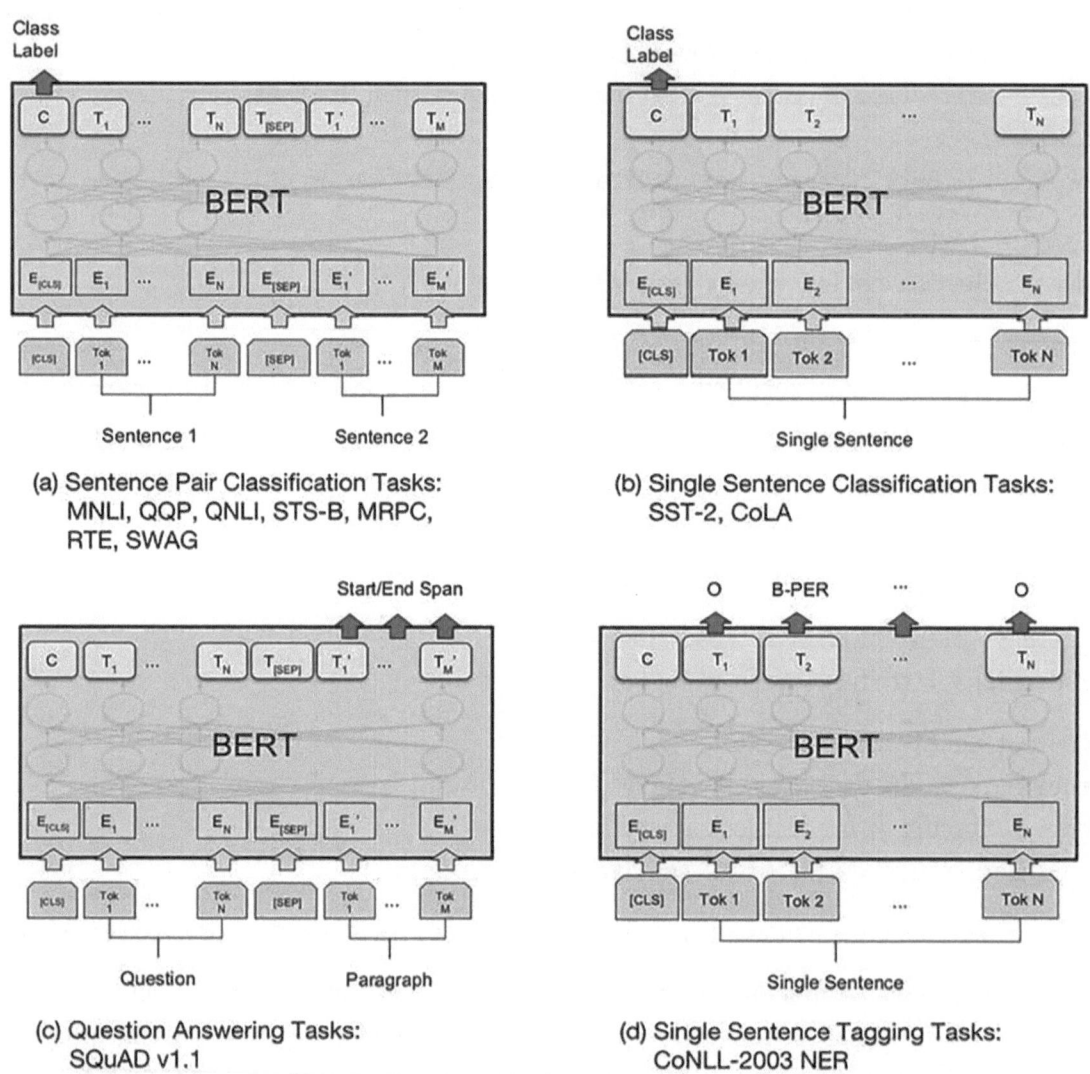

Figure 5-11. BERT fine-tuning[17] (source: https://arxiv.org/abs/1810.04805)

CHAPTER 5 LARGE LANGUAGE MODELS

Figure 5-12 illustrates the BERT fine-tuning process. The process begins with installing libraries essential for loading and fine-tuning the BERT model. This is followed by data loading. Then, we pre-process the data to prepare it for the model. Next, we load a pre-trained BERT model, establishing the foundation for fine-tuning.

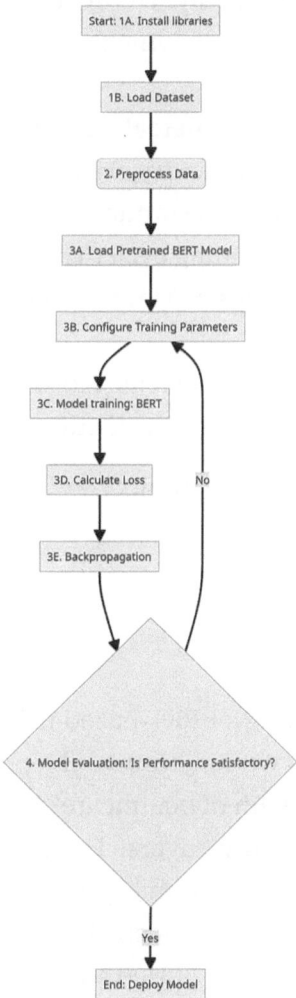

Figure 5-12. *BERT fine-tuning flowchart*

The next critical step is configuring training parameters. This involves setting up hyperparameters, such as the learning rate, batch size, and number of epochs. Once the settings are in place, we start the model training. During training, we evaluate the model's learning by calculating loss, which measures the predictions against actual outcomes. The loss calculation plays a crucial role. It guides the model's weight adjustments during backpropagation, allowing the model to learn from its mistakes and refine its understanding.

After each training iteration, we evaluate the model's performance on the hold-out test dataset. If the performance is satisfactory, we deploy the model. However, if the performance does not meet the required standards, we revise the training parameters based on feedback, reiterating the training cycle. This iterative process ensures the model is tuned until it achieves the desired level of accuracy and is ready for effective deployment.

We merge a few steps shown in the flowchart to simplify the process in our exercises. So, we will follow the four consistent steps in all our exercises.

1. Installing libraries and data loading
2. Data preprocessing
3. Model training
4. Model evaluation

Let's try our hands at fine-tuning a BERT-based model.

Have you ever wondered how we can harness the power of advanced technology to enhance patient safety in the realm of healthcare? Let's delve into the world of NLP and explore how it can help us identify Adverse Drug Events (ADEs)—instances where patients experience harm due to medication exposure.

ADEs are significant concerns in healthcare, but a substantial amount of information about these drug-related safety issues is buried within unstructured medical case reports. These reports are often challenging for machines to decipher, making them primarily accessible to human readers.

In this section, we will embark on an exciting journey to fine-tune a BERT-based model specifically designed to recognize ADEs within text. To achieve this, we'll leverage a comprehensive ADE dataset[18] available from the Hugging Face Hub. We'll teach our model to distinguish between ADE-related and non-ADE-related texts through fine-tuning.

But that's not all; we'll also fine-tune a BERT-based model for another task, named entity recognition, to extract information related to an adverse effect of a drug using the same ADE dataset from Hugging Face.

So, if you're ready to unlock the potential of NLP and make strides in patient safety, let's dive into this journey together and explore the fascinating world of fine-tuning BERT for a critical downstream task.

As we embark on this journey to fine-tune BERT for a critical downstream task, we can't help but reflect on the significant impact that ADEs have on our healthcare system. According to the Office of Disease Prevention and Health Promotion, ADEs wield severe consequences that touch the lives of countless patients.

In inpatient settings, ADEs are responsible for an estimated one in three of all hospital adverse events. They affect approximately 2 million hospital stays annually, adding 1.7 to 4.6 days to the already challenging hospitalization experience. On the outpatient front, ADEs lead to more than 3.5 million physician office visits, contribute to an estimated 1 million emergency department visits, and result in approximately 125,000 hospital admissions.

Now, imagine the potential impact of automating the identification of ADE-related phrases within unstructured text data using the power of AI. It's not just about streamlining processes; it's about enhancing patient safety. By leveraging AI to flag ADE-related content, we can prevent ADEs from occurring in the first place.

The benefits are substantial: safer and higher quality healthcare services, reduced healthcare costs, more informed and engaged consumers, and improved health outcomes. Together, we will explore the path to making this vision a reality as we fine-tune BERT to play a pivotal role in this essential endeavor.

Fine-Tuning for Sentence Classification

We will use the ADE-Corpus-V2 dataset publicly available on Hugging Face. What makes this dataset particularly intriguing is its versatility—it's divided into three insightful subsets:

- The ADE classification corpus
- The drug ADE relation corpus
- The drug dosage relation corpus

CHAPTER 5 LARGE LANGUAGE MODELS

For our first task, we'll focus on the classification subset. This subset is all about categorizing sentences, helping us determine whether they are ADE-related (label=1) or not (label=0).

Now, let's take a peek at what this dataset looks like; see Figure 5-13.

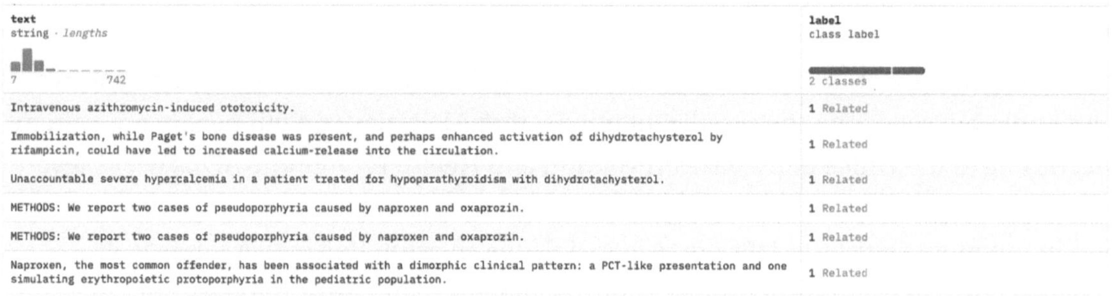

text string · lengths	label class label
7 ▬▬▬ 742	2 classes
Intravenous azithromycin-induced ototoxicity.	1 Related
Immobilization, while Paget's bone disease was present, and perhaps enhanced activation of dihydrotachysterol by rifampicin, could have led to increased calcium-release into the circulation.	1 Related
Unaccountable severe hypercalcemia in a patient treated for hypoparathyroidism with dihydrotachysterol.	1 Related
METHODS: We report two cases of pseudoporphyria caused by naproxen and oxaprozin.	1 Related
METHODS: We report two cases of pseudoporphyria caused by naproxen and oxaprozin.	1 Related
Naproxen, the most common offender, has been associated with a dimorphic clinical pattern: a PCT-like presentation and one simulating erythropoietic protoporphyria in the pediatric population.	1 Related

Figure 5-13. *ADE classification dataset*

Step 1: Installing Libraries and Data Loading

We will use the Hugging Face libraries datasets, accelerate and evaluate to load the data, train the model on a distributed configuration, and evaluate the fine-tuned model, respectively.

```
!pip install datasets
!pip install accelerate -U
!pip install evaluate
```

Next, we will load the ADE dataset from Hugging Face as shown here:

```
from datasets import load_dataset

ade = load_dataset("ade_corpus_v2", "Ade_corpus_v2_classification")
```

Here, we use the load_dataset function to load a specific dataset, namely, the ade_corpus_v2 dataset, with the Ade_corpus_v2_classification configuration.

```
Dataset({
    features: ['text', 'label'],
    num_rows: 23516
})
```

There are 23,516 records in this dataset with two features: text and label representing whether the drug had an adverse effect.

Step 2: Data Preprocessing

Now that we have loaded the data, we need to prepare it for fine-tuning. We will start the data processing step by splitting the data into train and test, as shown here:

```
# Splitting the dataset into training and test sets
train_test_split = ade['train'].train_test_split(test_size=0.2)

# Assigning the split datasets to train and test variables
ade['train'] = train_test_split['train']
ade['test'] = train_test_split['test']
print(f"Number of training examples: {ade['train'].num_rows}")
print(f"Number of test examples: {ade['test'].num_rows}")
```

In the previous code block, we are splitting the dataset into an 80/20 ratio. We will use 80% of the dataset for training our model and 20% to evaluate our model on an unseen dataset. Here is the output of the previous code block:

```
Number of training examples: 18812
Number of test examples: 4704
```

The next step in data preparation is to tokenize the dataset for fine-tuning a BERT-based model. We will initialize the tokenizer and apply it to our dataset as shown here:

```
from transformers import AutoTokenizer

text_tokenizer = AutoTokenizer.from_pretrained("distilbert-base-uncased")
```
[19]

```
def tokenize_texts(examples):
    return text_tokenizer(examples["text"], truncation=True)

tokenized_ade = ade.map(tokenize_texts, batched=True)
```

In the previous code block, we are carrying out these three steps:

1. First, we import `AutoTokenizer` from the transformers library. We then initialize it with `from_pretrained("distilbert-base-uncased")`. This tokenizer is designed to work with the `DistilBERT` model, a lighter version of BERT that retains most of its performance. The uncased version means that the tokenizer will treat uppercase and lowercase letters as the same, which is standard for English language models.

2. Next, we define a function named `tokenize_texts`. This function takes text examples and applies the tokenizer to them. The `truncation=True` parameter ensures that any text longer than the model can handle is appropriately truncated. This step is crucial because BERT-based models have a maximum sequence length limit.

3. Finally, we use the map method on the ade dataset to apply the `tokenize_texts` function to each example in the dataset. The `batched=True` parameter allows this process to happen in batches, which is more efficient than tokenizing each example individually. This results in `tokenized_ade`, which is our dataset with all the text converted into a format that our `DistilBERT` model can understand and process.

Now, we can again split our tokenized dataset into train and test using the following code:

```
# Splitting the dataset into training and test sets
tokenized_train_test_split = tokenized_ade['train'].train_test_split(test_size=0.2)

# Assigning the split datasets to train and test variables
tokenized_ade['train'] = tokenized_train_test_split['train']
tokenized_ade['test'] = tokenized_train_test_split['test']
print(f"Number of training examples: {tokenized_ade['train'].num_rows}")
print(f"Number of test examples: {tokenized_ade['test'].num_rows}")
```

By executing this code, we ensure that our model is trained on a robust set of examples and later evaluated on unseen data, a practice essential for testing the model's ability to generalize to new data. This step is critical to the machine learning workflow, especially in natural language processing tasks.

Finally, we will create a batch of examples using `DataCollatorWithPadding`. We will use this to dynamically pad the sentences to the longest length in a batch during collation instead of padding the whole dataset to the maximum length as shown here:

```
from transformers import DataCollatorWithPadding

padding_collator = DataCollatorWithPadding(tokenizer=text_tokenizer)
```

By incorporating this padding collator, we streamline our training process, ensuring that each batch of data fed into the model during training is properly formatted and padded. This step is vital for efficient and effective model training.

Step 3: Model Training

In this step, we will set up a BERT-based model for fine-tuning and evaluation. Let's begin by loading a DistilBERT model as shown here:

```
from transformers import AutoModelForSequenceClassification, TrainingArguments, Trainer

classification_model = AutoModelForSequenceClassification.from_pretrained("distilbert-base-uncased", num_labels=2)
```

In this code block, we're initializing our classification model for the task. We import the necessary components from the `transformers` library and then create `classification_model` using `AutoModelForSequenceClassification`. This model, specifically the `distilbert-base-uncased` version, is pre-trained and fine-tuned for sequence classification tasks, and we specify `num_labels=2` indicating it's a binary classification task. This step lays the groundwork for training our model on the dataset we've prepared.

Now, we will set up methods to evaluate the model. We will first import the evaluate library and load the accuracy method, as shown here:

```
import evaluate
accuracy = evaluate.load("accuracy")
```

Chapter 5 Large Language Models

Next, we will define a method to calculate accuracy, as shown here:

```python
import numpy as np

def calculate_metrics(eval_pred):
    predictions, labels = eval_pred
    predictions = np.argmax(predictions, axis=1)
    return accuracy.compute(predictions=predictions, references=labels)
```

This code block introduces the `calculate_metrics` function, which is crucial for assessing our model's effectiveness on the test set. The function retrieves both predictions and labels from `eval_pred`. Then, it leverages NumPy to identify the most probable prediction for each input. Finally, it calculates the accuracy of these predictions by comparing them to the actual labels. This evaluation process provides valuable insights into our model's strengths and weaknesses on unseen data.

Finally, let's fine-tune the DistilBERT model to classify and predict whether a sentence is ADE-related or not, as shown here:

```python
train_parameters = TrainingArguments(
    output_dir="./results",
    learning_rate=2e-5,
    per_device_train_batch_size=16,
    per_device_eval_batch_size=16,
    num_train_epochs=5,
    weight_decay=0.01,
)

trainer = Trainer(
    model=classification_model,
    args=train_parameters,
    train_dataset=tokenized_ade["train"],
    eval_dataset=tokenized_ade["test"],
    tokenizer=text_tokenizer,
    data_collator=padding_collator,
    compute_metrics=calculate_metrics,
)

trainer.train()
```

In this code block, we set up the training configuration and initiate the training of our model.

We first define `train_parameters` using `TrainingArguments`, specifying the output directory, learning rate, batch sizes, number of epochs, and weight decay. Then, we create a `Trainer` object, providing it with our classification model, training parameters, datasets, tokenizer, data collator, and custom metric calculation function. Finally, we start training the model with `trainer.train()`. This process is where our model learns from the data, adapting to perform the classification task effectively. Figure 5-14 shows the output of the training task.

[4705/4705 02:39, Epoch 5/5]

Step	Training Loss
500	0.297000
1000	0.204800
1500	0.147800
2000	0.111200
2500	0.077000
3000	0.060900
3500	0.039500
4000	0.034500
4500	0.024100

Figure 5-14. BERT classification fine-tuning

Finally, we can save our fine-tuned model locally, as shown here:

```
trainer.save_model("fine_tuned_classification_model")
```

This will ensure that the fine-tuned model is available on the disk, and we can load this model later to make predictions or even further fine-tune it.

Step 4: Model Evaluation

The last step in any ML/NLP pipeline is to evaluate your model after training it. But before we evaluate the model, let's test our model with an example inference on the holdout test dataset. Let's take an example record from our test dataset shown here:

CHAPTER 5 LARGE LANGUAGE MODELS

```
ade["test"][0]
{'text': 'In patients with swallowing dysfunction and pneumonia, a history
of mineral oil use should be obtained and a diagnosis of ELP should be
considered in the differential diagnoses if mineral oil use has occurred.',
 'label': 1}
```

We will make use of the `pipeline` method from the transformers library to make inferences on this example record, as shown here:

```
text = ade["test"][0]['text']

from transformers import pipeline

text_classifier = pipeline("text-classification", model="fine_tuned_
classification_model")
text_classifier(text)
```

First, we extract a text sample from the test dataset. Then, we initialize a text classification pipeline using our fine-tuned model and call this pipeline to classify the sample text, demonstrating how our model performs on individual examples. The fine-tuned model returns the following output:

```
[{'label': 'LABEL_1', 'score': 0.9991610050201416}]
```

This shows that it predicted that this record was ADE-related, with a confidence score of 99.9%. Now, we can run this inference for the entire dataset and record the predictions along with the actual labels in a Pandas dataframe, as shown here:

```
from transformers import pipeline
import pandas as pd

# Load the fine-tuned classification_model for text classification
text_classifier = pipeline("text-classification", model="fine_tuned_
classification_model")

# Assuming you have the 'ade["test"]' dataset
test_data = ade["test"]
```

```python
# Create empty lists to store results
texts = []
true_labels = []
predicted_labels = []

# Iterate through each record in the dataset
for record in test_data:
    text = record["text"]
    true_label = record["label"]

    # Make a prediction using the text_classifier
    prediction = text_classifier(text)
    predicted_label = prediction[0]["label"]

    # Append the results to the lists
    texts.append(text)
    true_labels.append(true_label)
    predicted_labels.append(predicted_label)

# Create a DataFrame to display the results
evaluation_results = pd.DataFrame({
    "Text": texts,
    "True Label": true_labels,
    "Predicted Label": predicted_labels
})

# Print the DataFrame
print(evaluation_results)
```

In this code block, we're using our fine-tuned classification model to evaluate its performance on the ade["test"] dataset. We load the model into a text classification pipeline, iterate through each record in the test dataset to predict labels for the texts, and then compile these texts, their true labels, and the predicted labels into a Pandas DataFrame. This DataFrame, evaluation_results, provides a clear and structured way to review and analyze the model's predictions, giving us valuable insights into its accuracy and effectiveness. Figure 5-15 shows the dataframe with the evaluation results.

	Text	True Label	Predicted Label
0	In patients with swallowing dysfunction and pn...	1	LABEL_1
1	Clonidine: a practical guide for usage in chil...	0	LABEL_0
2	When a CPS perforates the transverse foramen, ...	0	LABEL_0
3	Choanal atresia and athelia: methimazole terat...	1	LABEL_1
4	Transnasal fiberoptic laryngoscopy revealed a ...	0	LABEL_0

Figure 5-15. ADE classification prediction

Finally, we can calculate the model accuracy by comparing the predicted label with the true label, as shown here:

```
# Map the "LABEL_0" to 0 and "LABEL_1" to 1 in the 'Predicted Label' column
evaluation_results['Predicted Label'] = evaluation_results['Predicted Label'].map({"LABEL_0": 0, "LABEL_1": 1})

# Convert the 'True Label' column to integer type
evaluation_results['True Label'] = evaluation_results['True Label'].astype(int)

# Calculate accuracy by comparing 'True Label' and 'Predicted Label'
correct_predictions = (evaluation_results['True Label'] == evaluation_results['Predicted Label']).sum()
total_samples = len(evaluation_results)
accuracy = correct_predictions / total_samples

# Print the accuracy
print(f"Accuracy: {accuracy:.2%}")

# Save the DataFrame to a CSV file
evaluation_results.to_csv("accuracy_results.csv", index=False)
```

In this code block, we're processing and evaluating the accuracy of our model's predictions. We map the predicted labels from textual to numerical format, ensure the true labels are integers, and then calculate the accuracy by comparing the true and predicted labels. Finally, we print the accuracy percentage and save the evaluation results as a CSV file for further analysis or reporting. Here is the output of the previous code:

```
Accuracy: 94.58%
```

Conclusion

In our first fine-tuning exercise of a Language model, we successfully fine-tuned a BERT-based model, specifically the DistilBERT variant, to perform a sentence classification task, achieving a remarkable accuracy of 94.58%. Here's a brief overview of how we accomplished this:

1. **Dataset Preparation:** We started by selecting an appropriate dataset for our task. In this case, we used the `ade_corpus_v2` dataset from the Hugging Face repository, which is well-suited for sentence classification tasks. We then split this dataset into training and test sets, ensuring we had separate data for training our model and evaluating its performance.

2. **Tokenization and Data Preparation:** We used the `AutoTokenizer` from the `transformers` library to tokenize our dataset. This tokenizer is compatible with the DistilBERT model and ensures that our text data is in the correct format for the neural network. We also applied preprocessing, like padding, to ensure all input sequences were of uniform length.

3. **Model Selection and Training:** For the model, we chose `distilbert-base-uncased`—a lighter version of BERT that retains most of BERT's performance but is faster and less resource-intensive. We fine-tuned this pre-trained model on our dataset, adjusting parameters such as the learning rate, the batch size, and the number of epochs to optimize training.

4. **Evaluation and Metrics:** To evaluate our model, we used a custom function to calculate accuracy, comparing the predicted labels against the true labels in our test dataset. This helped us measure how well our model was performing.

5. **Achieving High Accuracy:** Our model boasts an impressive 94.58% accuracy, showcasing the effectiveness of our chosen approach. This stellar performance is attributed to utilizing the pre-trained DistilBERT model and fine-tuning it on a carefully selected, relevant dataset. This high accuracy suggests the model's ability to accurately classify sentences, making it a valuable tool for various sentence classification tasks.

Furthermore, this success story is a testament to the power of transfer learning and the efficiency of BERT-based models in tackling complex natural language processing tasks with remarkable accuracy. The key to achieving this outcome lies in our systematic approach, meticulously encompassing every step from dataset preparation to model evaluation.

Fine-Tuning for Named Entity Recognition

In this section of the book, we build upon our previous exploration of fine-tuning a BERT-based model for sentence classification. Having previously focused on identifying whether a sentence is related to ADE using the ADE dataset from Hugging Face, we now delve deeper into the nuances of NER. Our objective is to accurately identify specific entities—namely, drugs and their associated adverse effects—within the same dataset.

To achieve this, we will leverage a variation of the BERT model pre-trained on scientific and biological datasets. This specialized version of BERT is adept at understanding the complex vocabulary in medical and scientific texts. By fine-tuning this model on our targeted NER task, we aim to harness its advanced understanding of the corpus to recognize and categorize key entities with high precision.

This section is a progression from sentence-level classification to the more intricate entity-level recognition and a demonstration of how domain-specific pre-training can significantly enhance a model's performance in specialized tasks. As we navigate this process, we will uncover the intricacies of fine-tuning for NER and the impactful role of domain-specific knowledge in building effective models.

Step 1. Installing Libraries and Data Loading

Let's begin by installing the libraries as shown here:

```
!pip install datasets transformers seqeval
!pip install accelerate -U
```

In this step, we are installing crucial Python libraries that form the backbone of our NER project. The first command installs the `datasets` library for data handling, `transformers` for accessing powerful NLP models, and `seqeval` for evaluating our NER tasks, while the second command ensures that the `accelerate` library is up-to-date, enhancing our model's training efficiency across various hardware platforms.

We will import a set of libraries and modules to be used later in the notebook.

```
from datasets import Dataset, ClassLabel, Sequence, load_dataset, load_metric
import numpy as np
import pandas as pd
from spacy import displacy
import transformers
from transformers import (AutoModelForTokenClassification,
                          AutoTokenizer,
                          DataCollatorForTokenClassification,
                          pipeline,
                          TrainingArguments,
                          Trainer)
```

In the previous code block, we are importing libraries and modules essential for our NER task. We begin by importing various utilities from the `datasets` library, including functions to load datasets, metrics, and classes for handling different data types. Next, we bring in `numpy` and `pandas`, two foundational libraries for numerical and data frame operations. The SpaCy library's `displacy` module is included for visualization. Finally, we import several components from the transformers library: `AutoModelForTokenClassification` and `AutoTokenizer` for model and tokenizer setup, `DataCollatorForTokenClassification` for data preparation, `pipeline` for easy inference, and `Trainer` and `TrainingArguments` for model training and configuration. These imports equip us with the tools to efficiently process, train, and evaluate our NER model.

Now, it's time to load the dataset for further processing.

```
datasets = load_dataset("ade_corpus_v2", "Ade_corpus_v2_drug_ade_relation")
```

We are loading the `Ade_corpus_v2_drug_ade_relation` subset of the `ade_corpus_v2` dataset, which we used from Hugging Face's datasets in the previous section. This dataset contains text, drug, effect, and indexes for a total of 6,821 records, as shown here:

```
DatasetDict({
    train: Dataset({
        features: ['text', 'drug', 'effect', 'indexes'],
        num_rows: 6821
    })
})
```

Here is an example record from the dataset:

```
{'text': 'Intravenous azithromycin-induced ototoxicity.',
 'drug': 'azithromycin',
 'effect': 'ototoxicity',
 'indexes': {'drug': {'start_char': [12], 'end_char': [24]},
  'effect': {'start_char': [33], 'end_char': [44]}}}
```

Step 2: Data Pre-processing

Before diving into data preparation for the model, let's scrutinize the dataset more closely. We'll find that some sentences appear multiple times, each instance showcasing different combinations of drugs and their associated adverse effects. Here are some illustrative examples:

```
{'text': 'RESULTS: A 44-year-old man taking naproxen for chronic low back pain and a 20-year-old woman on oxaprozin for rheumatoid arthritis presented with tense bullae and cutaneous fragility on the face and the back of the hands.', 'drug': 'naproxen', 'effect': 'cutaneous fragility', 'indexes': { "drug": { "start_char": [ 34 ], "end_char": [ 42 ] }, "effect": { "start_char": [ 163 ], "end_char": [ 182 ] } }}
{'text': 'RESULTS: A 44-year-old man taking naproxen for chronic low back pain and a 20-year-old woman on oxaprozin for rheumatoid arthritis presented with tense bullae and cutaneous fragility on the face and the back of the hands.', 'drug': 'oxaprozin', 'effect': 'cutaneous fragility', 'indexes': { "drug": { "start_char": [ 96 ], "end_char": [ 105 ] }, "effect": { "start_char": [ 163 ], "end_char": [ 182 ] } }}
{'text': 'RESULTS: A 44-year-old man taking naproxen for chronic low back pain and a 20-year-old woman on oxaprozin for rheumatoid arthritis presented with tense bullae and cutaneous fragility on the face and the back of the hands.', 'drug': 'naproxen', 'effect': 'tense bullae', 'indexes': { "drug": { "start_char": [ 34 ], "end_char": [ 42 ] }, "effect": { "start_char": [ 146 ], "end_char": [ 158 ] } }}
{'text': 'RESULTS: A 44-year-old man taking naproxen for chronic low back pain and a 20-year-old woman on oxaprozin for rheumatoid arthritis presented with tense bullae and cutaneous fragility on the face and the
```

back of the hands.', 'drug': 'oxaprozin', 'effect': 'tense bullae', 'indexes': { "drug": { "start_char": [96], "end_char": [105] }, "effect": { "start_char": [146], "end_char": [158] } }}

This repetition is not conducive for NER tasks. If we were to assign token labels for each row as it is, the same tokens in identical sentences would receive different labels. Such inconsistency can lead to confusion in the model during the fine-tuning process. Therefore, it's necessary to amalgamate all provided ranges for each unique sentence. This consolidation will allow for a single pass to accurately label all known entities in the text.

```
merged_dataset = {}

for item in datasets["train"]:
    text = item["text"]
    if text not in merged_dataset:
        merged_dataset[text] = {
            "text": text,
            "drugs": [item["drug"]],
            "effects": [item["effect"]],
            "drug_starts": set(item["indexes"]["drug"]["start_char"]),
            "drug_ends": set(item["indexes"]["drug"]["end_char"]),
            "effect_starts": set(item["indexes"]["effect"]["start_char"]),
            "effect_ends": set(item["indexes"]["effect"]["end_char"])
        }
    else:
        merged_data = merged_dataset[text]
        merged_data["drugs"].append(item["drug"])
        merged_data["effects"].append(item["effect"])
        merged_data["drug_starts"].update(item["indexes"]["drug"]
        ["start_char"])
        merged_data["drug_ends"].update(item["indexes"]["drug"]
        ["end_char"])
        merged_data["effect_starts"].update(item["indexes"]["effect"]
        ["start_char"])
        merged_data["effect_ends"].update(item["indexes"]["effect"]
        ["end_char"])
```

CHAPTER 5　LARGE LANGUAGE MODELS

In the previous block, we're creating a merged_dataset dictionary to consolidate information from our training dataset. We check if the text is already in our dictionary for each record. If not, we add it along with the drug and effect information, including their start and end indices. If the text is already present, we append the new drug and effect details to the existing entry. This process helps us handle instances where the exact text mentions multiple drugs and effects, ensuring each instance is uniquely captured for our NER task.

With the consolidated dataset, we must assign per-token labels to each sentence. First, we re-define our Python data structure as a Hugging Face dataset object, as shown here:

```
df_ade = pd.DataFrame(
  list(merged_dataset.values()))
```

The transformed dataframe looks like Figure 5-16.

	text	drugs	effects	drug_starts	drug_ends	effect_starts	effect_ends
0	Intravenous azithromycin-induced ototoxicity.	[azithromycin]	[ototoxicity]	{12}	{24}	{33}	{44}
1	Immobilization, while Paget's bone disease was...	[dihydrotachysterol]	[increased calcium-release]	{91}	{109}	{143}	{168}
2	Unaccountable severe hypercalcemia in a patien...	[dihydrotachysterol]	[hypercalcemia]	{84}	{102}	{21}	{34}
3	METHODS: We report two cases of pseudoporphyri...	[naproxen, oxaprozin]	[pseudoporphyria, pseudoporphyria]	{58, 71}	{80, 66}	{32}	{47}
4	Naproxen, the most common offender, has been a...	[Naproxen]	[erythropoietic protoporphyria]	{0}	{8}	{134}	{163}

Figure 5-16. *ADE transformed dataset for NER*

As there are no overlapping spans, sorting will enable us to achieve one-to-one matching of index spans. Remember that sets do not maintain the order in which elements are inserted.

```
df_ade["drug_starts"] = df_ade["drug_starts"].apply(list).apply(sorted)
df_ade["drug_ends"] = df_ade["drug_ends"].apply(list).apply(sorted)
df_ade["effect_starts"] = df_ade["effect_starts"].apply(list).apply(sorted)
df_ade["effect_ends"] = df_ade["effect_ends"].apply(list).apply(sorted)
```

In this code block, we are processing the df_ade DataFrame by converting the start and end indices of both drug and effect entities into lists and then sorting them. This ensures that the indices are in a sequential and organized format, which is crucial for our subsequent NER tasks.

We will store the data in JSON format for subsequent import into a Dataset object, as shown here:

```
df_ade.to_json("dataset.jsonl", orient="records", lines=True)
ade_dataset = load_dataset("json", data_files="dataset.jsonl")
```

It is time to create a train test split of the transformed dataset as shown here:

```
ade_train_test = ade_dataset["train"].train_test_split()
```

The whole dataset has been divided into train and test datasets:

```
DatasetDict({
    train: Dataset({
        features: ['text', 'drugs', 'effects', 'drug_starts', 'drug_ends',
        'effect_starts', 'effect_ends'],
        num_rows: 3203
    })
    test: Dataset({
        features: ['text', 'drugs', 'effects', 'drug_starts', 'drug_ends',
        'effect_starts', 'effect_ends'],
        num_rows: 1068
    })
})
```

Now, the most important data preparation step is token labeling. We employ the BIO (Beginning, Inside, Outside) tagging scheme for two entities: DRUG and EFFECT. This approach leads to five distinct label classes for each token:

1. O: Outside any entity of interest
2. B-DRUG: The beginning of a DRUG entity
3. I-DRUG: Inside a DRUG entity
4. B-EFFECT: The beginning of an EFFECT entity
5. I-EFFECT: Inside an EFFECT entity

CHAPTER 5   LARGE LANGUAGE MODELS

```
entity_label_list = ['O', 'B-DRUG', 'I-DRUG', 'B-EFFECT', 'I-EFFECT']

custom_sequence = Sequence(feature=ClassLabel(num_classes=5,
                                    names=entity_label_list,
                                    names_file=None, id=None),
                                    length=-1, id=None)

ade_train_test["train"].features["ner_tags"] = custom_sequence
ade_train_test["test"].features["ner_tags"] = custom_sequence
```

In this code block, we first define entity_label_list, which contains the labels for our NER task, including both drug and effect entities with the BIO tagging scheme. We then create a custom_sequence, specifying it as a sequence of class labels for NER tags. Finally, we update the features of both the train and test sets in our ade_train_test dataset to include these NER tags, ensuring our dataset is properly formatted for training and evaluating our NER model.

Let's create a utility function that takes a row from the ADE dataset and creates tags/labels for drug and effect entities.

```
def create_bio_tags_for_entities(data_row, display_log=False):
    """ This function takes a row from the 'Ade_corpus_v2_drug_ade_
    relation' dataset
    and creates BIO tags for 'drug' and 'effect' entities.
    """

    sentence = data_row["text"]

    bio_tags = []
    current_label = "O"
    tag_prefix = ""

    # Iterate through tokens and navigate through all spans of drug
    and effect
    drug_span_index = 0
    effect_span_index = 0

    tokenized_data = tokenizer(sentence, return_offsets_mapping=True)

    for index in range(len(tokenized_data["input_ids"])):
        token_start, token_end = tokenized_data["offset_mapping"][index]
```

```python
        # Handling [CLS] and [SEP] tokens
        if token_end - token_start == 0:
            bio_tags.append(-100)
            continue

        if drug_span_index < len(data_row["drug_starts"]) and token_start
        == data_row["drug_starts"][drug_span_index]:
            current_label = "DRUG"
            tag_prefix = "B-"

        elif effect_span_index < len(data_row["effect_starts"]) and token_
        start == data_row["effect_starts"][effect_span_index]:
            current_label = "EFFECT"
            tag_prefix = "B-"

        bio_tags.append(entity_label_list.index(f"{tag_prefix}{current_
        label}"))

        if drug_span_index < len(data_row["drug_ends"]) and token_end ==
        data_row["drug_ends"][drug_span_index]:
            current_label = "O"
            tag_prefix = ""
            drug_span_index += 1

        elif effect_span_index < len(data_row["effect_ends"]) and token_end
        == data_row["effect_ends"][effect_span_index]:
            current_label = "O"
            tag_prefix = ""
            effect_span_index += 1

        # Switch to 'inside' tag after entering an entity
        if tag_prefix == "B-":
            tag_prefix = "I-"

    if display_log:
        print(f"{data_row}\n")
        original_tokens = tokenizer.convert_ids_to_tokens(tokenized_
        data["input_ids"])
```

Chapter 5 Large Language Models

```
        for index in range(len(bio_tags)):
            print(original_tokens[index], bio_tags[index])
    tokenized_data["labels"] = bio_tags

    return tokenized_data
```

In this code block, we've designed a function `create_bio_tags_for_entities` to assign BIO (Beginning, Inside, Outside) tags for drug and effect entities in each data row from the `Ade_corpus_v2_drug_ade_relation` dataset. The function tokenizes the sentence and iterates through each token, applying the appropriate BIO tag based on drug and effect entities' start and end indices. Special tokens like [CLS] and [SEP] are handled separately. If the `display_log` option is enabled, it prints the original tokens alongside their BIO tags for a detailed view. Finally, the function appends these BIO tags to the tokenized data, making it ready for use in our NER model.

We can test our tagging function on one sample record, as shown here:

```
create_bio_tags_for_entities(ade_train_test["train"][2], display_log=True)
```

The output of the previous code is displayed here:

```
{'text': 'The changes were progressive regardless of discontinuation of cyclophosphamide and led to severe restrictive ventilatory defect.', 'drugs': ['cyclophosphamide'], 'effects': ['severe restrictive ventilatory defect'], 'drug_starts': [62], 'drug_ends': [78], 'effect_starts': [90], 'effect_ends': [127]}

[CLS] -100
the 0
changes 0
were 0
progressive 0
regardless 0
of 0
discontinuation 0
of 0
cyclo 1
##phosph 2
##amide 2
```

```
and 0
led 0
to 0
severe 3
restrictive 4
ventilator 4
##y 4
defect 4
. 0
[SEP] -100
{'input_ids': [102, 111, 1334, 267, 8381, 7161, 131, 21710, 131, 14448,
5952, 5659, 137, 4030, 147, 3167, 17843, 27879, 30126, 7465, 205, 103],
'token_type_ids': [0, 0, 0, 0, 0, 0, 0, 0, 0, 0, 0, 0, 0, 0, 0, 0, 0, 0,
0, 0, 0], 'attention_mask': [1, 1, 1, 1, 1, 1, 1, 1, 1, 1, 1, 1, 1, 1, 1,
1, 1, 1, 1, 1, 1], 'offset_mapping': [(0, 0), (0, 3), (4, 11), (12, 16),
(17, 28), (29, 39), (40, 42), (43, 58), (59, 61), (62, 67), (67, 73), (73,
78), (79, 82), (83, 86), (87, 89), (90, 96), (97, 108), (109, 119), (119,
120), (121, 127), (127, 128), (0, 0)], 'labels': [-100, 0, 0, 0, 0, 0, 0,
0, 0, 1, 2, 2, 0, 0, 0, 3, 4, 4, 4, 4, 0, -100]}
```

After testing it on sample record, we can apply the utility function on the entire dataset using the following code:

```
labeled_dataset = ade_train_test.map(create_bio_tags_for_entities)
```

In this code block, we apply the `create_bio_tags_for_entities` function to each entry in our `ade_train_test` dataset. This is done using the `map` method, which efficiently processes each dataset row to generate BIO tags for the drug and effect entities. The result is a new dataset, `labeled_dataset`, where each data point now includes these tags, essential for training our NER model.

Step 3: Model Training

Let's begin this step by loading our model and setting it up for the right task. You can use any BERT-based model that has been fine-tuned on scientific (specifically medical) corpus so that it has learned the medical vocabulary. We will fine-tune the already fine-tuned BERT-based model for a downstream task of NER.

CHAPTER 5   LARGE LANGUAGE MODELS

```
task = "ner" # Should be one of "ner", "pos" or "chunk"
# model_checkpoint = "allenai/scibert_scivocab_uncased"
model_checkpoint = "dmis-lab/biobert-v1.1"
# model_checkpoint = "alvaroalon2/biobert_diseases_ner"
batch_size = 16
```

We're setting up the initial configuration for our NER task in this code block. We specify that the task is ner, which could be pos (part-of-speech tagging) or chunk (chunking). We then choose a pre-trained model for this task, opting for the dmis-lab/biobert-v1.1 model, a BioBERT variant specifically trained on biomedical literature. It is ideal for our dataset focused on drugs and their effects. Additionally, we set the batch size for training to 16, balancing efficiency and resource usage.

Note Feel free to play with other models mentioned here or in the open source that are specific to the biomedical domain.

Let's set up the model and training configuration for the NER task using the following code:

```
model = AutoModelForTokenClassification.from_pretrained(model_checkpoint, num_labels=len(entity_label_list))
model_name = model_checkpoint.split("/")[-1]
args = TrainingArguments(
    f"{model_name}-finetuned-{task}",
    evaluation_strategy = "epoch",
    learning_rate=1e-5, per_device_train_batch_size=batch_size, per_device_eval_batch_size=batch_size, num_train_epochs=5, weight_decay=0.05, logging_steps=1 )
data_collator = DataCollatorForTokenClassification(tokenizer)
```

We are setting up our model and training configuration for the NER task in the previous code block. First, we initialize the model using AutoModelForTokenClassification from our chosen pre-trained model_checkpoint and specify the number of labels based on our entity_label_list. Next, we define model_name for easier reference, derived from the model_checkpoint.

We then set up the training arguments in args using TrainingArguments. This includes setting the name for the training session, the evaluation strategy, the learning rate, the batch sizes for training and evaluation, the number of training epochs, the weight decay, and the logging frequency.

Finally, we create a data_collator using DataCollatorForTokenClassification and pass our tokenizer to it. This data collator will handle the batching and encoding of our data during training and evaluation, ensuring it is in the correct format for the model.

Let's start the training process by setting up the trainer class as shown here:

```
trainer = Trainer(
    model,
    args,
    train_dataset=labeled_dataset["train"],
    eval_dataset=labeled_dataset["test"],
    data_collator=data_collator,
    tokenizer=tokenizer,
    compute_metrics=evaluate_model_performance,
)
trainer.train()
```

In this code block, we are setting up and initiating the training process for our NER model. We create a Trainer object, supplying it with our configured model, training arguments (args), the training and evaluation datasets from labeled_dataset, the data_collator for batch processing, and the tokenizer. We also specify the compute_metrics function, which we defined earlier, to evaluate the model's performance. Finally, by calling trainer.train(), we kick off the training process, where our model will learn to recognize and classify named entities based on the data provided. This is a crucial step in fine-tuning our model to achieve high accuracy in NER tasks. You can observe the output of the training process (Figure 5-17) and see that the model can improve its F1 score in each epoch. It has an excellent accuracy and F1 score with only five training epochs.

Epoch	Training Loss	Validation Loss	Precision Score	Recall Score	F1 Score	Accuracy Score
1	0.121600	0.142491	0.782477	0.865256	0.821787	0.949080
2	0.163100	0.129841	0.811828	0.896808	0.852205	0.955228
3	0.017600	0.136384	0.819044	0.897179	0.856333	0.957217
4	0.158500	0.135221	0.828601	0.899035	0.862382	0.958410
5	0.091400	0.138306	0.828807	0.894952	0.860610	0.957832

Figure 5-17. NER fine-tuning

Step 4: Model Evaluation

We are covering the model evaluation in this step, but we configured the mechanism to evaluate the fine-tuned model before we defined the Trainer object in the previous step. We create a method called evaluate_model_performance to calculate accuracy and use this method in the compute_metrics argument while training the model. The method evaluate_model_performance is defined here:

```
metric = load_metric("seqeval")

def evaluate_model_performance(prediction_data):
    predicted_values, actual_labels = prediction_data
    predicted_values = np.argmax(predicted_values, axis=2)

    # Filter out the special token indices
    refined_predictions = [
        [entity_label_list[pred] for pred, true_label in zip(pred_row,
        label_row) if true_label != -100]
        for pred_row, label_row in zip(predicted_values, actual_labels)
    ]
    refined_labels = [
        [entity_label_list[true_label] for pred, true_label in zip(pred_
        row, label_row) if true_label != -100]
        for pred_row, label_row in zip(predicted_values, actual_labels)
    ]
    metric_results = metric.compute(predictions=refined_predictions,
    references=refined_labels)
    return {
        "precision_score": metric_results["overall_precision"],
```

```
        "recall_score": metric_results["overall_recall"],
        "f1_score": metric_results["overall_f1"],
        "accuracy_score": metric_results["overall_accuracy"],
    }
```

We're preparing to evaluate the performance of our NER model in this code block. We start by loading the seqeval metric, a standard for sequence evaluation tasks. Then, we define a function called evaluate_model_performance to process the predictions from our model.

The function takes predicted and actual labels, and for each prediction, it chooses the label with the highest probability (using np.argmax). We then filter out special token indices marked with -100. Next, we convert the numerical predictions and labels to their corresponding textual labels as defined in entity_label_list.

Finally, we calculate the evaluation metrics—precision, recall, F1 score, and accuracy—using the seqeval metric on these refined predictions and actual labels. This function comprehensively evaluates our model's performance in terms of its ability to identify and classify named entities correctly.

Now it's time to run the fine-tuned model on the hold-out test dataset and evaluate the results, as shown here:

```
predicted_results, actual_labels, _ = trainer.predict(labeled_
dataset["test"])
predicted_results = np.argmax(predicted_results, axis=2)

# Exclude special token indices
processed_predictions = [
    [entity_label_list[each_pred] for each_pred, each_label in zip(single_
    prediction, single_label) if each_label != -100]
    for single_prediction, single_label in zip(predicted_results,
    actual_labels)
]
processed_labels = [
    [entity_label_list[each_label] for each_pred, each_label in zip(single_
    prediction, single_label) if each_label != -100]
    for single_prediction, single_label in zip(predicted_results,
    actual_labels)
]
```

CHAPTER 5   LARGE LANGUAGE MODELS

```
evaluation_results = metric.compute(predictions=processed_predictions,
references=processed_labels)
evaluation_results
```

In the previous code block, we use our trained model to predict named entities on the test dataset. We use the `predict` method of the `trainer` to obtain predictions, which we then process to find the most likely label for each token using `np.argmax`.

Since our model outputs predictions for every token, including special ones like [CLS] and [SEP], we filter these out (marked with an index of -100) to focus only on meaningful predictions. We then convert the numerical indices of our predictions and actual labels into their textual form using `entity_label_list`.

Finally, we compute the evaluation metrics using the `metric` object with our processed predictions and actual labels. This gives us a comprehensive view of our model's performance on the test set, including its precision, recall, F1 score, and overall accuracy. This evaluation is essential to understand how well our model has learned to identify and classify named entities.

The evaluation results look promising. Even with only five epochs of fine-tuning, the model can predict the right entities with an F1 score of 0.86.

```
{'DRUG': {'precision': 0.915680473372781, 'recall': 0.9626749611197511,
'f1': 0.9385898407884761, 'number': 1286}, 'EFFECT': {'precision':
0.7533718689788054, 'recall': 0.8330965909090909, 'f1': 0.7912310286677909,
'number': 1408}, 'overall_precision': 0.8288071502234445, 'overall_recall':
0.8949517446176689, 'overall_f1': 0.8606103872925219, 'overall_accuracy':
0.9578315431629959}
```

Let's visualize the prediction from our fine-tuned model on some unseen dataset. First, we will create a method for visualization using the following function:

```
def display_entity_annotations(text):
    annotated_tokens = ade_ner_model(text)
    entity_annotations = []

    for token in annotated_tokens:
        entity_type = int(token["entity"][-1])
        if entity_type != 0:
            token["label"] = entity_label_list[entity_type]
            entity_annotations.append(token)
```

```
    render_params = [{"text": text, "ents": entity_annotations,
    "title": None}]

    rendered_html = displacy.render(render_params, style="ent",
    manual=True, options={
        "colors": {
                "B-DRUG": "#00FF00",
                "I-DRUG": "#00FF00",
                "B-EFFECT": "#ff0000",
                "I-EFFECT": "#ff0000",
            },
    })
    display(HTML(rendered_html))
```

In this code block, we have defined a function called `display_entity_annotations` to visualize the entity annotations predicted by our `ade_ner_model` for a given text. The function processes the tokens the model identifies, assigning a label from `entity_label_list` to each token based on its entity type. We then compile these annotations into a format suitable for rendering with SpaCy's `displacy` tool.

The function utilizes `displacy.render` to create an HTML representation of the text with highlighted entities, where each entity type is color-coded for easy differentiation (B-DRUG and I-DRUG in green, B-EFFECT and I-EFFECT in red). Finally, we display this rendered HTML, offering a clear and interactive way to view the model's predictions on the text, which is particularly helpful for understanding and demonstrating the model's capabilities in NER tasks.

Let's take a few examples of adverse effects from the Wikipedia page on adverse effects and visualize them with the displaCy library (https://en.wikipedia.org/wiki/Adverse_effect#Medications):

```
examples = [
"Rhabdomyolysis associated with statins (anticholesterol drugs)",
"Seizures caused by withdrawal from benzodiazepines",
"Drowsiness or increase in appetite due to antihistamine use. Some
antihistamines are used in sleep aids explicitly because they cause
drowsiness",
"Stroke or heart attack associated with sildenafil (Viagra), when used with
nitroglycerin",
```

CHAPTER 5 LARGE LANGUAGE MODELS

"Suicide, increased tendency associated to the use of fluoxetine and other selective serotonin reuptake inhibitor (SSRI) antidepressants",
"Tardive dyskinesia associated with use of metoclopramide and many antipsychotic medications"
]

```
from IPython.core.display import HTML
import spacy
from spacy import displacy

# for example in examples:
#     visualize_entities(example)
#     print(f"{'*' * 50}\n")
for sample in examples:
    display_entity_annotations(sample)
    print(f"{'=' * 50}\n")
```

The fine-tuned model predicts the drug and effects in the sentences as shown in Figure 5-18.

Figure 5-18. NER output

The predictions from the fine-tuned model demonstrate its ability to effectively identify and classify entities related to adverse drug effects and the drugs themselves.

Conclusion

As we conclude this section on fine-tuning a BERT-based Language model for the NER task on the ADE dataset, it's evident that the advanced capabilities of BERT and its variants offer significant advantages in understanding and extracting meaningful information from complex datasets. Our journey through fine-tuning this model has demonstrated the model's robustness in accurately identifying drugs and their adverse effects and the versatility and effectiveness of BERT in handling domain-specific language in biomedical texts.

The fine-tuned model's proficiency in distinguishing between intricate entity types like drugs and their effects showcases the potential of BERT-based models in transforming the landscape of text analysis in specialized fields. The successful application of these models on the ADE dataset reaffirms their suitability for biomedical NER tasks, paving the way for further exploration and application in other specialized domains.

The insights gained from this exercise emphasize the importance of choosing appropriate pre-trained models and fine-tuning strategies tailored to specific datasets and tasks. As we continue to advance in the field of NLP, the ability to effectively fine-tune such models will become increasingly crucial in harnessing their full potential for diverse and challenging NER tasks.

This section underscores the power of BERT-based models in NER, highlighting their adaptability, accuracy, and potential for significant contributions to biomedical text analysis and beyond.

Decoder-Only Models (Generative Pre-trained Transformer)

Decoder frameworks exclusively deploy the decoder segment of a transformer model. Within these stages, the attention mechanism is designed to consider only the preceding words in the sentence when assessing a specific word, lending these models the designation of auto-regressive models.

The preliminary training phase for decoder models primarily focuses on forecasting the subsequent word in a sequence. Such frameworks are optimally aligned with text generation tasks, given their predictive nature. Notable examples of these models encompass the following:

- CTRL
- GPT series
- Transformer XL

Let's consider their application in crafting stories or generating engaging content to further illuminate the capabilities of decoder-only models, especially in creative text generation. These models, such as the renowned GPT series developed by OpenAI, have revolutionized how we approach creative writing with AI. By leveraging a decoder-only architecture, these models generate coherent and contextually rich narratives, often indistinguishable from human-authored text.

Imagine you are tasked with generating a short story about a lost civilization. By inputting a simple prompt into a GPT model, such as "In a world forgotten by time, an ancient city lies hidden," the model takes this seed and weaves a narrative, expanding on the prompt with surprising complexity and creativity. The process is akin to planting a creative seed and watching as the model cultivates a text garden, each sentence blooming from the last.

Here's a brief code snippet to showcase how one might utilize a GPT[20] model for this purpose:

```
from transformers import pipeline

# Load the GPT model using the pipeline API
text_generator = pipeline("text-generation", model="gpt2")
# Define the prompt
prompt = "In a world forgotten by time, an ancient city lies hidden"

# Generate text based on the prompt
generated_text = text_generator(prompt, max_length=100, num_return_sequences=1)

print(generated_text[0]['generated_text'])
```

The output of the previous code snippet is shown here:

```
In a world forgotten by time, an ancient city lies hidden somewhere on the
horizon and a lost civilisation is trying to catch up with one. In A View
from Earth, writer Paul Rynne explores the legacy of those characters for
more than six hours, interviewing their heroes and their creators. To read
Part Two of Rynne's fascinating work, click here.
A view from earth, or in other media
You're not living on a planet when you start visiting your relatives.
```

This example demonstrates the practical use of decoder-only models in text generation and underscores their potential to act as a catalyst for creativity. Through such applications, these models serve not merely as tools for automated writing but as collaborators in the creative process, offering new possibilities for storytelling and content creation. We will also dive into many more examples using these models in the upcoming chapters.

Autoregressive Modeling: The Technical Backbone

At the heart of decoder-only models lies the autoregressive property, a method where the prediction of each subsequent token is conditioned on the tokens that precede it. Unlike their encoder-decoder counterparts that juggle input comprehension and output generation, decoder-only models are dedicated artisans of output, meticulously crafting token by token, each word a successor of the previous, relying solely on the partial input sequence provided to them.

This focus on sequential token generation enables these models to traverse the language landscape with fluidity and creativity that has been pivotal in tasks such as text completion, where each new word is a calculated prediction based on all the previous words.

The Mechanics of Sequence Generation

The technical prowess of decoder-only models is in creating sequences and in the probabilistic finesse with which they approach this task. They are adept at forecasting the likelihood of each possible next token, drawing from a well of past knowledge encapsulated in their internal representations. By employing activation functions like softmax, they convert these representations into probability distributions, from which the next token is sampled.

The architecture of these models, whether transformer-based with their multiheaded self-attention mechanisms or recurrent-based with their looping feedback pathways, is engineered to capture the subtleties and contexts of language. Each model iteration, from the pioneering GPT to the latest GPT-4, is a testament to the evolution of this capability.

Real-World Implications and Technical Applications

In practical applications, decoder-only models have demonstrated remarkable versatility. They serve as the backbone of machine translation services without an encoder, directly generating translations in an end-to-end manner. In text summarization, they distill expansive content into succinct representations. When it comes to synthesizing dialogue, they are the virtual puppeteers, pulling strings of conversation with lifelike ease.

Moreover, the domain of language modeling itself has been transformed by decoder-only models. They have redefined benchmarks and expectations, producing syntactically, semantically sound, and contextually nuanced text.

The Road Ahead: Challenges and Opportunities

As we venture further into the age of language model sophistication, the journey of decoder-only models is both exhilarating and fraught with technical challenges. Training these behemoths is a Herculean task, demanding extensive computational resources and large datasets. Fine-tuning them for specific tasks requires precision and understanding the subtle interplay between the model's parameters and the task at hand.

Yet, the innovation potential is boundless. The future holds promise for even more advanced iterations of decoder-only models, with improved efficiency, reduced bias, and heightened creativity. As architects of text, they are not just shaping the sentences of tomorrow but also redefining the relationship between humans and the written word.

We stand at the threshold of a new frontier in NLP, with decoder-only models as our guides. Their capacity to weave complex text tapestries from simple data threads is a technological marvel and a gateway to uncharted intellectual territories.

Encoder-Decoder Models

Encoder-decoder frameworks, also known as *sequence-to-sequence models*, integrate both the encoder and decoder components of the transformer structure. In these models, while the encoder's attention layers have complete visibility over all words in the initial input, the decoder's attention layers are limited to words that precede the current focus in the sequence.

Training these models often involves more intricate strategies beyond the primary objectives of encoder or decoder models. For example, the T5 model is trained by substituting various text lengths with a unique mask token and then teaching the model to infer the masked text. These sequence-to-sequence models excel in tasks that require generating new text based on provided input, including summarization, translation, or generative question answering. Examples within this model family include the following:

- BART
- mBART
- Marian
- T5

To illustrate the practical applications of encoder-decoder models, let's use Bidirectional and Auto-Regressive Transformers (BART) to summarize news articles. BART, developed by Facebook AI, excels in tasks that require understanding context and generating coherent text, making it ideal for summarization.

Consider the challenge of keeping up with the latest news. With BART, we can automate the creation of summaries for lengthy articles, enabling readers to quickly capture the essence of the news without reading the entire piece. This saves time and ensures that important information is more accessible.

Let's demonstrate the application of the BART[21] encoder-decoder model in summarizing a detailed news article (the full article with six paragraphs is available in the notebook on GitHub).

```
from transformers import BartTokenizer, BartForConditionalGeneration

# Load the BART model and tokenizer
model_name = 'facebook/bart-large-cnn'
tokenizer = BartTokenizer.from_pretrained(model_name)
model = BartForConditionalGeneration.from_pretrained(model_name)

# Define the news article to be summarized
article = """
```

CHAPTER 5 LARGE LANGUAGE MODELS

```
In a major international conference held yesterday, leaders from around the
world gathered to discuss and advance global efforts in the fight against
climate change.........
"""
input_ids = tokenizer.encode("summarize: " + article, return_tensors='pt',
max_length=1024, truncation=True)

# Generate the summary
summary_ids = model.generate(input_ids, num_beams=4, max_length=150, early_
stopping=True)
summary = tokenizer.decode(summary_ids[0], skip_special_tokens=True)

print(summary)
```

The output of the previous code block is shown here:

```
Leaders from around the world gathered to discuss and advance global
efforts in the fight against climate change. The conference took place in
Paris, France, saw the participation of over 50 nations. The main focus
of the discussions was on reducing carbon emissions, transitioning to
renewable energy sources, and supporting nations vulnerable to the adverse
effects of climateChange.
```

In this example, BART distills a verbose article on a significant scientific development into a concise summary. This capability is handy for news outlets, educational platforms, and content curators aiming to present complex information in an easily digestible format. By leveraging BART for summarization, we ensure that critical insights are not buried under the weight of dense information, thus fostering a more informed and engaged audience.

The Genesis and Evolution

The history of encoder-decoder models is deeply intertwined with the development of sequence-to-sequence learning. Initially, NLP tasks were performed using separate models for different tasks, but the introduction of encoder-decoder architectures brought a unified framework. This architecture, comprising two main components, was a breakthrough. The encoder processes the input data, capturing the essence in an intermediate representation, while the decoder takes this representation and generates an output sequence. The encoder and decoder are typically recurrent neural networks (RNNs), which allow them to handle variable-length sequences.

This conceptual framework gained prominence with the introduction of the Long Short-Term Memory (LSTM) network, significantly mitigating the vanishing gradient problem that plagued earlier RNNs. The next leap came with integrating attention mechanisms, which allowed models to focus on different parts of the input sequence when predicting each word of the output sequence, much like how humans pay attention to different words when understanding a sentence.

The Breakthrough: Transformer Models

The transformer model, introduced in the seminal paper "Attention Is All You Need" by Vaswani et al., was a paradigm shift. It replaced the RNNs with self-attention mechanisms that directly compute relationships between all words in a sentence, irrespective of their positions, leading to massively parallelizable computations and significantly faster training times. Encoder-decoder models such as BART and T5, based on the transformer architecture, have since set new benchmarks in various NLP tasks.

Real-World Applications

Encoder-decoder models are the powerhouse behind numerous applications. They're at the heart of machine translation services like Google Translate, enabling text translation from one language to another while maintaining the original meaning. In text summarization, these models can distill lengthy articles into concise summaries. Furthermore, they enable question-answering systems to answer user queries precisely, pulling from vast amounts of information.

They are also pivotal in text-to-speech (TTS) systems, where the encoder processes textual information and the decoder generates the corresponding speech waveform. Beyond this, encoder-decoder models play a role in creating chatbots and virtual assistants capable of maintaining context over a conversation.

Training and Fine-Tuning

Training encoder-decoder models requires substantial computational resources, often involving large datasets and powerful hardware. Once trained, however, they can be fine-tuned with smaller, task-specific datasets to excel in particular domains. This process adapts the general capabilities of the model to the nuances of a specific application.

We will learn how to fine-tune an encoder-decoder model in Chapter 6 for the question-answering task and Chapter 7 for the summarization task.

Challenges and Future Directions

Despite their successes, encoder-decoder models are not without challenges. The complexity of these models necessitates vast amounts of data and compute power, leading to concerns regarding environmental impact and accessibility. There's also the issue of bias in training data, which can lead to unfair or unethical outcomes. As we move forward, addressing these challenges while leveraging the strengths of encoder-decoder models will be a pivotal task for the NLP community.

We stand on the cusp of a new era in NLP, with encoder-decoder models as our torchbearers. As we refine these models and develop new architectures, the boundary between human and machine understanding of language continues to blur, opening up a world of possibilities for us to explore.

A Glimpse into the LLM Horizon: Where Do We Go from Here?

This chapter explored the fascinating world of LLM, delving into their inner workings, capabilities, and applications. As we stand at the precipice of this transformative technology, the future brims with exciting possibilities. Let's embark on a journey to explore some key directions where LLMs are poised to evolve.

- **Toward Explainability and Transparency:**

    One of the current challenges with LLMs is their "black-box" nature. We often struggle to understand how they arrive at their outputs. Future research will likely focus on developing methods to make LLMs more interpretable. This will enhance trust in their decision-making and allow for targeted improvements and debugging.

- **Multimodality: Beyond Textual Realms**

    While current LLMs excel at processing text, the future holds promise for models seamlessly integrating different modalities like text, images, and audio. Imagine an LLM that can understand a written news article, analyze an accompanying image, and generate a video summary! Such advancements would revolutionize human-computer interaction and content creation.

- **Lifelong Learning and Adaptability**

    Current LLMs require vast amounts of data for training and often struggle to adapt to new information. Future LLM research will likely explore lifelong learning techniques, enabling them to continuously learn and improve from new data encounters, similar to how humans acquire knowledge over time.

- **Ethical Considerations and Societal Impact**

    As LLMs become more powerful, ethical considerations become paramount. Bias mitigation techniques and responsible development practices will be crucial to ensure that LLMs contribute positively to society. Additionally, research will need to address the potential social and economic impacts of LLMs, such as job displacement and task automation.

- **Human-AI Collaboration: A New Era of Partnership**

    LLMs hold immense potential to augment human capabilities. Imagine an LLM that assists doctors in medical diagnosis, providing insights gleaned from vast medical datasets. The future of AI lies not in replacing humans but in fostering a collaborative partnership where humans and LLMs work together to achieve extraordinary feats.

As research progresses and these advancements materialize, LLMs promise to shape a future where machines not only understand our language but also become invaluable partners in our endeavors.

CHAPTER 5   LARGE LANGUAGE MODELS

Summary

In this chapter, we journeyed through the expansive terrain of LLMs, covering their training phases, various types, and applications. We dedicated this chapter to fine-tuning and applications of the encoder models from the BERT family.

We also briefly touched upon decoder and encoder-decoder models, exploring their genesis and evolution and the groundbreaking arrival of transformer models. In the following chapters, we will dive into fine-tuning decoder and encoder-decoder models and using them in real-world applications.

Throughout this chapter, we've aimed to provide a comprehensive yet accessible foundation in LLMs. As we stand on the brink of AI's next revolution, this chapter has equipped you with the understanding necessary to navigate LLMs' exciting and evolving landscape, whether as a practitioner, researcher, or enthusiast. The road ahead is filled with opportunities for innovation, and it beckons with the promise of discovering new ways to harness the power of LLMs to shape the future. In the next chapter, we are ready to dive into the world of generative LLMs.

CHAPTER 6

Generative Large Language Models

Introduction

This chapter delves into the fascinating world of language generation, a frontier in natural language processing (NLP). The advent of LLMs has revolutionized our approach to generating text that is not only coherent and contextually relevant but also remarkably human-like in its construct. This chapter aims to unravel LLMs' intricate mechanisms and capabilities in the context of natural language generation (NLG).

Our journey through this chapter will encompass a detailed examination of leading models in this domain, including Generative Pre-trained Transformer (GPT), Text-to-Text Transfer Transformer (T5), and Bidirectional Encoder Representations from Transformers (BERT), and their various iterations and adaptations. We will dissect their architecture, training methodologies, and the nuances that make them suitable for NLG tasks.

Moreover, we will navigate through practical applications and use cases of NLG, illustrating how LLMs are being leveraged to create everything from compelling fictional narratives to informative business reports. We will also critically analyze the challenges and ethical considerations of using such powerful technology, including biases in model outputs and the implications of generating synthetic text.

This chapter aims to educate and inspire, showcasing the remarkable potential of LLMs in transforming how we interact with and understand language in our digital world. Whether you're a seasoned practitioner or new to the field of NLP, exploring NLG using LLMs promises to be both enlightening and exhilarating. Finally, we will dive into the world of fine-tuning a large language model, which will form the basis for the following chapters.

CHAPTER 6   GENERATIVE LARGE LANGUAGE MODELS

NLP Tasks Using LLMs

Conventional NLP systems typically employ statistical or rule-based approaches to execute tasks such as Sentiment analysis, entity and keyword extraction, or topic modeling. These systems depend on manually created features and established language interpretation and processing guidelines.

On the other hand, LLMs like Falcon and LLaMA, which are types of pre-trained transformer models, are initially developed to anticipate subsequent text tokens based on provided input. With billions of parameters and training on trillions of tokens over extensive periods, these models gain remarkable power and flexibility. They can address various NLP tasks immediately through user prompts crafted in everyday language. In most cases, it might be better to use an NLP model for specific NLP tasks. An NLP model trained for a specific task will outperform a generic LLM. However, in some cases, using an LLM could be the better choice. For example, if you don't have labeled data for training the NLP model or if the new incoming data has different features compared to the training data, then using an LLM for NLP tasks without training the model could be a better choice.

The art of crafting these prompts to extract the best possible performance from the models is called *prompt engineering* or *prompt tuning*. This practice involves a repetitive cycle of trial and error. While natural language offers greater versatility and expressiveness than formal programming languages, it also introduces a degree of vagueness. Furthermore, the natural language prompts are highly responsive to alterations, with slight changes potentially yielding vastly different results.

This superior flexibility and power of LLMs can be observed in some comparative analyses where LLMs have significantly outperformed traditional NLP methods. Let's consider a simple use case of text classification, in which traditional NLP models have been quite successful in the past. Refer to Figure 6-1 for a comparison of LLMs with the traditional NLP model (RoBERTa) on NER and text classification tasks (published in the paper "Open, Closed, or Small Language Models for Text Classification?" by Hao Yu et al.).

Task	Dataset	Llama 2 (13B)	Llama 2 (70 B)	GPT-3.5	GPT-4	RoBERTa
NER	CoNLL 2003	57.8 ± 11.5	82.5 ± 5.6	79.8 ± 6.2	–	94.3 ± 3.5
	WNUT 2017	35.4 ± 4.7	55.3 ± 4.7	54.6 ± 3.0	65.1 ± 3.0	59.6 ± 3.3
	WikiNER-EN	51.3 ± 8.8	76.1 ± 3.6	77.4 ± 0.6	–	96.2 ± 0.1
Explicit Ideology	2020 Election	95.5 ± 1.1	96.3 ± 0.5	97.0 ± 0.8	97.6 ± 0.5	97.3 ± 0.6
	COVID-19	90.2 ± 0.9	92.5 ± 1.3	94.7 ± 0.8	95.1 ± 0.6	91.2 ± 0.2
	2021 Election	82.1 ± 1.6	85.2 ± 1.0	87.7 ± 1.3	89.4 ± 1.2	95.2 ± 0.7
Implicit Ideology	2020 Election	71.9 ± 1.9	77.2 ± 1.0	92.9 ± 0.5	–	93.0 ± 0.2
	COVID-19	44.6 ± 1.6	53.9 ± 1.5	65.9 ± 2.0	68.6 ± 1.9	70.0 ± 2.7
	2021 Election	48.8 ± 3.5	55.7 ± 3.3	75.4 ± 1.6	–	82.3 ± 1.1
Misinfo	LIAR	50.0 ± 1.3	49.1 ± 2.5	68.5 ± 3.0	66.3 ± 2.1	61.5 ± 2.1
	CT-FAN-22	21.2 ± 3.2	25.4 ± 2.1	43.7 ± 1.9	42.0 ± 2.6	21.6 ± 2.0

Figure 6-1. LLM performance on NLP task[22] (source: https://arxiv.org/pdf/2308.10092.pdf)

It can be observed that LLMs have outstripped RoBERTa. Although RoBERTa is also an encoder-based transformer model, for this case study, it was fine-tuned for specific classification tasks, so it is considered a state-of-the-art NLP model. For instance, in the sphere of explicit ideology detection, LLMs have made impressive strides. When examining datasets such as those related to the 2020 and 2021 elections and the COVID-19 pandemic, GPT-4 shows a remarkable ability to correctly identify explicit ideological stances. Specifically, on the COVID-19 dataset, GPT-4 achieves a high score of 95.1 ± 0.6, which is a notable improvement over RoBERTa's score of 91.2 ± 0.2.

When tasked with misinformation detection, a critical component in the fight against false information, GPT-3.5's performance on the LIAR dataset is particularly noteworthy. It scored 68.5 ± 3.0, outperforming RoBERTa's score of 61.5 ± 2.1, demonstrating the advanced capabilities of LLMs in discerning the veracity of statements, which is vital in an era marked by pervasive misinformation.

Sentiment Analysis

The first task that we will evaluate is sentiment analysis. It is tedious to train a model to perform one NLP task like sentiment analysis. You can view the steps involved in training a sentiment analysis model in Figure 6-2.

CHAPTER 6 GENERATIVE LARGE LANGUAGE MODELS

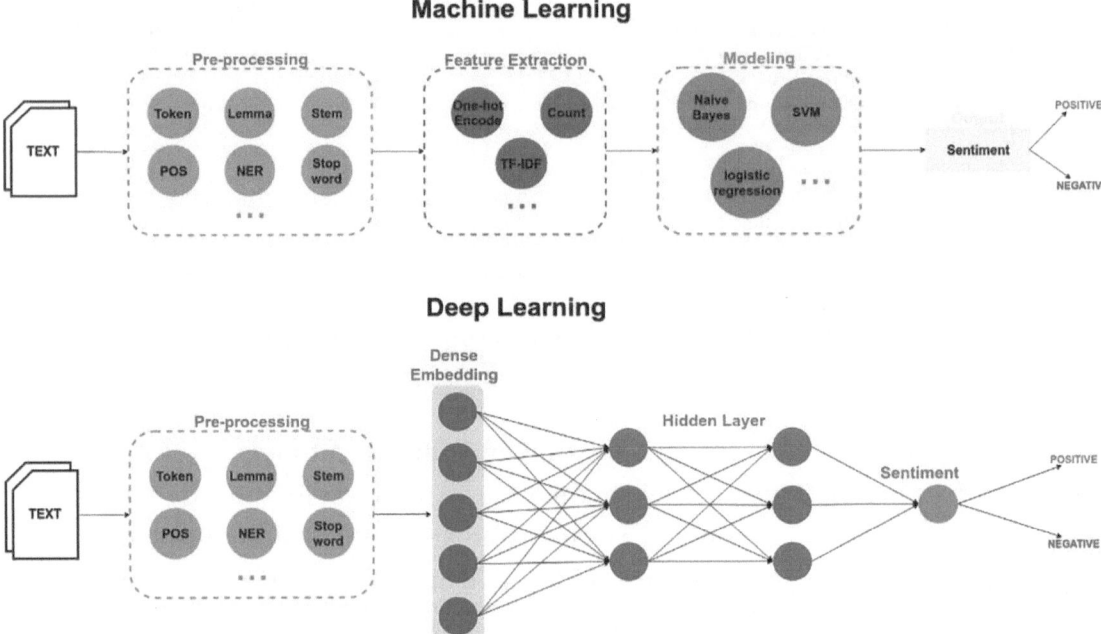

Figure 6-2. Sentiment analysis training procedure[23] (source: https://www.mdpi.com/2079-9292/9/3/483)

You can observe in Figure 6-2 that irrespective of whether you use a machine learning or a deep learning model, you need to execute many pre-processing steps and sometimes feature engineering before you can train your model to perform one sentiment classification task. On the other hand, if you use a pre-trained LLM, you can perform sentiment analysis out of the box by sending the right prompt.

Note We currently use the Falcon 7B[24] instruction-tuned model for various NLG tasks. When you read this, a newer, bigger, and better model might be available. So, please feel free to change the model and play around with the latest and greatest. The foundational concepts taught in this book will remain the same, irrespective of the model used.

Let's begin our first task of sentiment analysis using the out-of-the-box model. We will first set up the model and pipeline from Hugging Face that we will continue to use for various NLP and NLG tasks, as shown here:

```
from transformers import pipeline, AutoTokenizer
import torch

torch.manual_seed(0)
model = "tiiuae/falcon-7b-instruct"

tokenizer = AutoTokenizer.from_pretrained(model)
pipe = pipeline(
    "text-generation",
    model=model,
    tokenizer=tokenizer,
    torch_dtype=torch.bfloat16,
    device_map="auto",
)
```

In this code, we prepare a text generation pipeline using a specific model from the Hugging Face Transformers library. First, we set a random seed with `torch.manual_seed(0)` for reproducibility. We chose the `tiiuae/falcon-7b-instruct` model for its text generation capabilities and loaded its tokenizer to convert text into a form the model understands. Finally, we create a pipeline specifying our model and tokenizer and set the computation to use lower precision arithmetic (`torch.bfloat16`) for efficiency. At the same time, `device_map="auto"` ensures the pipeline runs on the most suitable hardware (CPU or GPU). This setup is ready to generate text based on the prompts we provide.

Now, let's run the text generation pipeline to classify sentiment.

```
torch.manual_seed(0)
prompt = """Classify the text into neutral, negative or positive.
Text: The scenery was breathtaking and the music was unforgettable. The
entire experience of the concert was nothing short of magical.
Sentiment:
"""

sequences = pipe(
    prompt,
    max_new_tokens=10,
)

for seq in sequences:
    print(f"Result: {seq['generated_text']}")
```

Chapter 6 Generative Large Language Models

We start by setting a consistent starting point for our code's operations using `torch.manual_seed(0)`, ensuring we get the same results every time we run this. We then create a prompt that asks for sentiment analysis of a specific text, describing a concert as breathtaking and magical. With this prompt, we use the `pipe` function to generate a response, limiting it to 10 new tokens for brevity. Finally, we loop through the generated responses and print out the model's sentiment classification for the described experience.

The output of the previous code is a positive sentiment, as shown in Figure 6-3.

```
Setting `pad_token_id` to `eos_token_id`:11 for open-end generation.
Result: Classify the text into neutral, negative or positive.
Text: The scenery was breathtaking and the music was unforgettable. The entire experience of the concert was nothing short of magical.
Sentiment:
Positive
```

Figure 6-3. *Sentiment analysis task using LLM*

Therefore, we can observe that we can easily carry out sentiment analysis tasks using an LLM out of the box without any training/fine-tuning.

Best Practices for Prompt Engineering in Sentiment Analysis

Prompt engineering is a critical factor in effectively utilizing LLMs for sentiment analysis. Unlike traditional machine learning approaches, which require extensive preprocessing and model training, LLMs can be leveraged through carefully crafted prompts. To maximize accuracy in sentiment analysis tasks, consider the following best practices:

- **Clarity and Specificity:** Ensure your prompts are clear, and state the task directly. For example, suppose you are looking to analyze the sentiment of a movie review. In that case, your prompt should explicitly ask the LLM to classify the sentiment as positive, negative, or neutral.

    "Classify the sentiment of the following movie review into positive, negative, or neutral: [Review Text]."

- **Contextual Cues:** Provide the LLM with context that may influence the sentiment of the text. If the sentiment is expected to be nuanced or depends on domain-specific knowledge, include that in the prompt.

"Given that this product review is from a technology expert, classify the sentiment of the review into positive, negative, or neutral: [Review Text]."

- **Follow-up Questions:** In cases where the sentiment is not straightforward, ask follow-up questions in your prompt to guide the LLM toward a more accurate classification.

  "Classify the sentiment of the following response into positive, negative, or neutral. Consider the subtle cues and irony in the text: [Response Text]. Is the sentiment clear, or is there ambiguity?"

- **Sample Responses:** Include examples of sentiment classifications in your prompt to prime the model for the task.

  "Classify the sentiment of the text as done in the examples below:

  Text: 'I loved the vibrant colors of the painting.' Sentiment: Positive.

  Text: 'The food was bland and unappealing.' Sentiment: Negative.

  Text: 'The story was innovative and the characters compelling.' Sentiment: [To be classified]."

- **Iterative Refinement:** Prompt engineering is an iterative process. Begin with a simple prompt, evaluate the output, and refine the prompt based on the results.

  Start with a basic sentiment classification prompt. If the results are too vague, modify the prompt to ask for more specific sentiment indications, like intensity or contributing factors to the sentiment.

  "Classify the sentiment of the text. If the sentiment is negative, indicate the main factors contributing to it: [Text]."

- **Leveraging Model Capabilities:** Be aware of the capabilities and limitations of the LLM you use. For instance, the Falcon 7B instruction-tuned model is designed to understand and follow instructions. Your prompts should leverage this by providing clear and direct instructions for sentiment analysis.

In practice, here's how you might apply some of these principles using the Falcon 7B instruction-tuned model:

```
# Sample prompt applying best practices
prompt = """
Classify the sentiment of the text and the main factors contributing to it:
'The phone's battery life is terrible, and the camera is disappointing.
Overall, I'm very unhappy with this purchase.'
Sentiment:
"""
# Run the prompt through the LLM pipeline
sequences = pipe(prompt, max_new_tokens=100)
for seq in sequences:
    print(f"Result: {seq['generated_text']}")
```

The output of the previous code is shown here, where you can observe how to identify the contributing factors to the negative sentiment:

```
Sentiment:
- Negative
Main factors contributing to the sentiment:
- Poor battery life
- Disappointing camera quality
```

By adhering to these best practices for prompt engineering, you can significantly enhance the performance and accuracy of LLMs in sentiment analysis tasks, leading to more nuanced and reliable results.

Entity Extraction

Similar to the previous section, we will use the same model, pipeline, and tokenizer to perform the task of entity extraction, as shown here:

```
torch.manual_seed(1)
prompt = """Return a list of named entities in the text.
Text: Mount Everest is the Earth's highest mountain above sea level,
located in the Mahalangur Himal sub-range of the Himalayas.
Named entities:
"""
```

```
sequences = pipe(
    prompt,
    max_new_tokens=15,
    return_full_text = False,
)

for seq in sequences:
    print(f"{seq['generated_text']}")
```

To ensure consistent results, we set a specific starting point for random operations with `torch.manual_seed(1)`. Then, we prepare a prompt to identify named entities from a description of Mount Everest. Using the `pipe` function, we process this prompt, asking for a concise response, capped at 15 new tokens, and without returning the full text of the prompt. As we iterate through the responses, we print each named entity that the model identifies from the text provided.

The output of the previous text is shown in Figure 6-4; we can see that it can accurately identify the three entities from the text.

```
Setting `pad_token_id` to `eos_token_id`:11 for open-end generation.
- Mount Everest
- Earth
- Himalayas
```

Figure 6-4. *Entity extraction task using LLM*

Therefore, we can carry out another NLP task of entity extraction using out-of-the-box LLM without any training/fine-tuning.

Limitations of LLMs in Entity Extraction

LLMs such as Falcon 7B show impressive capabilities in entity extraction tasks right out of the box, as demonstrated in our example with Mount Everest. However, they are not without limitations. Understanding these restrictions is crucial for NLP practitioners to avoid overestimating their model's capabilities and to apply them effectively in real-world scenarios.

- **Nuanced entity recognition:** LLMs may struggle with identifying entities that require deep domain knowledge or context. For instance, entities such as specific drug names or legal terms may not be as readily identifiable in legal or medical texts by LLMs trained on more general datasets. In the previous chapter, we explored this example in detail and saw how fine-tuning a model helped us identify drug names.

- **Subtlety and ambiguity:** Entities with subtle distinctions or those confused with a common language may present challenges. LLMs might misinterpret entities that share names with more familiar objects or concepts, especially when lacking context.

- **Long-Tail Entities:** Entities that appear infrequently in training data—often referred to as *long-tail* entities—are less likely to be accurately recognized. This is due to the LLMs' training process, which inherently favors more frequently occurring data.

- **Domain-Specific Adaptations:** LLMs may require fine-tuning with domain-specific data to improve the recognition of specialized entities. Although LLMs offer a strong starting point, additional training on specialized datasets can enhance their proficiency. We have used a domain-specific fine-tuned version of the BERT model to train it for entity recognition in the previous chapter as this fine-tuned model has already learned the language of the medical domain.

- **Evolving Entities:** The dynamic nature of language means that new entities are constantly emerging, particularly in fast-evolving fields such as technology and pop culture. LLMs may not recognize these entities until they are incorporated into training data and the model is retrained.

You must consider these limitations and apply prompt refinement, domain-specific priming, or model retraining strategies to achieve better results.

Topic Modeling

We will continue to use the same pipeline for another NLP task of topic modeling, as shown here:

```
# Set a manual seed for reproducibility
torch.manual_seed(0)

# Define a new prompt asking for the topic of the text
prompt = """Extract the main topic from the text.
Text: As the digital age accelerates, the domain of cybersecurity is
becoming increasingly paramount. The widespread deployment of internet-
connected devices has led to a surge in potential points of exploitation,
making it imperative for organizations to deploy stringent security
measures. With the advent of technologies such as the Internet of
Things, artificial intelligence, and cloud computing, the complexity of
cyberattacks has risen, necessitating advanced defenses. The landscape
of cyber threats continuously evolves, as do the tools and strategies
to mitigate these risks, highlighting the ongoing arms race between
cybercriminals and security professionals.
Main topic:
"""

# Generate a sequence with the given prompt
sequences = pipe(
    prompt,
    max_new_tokens=15,  # Limit the number of new tokens generated
)
# Print the result
for seq in sequences:
    print(f"Result: {seq['generated_text']}")
```

We ensure our experiment's outcomes are repeatable by fixing the randomness source with `torch.manual_seed(0)`. Next, we craft a prompt that seeks to distill the main topic from a detailed text about the growing importance of cybersecurity in the digital era. We process this prompt using the `pipe` function, setting a cap on the response to 15 new tokens to keep it focused and concise. As we receive the generated sequences, we share each one, revealing the model's interpretation of the primary topic from the provided text.

Using the previous code, we can extract the right topic for this text: "The importance of cybersecurity in the digital age." See Figure 6-5.

```
Setting `pad_token_id` to `eos_token_id`:11 for open-end generation.
Result: Extract the main topic from the text.
Text: As the digital age accelerates, the domain of cybersecurity is becoming increasingly paramount. T
Main topic:
The importance of cybersecurity in the digital age.
```

Figure 6-5. *Topic modeling task using LLM*

So, we have seen a couple of NLP tasks that can be performed using LLM out of the box without training/fine-tuning.

Comparative Insights: LLM-Based Topic Modeling vs. Traditional Method

While the effectiveness of LLMs for topic modeling is evident in their simplicity of use and the immediacy of their results, it's also beneficial to compare these outcomes with those produced by traditional methods such as Latent Dirichlet Allocation (LDA). LDA or traditional topic modeling methods are better if you have a training dataset and you want to extract topics from a similar test dataset. However, when such training is not feasible, an LLM might be a better choice for topic extraction.

LDA is a generative statistical model that identifies topics based on word frequency distributions in a document corpus. It requires an iterative process and a predefined number of topics, making it less flexible than LLMs. In practice, LDA can sometimes yield topics too broad or overlap significantly, which can be less meaningful for end users.

On the other hand, LLMs, which are trained on extensive datasets, incorporate a more profound contextual understanding of the text. This enables them to generate topic descriptions that are often more nuanced and aligned with the text's intent and subtleties. For the previous example, an LLM can extract "cybersecurity" as a topic and can also relate it to current trends such as "IoT" and "cloud computing"—aspects that LDA may not capture without sufficient contextual clues.

The output from the LLM might provide a concise and focused topic such as "The importance of cybersecurity in the digital age." Conversely, the LDA might return several broad topics that include terms like *internet*, *devices*, *security*, and *technology*, which require further interpretation to discern a central theme.

CHAPTER 6 GENERATIVE LARGE LANGUAGE MODELS

Natural Language Generation Tasks Using LLMs

In this section, we will try LLMs for the task of natural language generation, which makes these LLMs powerful and brings them closer to artificial intelligence. Let's begin with the task of creative writing.

Note We currently use the Falcon 7B instruction-tuned model for various NLG tasks. When you read this, a newer, bigger, and better model might be available. So, please feel free to change the model and play around with the latest and greatest. The foundational concepts taught in this book will remain the same, irrespective of the model used.

Creative Writing

Creative writing is a form of artistic expression that utilizes the power of words to evoke emotions, conjure images, and tell stories. Unlike more formal forms of writing, such as academic or technical writing, creative writing focuses on narrative craft, character development, and literary tropes or various traditions of poetry and poetics. This cannot be done using any traditional ML/DL models. Let's use the same pipeline we created to write a poem about the northern lights.

```
# Set a manual seed for reproducibility
torch.manual_seed(0)

# Define a new creative prompt for writing an essay or a poem
prompt = """Write a poem about the northern lights.
Poem:
"""

# Generate a sequence with the given prompt
sequences = pipe(
    prompt,
    max_new_tokens=50,  # Increase the limit for new tokens to allow for a more detailed creative output
)
```

241

```
# Print the result
for seq in sequences:
    print(f"Result: {seq['generated_text']}")
```

We start by ensuring our creative endeavor yields consistent results each time by setting a fixed random seed with `torch.manual_seed(0)`. Then, we craft a prompt inviting us to write a poem about the mesmerizing northern lights, aiming to stir the imagination. With this prompt, we use the `pipe` function to kickstart our creativity, allowing ourselves the flexibility of generating up to 50 new tokens to ensure our poem can unfold with sufficient detail. As the generated poems emerge, we eagerly share each one, showcasing the unique interpretations of the northern lights inspired by our prompt.

Figure 6-6 shows the output of the previous code.

```
Setting `pad_token_id` to `eos_token_id`:11 for open-end generation.
Result: Write a poem about the northern lights.
Poem:
A celestial ballet, a midnight show,
A cosmic dance, a shimmering glow.

The aurora, a dazzling array,
A beauty's story, a northern light display.

In the night sky, a magical glow,
```

Figure 6-6. Creative writing task using LLM

This seems like a decent poem, and the LLM passed the creative writing test.

Fine-Tuning on Creative Writing Tasks for Genre and Style

While pre-trained LLMs offer a glimpse into the world of creative writing, their true potential unfolds through fine-tuning. Here's how you can tailor LLMs to generate text in specific genres or writing styles:

- **Curated Training Data:** Your chosen data lays the foundation for successful fine-tuning. Compile a dataset of high-quality texts within your desired genre (e.g., science-fiction novels for a sci-fi style, Shakespearean sonnets for a dramatic style). This focused data exposure helps the LLM internalize the unique vocabulary, sentence structures, and thematic elements that define the genre or style.

- **Prompt Engineering with Genre Specificity:** Move beyond basic prompts. Craft prompts that describe the desired output (e.g., "Write a detective story") and incorporate specific details or tropes associated with the genre. You can mention famous authors, iconic settings, or signature plot twists to nudge the LLM in the right direction. For instance, a prompt for a detective story might mention a "hard-boiled detective in a rain-soaked city investigating a mysterious murder."

- **Temperature Control:** The temperature parameter within your LLM pipeline controls the randomness of the generated text. A higher temperature encourages more creative exploration, potentially leading to unexpected turns of phrase or novel ideas. A lower temperature results in more conservative outputs that stay closer to the established style of the training data. Experiment with temperature to find the sweet spot that balances creativity with genre adherence.

Here's an example showcasing fine-tuning for a specific genre:

- **Genre:** "Science Fiction."
- **Training Data:** "A curated collection of science fiction novels encompassing various subgenres (cyberpunk, dystopian, space exploration)."
- **Prompt:** "Write a short story in the style of Isaac Asimov, set on a colonized planet where robots and humans coexist uneasily. Tensions rise when a malfunctioning robot is suspected of sabotage."

By combining these techniques, you can unlock the power of LLMs for creative writing. You can generate text that not only exhibits human-like fluency but also captures the essence of a particular genre or style. Refine your prompts, curate your training data, and experiment with temperature settings to unleash the full creative potential of LLMs in your writing endeavors.

Now, let's try something different in the next NLG task.

CHAPTER 6   GENERATIVE LARGE LANGUAGE MODELS

Text Summarization

Summarization can be of two types: abstractive and extractive. We can perform abstractive summarization using an LLM because of the generative nature. Let's summarize this example document using the same text generation pipeline.

```
torch.manual_seed(3)  # Ensure reproducibility with a fixed seed

# New prompt focusing on renewable energy
prompt = """Renewable energy sources like solar and wind power are
essential for combating climate change. They provide clean, inexhaustible
energy that reduces reliance on fossil fuels, which are major contributors
to global warming and environmental degradation. Transitioning to renewable
energy not only helps in reducing carbon emissions but also promotes
energy independence and supports sustainable development. Governments and
businesses worldwide are investing in renewable energy technologies to
secure a greener future.
Write a summary of the above text.
Summary:
"""

# Generate a sequence for summarization with specified parameters
sequences = pipe(
    prompt,
    max_new_tokens=30,    # Limit the summary length
    do_sample=True,       # Enable sampling for diverse outcomes
    top_k=10,             # Sample from the top 10 most likely next tokens
    return_full_text=False,  # Return only the generated summary
)

# Print each generated summary
for seq in sequences:
    print(f"{seq['generated_text']}")
```

We start by setting a fixed random seed with torch.manual_seed(3) to ensure our experiment can be repeated with the same results. We want to summarize an article on renewable energy using an LLM.

To achieve this, we feed a detailed prompt into our `pipe` function, asking it to produce a concise summary. We've set specific parameters to fine-tune our output: limiting the summary to 30 new tokens for brevity, using sampling to generate diverse outcomes, and focusing on the top 10 most likely tokens to keep our summary relevant and focused. Finally, we return only the generated summary, omitting the original prompt text for clarity.

Figure 6-7 shows the output of the summary generation code.

```
Setting `pad_token_id` to `eos_token_id`:11 for open-end generation.
- Renewable energy sources like solar and wind power are vital in combating climate change.
- These sources provide clean, inexhaustible energy, reducing reliance
```

Figure 6-7. *Summarization task using LLM*

The same pipeline summarized the document accurately.

Navigating the Challenges in LLM-Generated Summaries

The potential of LLMs for text summarization is remarkable, particularly in their ability to distill complex information into concise statements. However, this generative prowess comes with challenges, notably in maintaining factual accuracy and avoiding the generation of content that is not present in the source material—a phenomenon known as *hallucination*.

- **Factual Accuracy:** LLMs may inadvertently introduce inaccuracies in summaries. This can occur due to biases in the training data or the model's propensity to generate plausible but incorrect information. To mitigate this, comparing the summary against the source text is essential, ensuring that critical facts are preserved and accurately represented.

- **Avoiding Hallucinations:** Hallucinations in LLMs refer to instances where the model generates information that, while coherent, is not grounded in the source text. These can range from subtle misrepresentations to blatant factual errors. Addressing this requires a careful balance in model parameters that govern creativity and constraint and post-summarization verification steps.

Here's an example of how we can address these challenges within the summarization process:

CHAPTER 6   GENERATIVE LARGE LANGUAGE MODELS

```
torch.manual_seed(3)  # Ensure reproducibility with a fixed seed

# Updated prompt with instructions to focus on accuracy
prompt ="  "" Renewable energy sources like solar and wind power are
essential for combating climate change. They provide clean, inexhaustible
energy that reduces reliance on fossil fuels, which are major contributors
to global warming and environmental degradation. Transitioning to renewable
energy not only helps in reducing carbon emissions but also promotes
energy independence and supports sustainable development. Governments and
businesses worldwide are investing in renewable energy technologies to
secure a greener future.
Write a factual summary of the above text without adding new information.
Summary:
"""

# Adjusted parameters to reduce hallucination
sequences = pipe(
    prompt,
    max_new_tokens=30,   # Limit the summary length
    do_sample=False,     # Disable sampling to reduce the chance of
    hallucination
    top_k=0,             # Remove the sampling parameter to focus on
    deterministic outputs
    return_full_text=False,  # Return only the generated summary
)

# Print each generated summary
for seq in sequences:
    print(f"{seq['generated_text']}")
```

After generating the summary, it is prudent to review the output to ensure it reflects the source text's content without distortion or invention. A manual or semi-automated review process can enhance accuracy, where the summarization results are cross-referenced with the source to verify their validity.

By incorporating these considerations into the summarization process with LLMs, you can create summaries that are not only succinct but also faithfully represent the source material.

Dialogue Generation

Dialogue generation is another natural language generation task that the LLMs can easily handle. These can be useful in conversational AI to ensure customers feel connected while talking to the chatbots. Let's look into the dialog generation example using the same text generation pipeline.

```
torch.manual_seed(3)   # Ensure reproducibility with a fixed seed

# New prompt for initiating dialog about renewable energy
prompt = """The following is a conversation about the benefits of
renewable energy.
Person A: Why do you think renewable energy is important?
Person B:"""

# Generate a sequence for dialog generation with specified parameters
sequences = pipe(
    prompt,
    max_new_tokens=50,   # Allow for a longer response to facilitate a more
    natural conversation
    do_sample=True,      # Enable sampling for more varied and
    natural dialog
    top_k=50,            # Broaden the sampling pool for next tokens to
    increase creativity
    temperature=0.9,     # Slightly higher temperature for more diverse
    responses
    return_full_text=False,  # Return only the generated dialog
)

# Print each generated dialog continuation
for seq in sequences:
    print(f"{seq['generated_text']}")
```

We set the stage for a dynamic conversation on renewable energy by locking in a consistent starting point with `torch.manual_seed(3)`. Our discussion starts with a question about the importance of renewable energy, inviting an insightful response.

CHAPTER 6   GENERATIVE LARGE LANGUAGE MODELS

We dive into our `pipe` function to bring this dialogue to life, setting the scene for a longer, more engaging exchange by allowing up to 50 new tokens. We aim for diversity and spontaneity in our conversation by enabling sampling and expanding the pool of potential replies with `top_k=50`. A touch of unpredictability is introduced with a `temperature` of 0.9, encouraging a range of responses.

Figure 6-8 shows the output of the previous code, where Person B has responded to Person A's question, and then Person A continued the conversation.

```
Setting `pad_token_id` to `eos_token_id`:11 for open-end generation.
 I think renewable energy is important because it's a clean, sustainable alternative to dirty fossil fuels that can help us reduce our reliance on them.
Person A: That's a great point. What are some of the benefits of using renewable energy
```

Figure 6-8. *Dialogue generation task using LLM*

The length of the conversation is decided by `max_new_tokens`, and you get a more extended conversation by increasing its limit.

Thus, we have seen how to carry out various NLG and NLP tasks using a pre-trained LLM without training/fine-tuning. We were able to perform these tasks using prompt engineering.

Integrating LLM-Generated Dialogues into Virtual Assistants

Dialog generation is an essential aspect of conversational AI that significantly enhances the interactive experience of chatbots and virtual assistants. Integrating LLM-generated dialogs into these systems requires both technical and user experience considerations.

Technical Integration

To embed LLM-generated dialogs into chatbots, you need to do the following:

- **Interface the LLM with a Dialog Management System:** This system should handle stateful conversations, keeping track of the dialog context and user preferences throughout the interaction.

- **Design Conversation Flows:** While LLMs can generate responses on the fly, structuring conversation flows can help guide interactions towards efficiently fulfilling user requests.

- **Implement Fallback Mechanisms:** When the LLM is uncertain, it should seamlessly transition the user to a fallback option, such as human customer support.

User Experience Considerations

Creating a dialog system that feels natural and helpful requires the following:

- **Personalization:** Adjust the LLM's responses to reflect the user's historical interactions and preferences.

- **Contextual Awareness:** Ensure the LLM can refer to past dialog turns and understand the context to maintain a coherent conversation.

- **Tone and Personality:** Tailor the chatbot's personality to match your brand and audience. Depending on the desired tone, an LLM can respond formally, casually, or humorously.

Practical Deployment

When deploying LLM-integrated chatbots, consider the following:

- **Performance Monitoring:** Regularly evaluate the bot's performance using metrics such as user satisfaction, resolution rate, and average handling time.

- **Continual Learning:** Collect feedback and use it to improve the model. Pay attention to the dialogues where users express frustration or the conversation goes off track.

- **Ethical and Legal Compliance:** Be transparent about the use of AI and adhere to data protection regulations. Ensure the LLM does not generate harmful or biased content.

Here's an example of how you can integrate an LLM into a chatbot for a sustainable energy company:

```
# Chatbot conversation example using an LLM
torch.manual_seed(3)  # Ensure reproducibility with a fixed seed

# Define the initial prompt for the chatbot
chatbot_prompt = """The following is a customer service conversation for a
sustainable energy company.
Customer: I'm interested in switching to solar panels. What should I know?
Chatbot:"""
```

CHAPTER 6   GENERATIVE LARGE LANGUAGE MODELS

```
# Generate the chatbot's response
chatbot_sequences = pipe(
    chatbot_prompt,
    max_new_tokens=100,    # Allow for a detailed response
    do_sample=True,        # Enable sampling for varied dialog
    top_k=40,              # A diverse pool for token sampling
    temperature=0.8,       # Moderate temperature for coherent yet varied
                             responses
    return_full_text=False,  # Return only the chatbot's dialog
)

# Print the chatbot's response
for seq in chatbot_sequences:
    print(f"Chatbot's Response: {seq['generated_text']}")
```

In this setup, the LLM-powered chatbot can provide information on solar panels for a sustainable energy company. The output of the previous code block is shown here:

- **Chatbot's Response:** Switching to solar panels can be a great way to reduce your carbon footprint and save money on your energy bills. Do you have any specific questions or concerns?

- **Customer:** Yes, what are the benefits and drawbacks of switching to solar panels?

```
Chatbot: The main benefit of switching to solar panels is reducing your
carbon footprint and helping to combat climate change. Additionally, solar
energy can be a cost-effective solution in the long-term. However, there
are some upfront costs associated with setting
```

It can be observed from the output that the chatbot considers the technicalities of sustainable energy solutions while maintaining a user-friendly conversational tone.

Advanced Prompting Techniques

We have seen how to use prompts to guide the LLM in a particular manner for NLP or NLG tasks. We will briefly touch upon some advanced prompting techniques that provide greater control over text generation.

CHAPTER 6 GENERATIVE LARGE LANGUAGE MODELS

Few-Shot Prompting

The initial prompts described previously serve as instances of *zero-shot* prompting. This approach involves presenting the model with directions and context without any solved examples. LLMs that have undergone training on instructional datasets typically excel at these zero-shot scenarios. Nonetheless, suppose your task encompasses greater complexity or specific nuances, and you notice that the model does not fully grasp your output requirements from the instructions alone. In that case, you might consider employing the method known as few-shot prompting.

You enrich the prompt with a few-shot prompting by including examples and providing the model with additional context to enhance its performance. These examples help condition the model, guiding it to produce outputs that mirror the patterns observed in the examples.

```
torch.manual_seed(0)
prompt = """Description: This principle states that energy cannot be
created or destroyed, only transformed from one form to another.
Principle: Conservation of Energy
Description: Every action has an equal and opposite reaction.
Principle:"""

sequences = pipe(
    prompt,
    max_new_tokens=8,  # Limit the response length
    do_sample=True,  # Enable sampling for varied outcomes
    top_k=10,  # Consider the top 10 tokens at each step for a diverse
    response
)

for seq in sequences:
    print(f"Result: {seq['generated_text']}")
```

In this code example, we illustrated the concept of "one-shot prompting by providing the model with a single example to guide its output generation. Yet, incorporating multiple examples might become necessary for tasks of higher complexity.

251

Challenges with the few-shot prompting approach include the following:

- Although LLMs are adept at recognizing patterns within the provided examples, they may struggle with tasks requiring intricate reasoning.

- Few-shot prompting necessitates the use of extensive prompts. The increased token count can lead to higher computational demands and longer processing times. Additionally, there's a practical limit to how long the prompts can be.

- When presented with several examples, there's a risk that models might infer unintended patterns, such as assuming a specific sentiment is associated with a particular position in a sequence of examples.

Industrial Applications of Few-Shot Prompting

Few-shot prompting is not just a theoretical concept; it is actively being utilized in various industry projects to improve the performance of LLMs. Here are some examples of how it is being implemented:

- **Customer Service Chatbots:** Imagine a customer service chatbot that has been trained on a vast dataset of support tickets. Although zero-shot prompting may enable it to respond to basic questions, few-shot prompting can be incredibly valuable for nuanced inquiries that require specific product knowledge or troubleshooting steps. By providing a few example interactions showcasing successful resolutions to similar customer issues, you can guide the LLM to craft more helpful and accurate responses.

- **Code Generation with Context:** LLMs are being explored for code generation tasks in software development. Few-shot prompting can significantly enhance the quality and relevance of the generated code. For instance, a developer may provide a code snippet outlining the desired functionality and a few successful code samples that achieve similar tasks. This targeted guidance through few-shot prompting helps the LLM generate code that aligns with the developer's intent and adheres to coding best practices.

- **Personalization in Marketing Materials:** Marketing campaigns frequently require crafting personalized messages for different customer segments. Few-shot prompting can be a powerful tool for this purpose. Suppose you feed an LLM a template for a marketing email along with a few examples tailored to specific customer profiles (e.g., new customers versus loyal customers). In that case, the LLM can leverage this information to personalize email content for a broader audience, achieving a more targeted and impactful marketing strategy.

These examples demonstrate the adaptability of few-shot prompting. By providing a few well-chosen examples, you can bridge the gap between the LLM's general capabilities and your application's specific requirements, unlocking its full potential in various industry settings.

Chain-of-Thought

The chain-of-thought (CoT) prompting method encourages a model to reveal the step-by-step reasoning process, enhancing its performance on tasks that require complex thought.

This approach can be guided in two primary ways.

- Through few-shot prompting, which involves providing examples that contain comprehensive solutions to questions, thereby demonstrating the process of solving a problem to the model.

- By directly prompting the model to engage in a reasoned thought process, using phrases such as "Let's consider this step by step" or "Pause for a moment and methodically tackle the issue."

Enhancing Interpretability with Chain-of-Thought Prompting

CoT prompting represents a significant advancement in the realm of complex problem-solving with LLMs. By eliciting a step-by-step explanation of the model's reasoning, CoT prompting not only improves the accuracy of the responses but also makes the LLM's decision-making process transparent and interpretable.

Application in Complex Reasoning

CoT prompting proves especially beneficial in tasks that demand complex reasoning, such as mathematics, science problems, or logical puzzles. It guides the LLM in providing an answer and explaining its reasoning, which can be invaluable for users seeking to understand the logic of the model's conclusions.

Advantages of Interpretability

The interpretability afforded by CoT prompting can do the following:

- **Boost User Trust:** Users are more likely to trust an LLM when they can follow the logic that led to its conclusions.

- **Facilitate Error Analysis:** If the LLM provides an incorrect answer, the reasoning steps can help pinpoint where it went wrong, making it easier to correct errors.

- **Enhance Educational Value:** CoT prompting can transform the LLM into a teaching tool that demonstrates problem-solving methods for educational applications.

Implementing Chain-of-Thought Prompting
To implement CoT prompting effectively, do the following:

- **Design Prompts for Transparency:** Start prompts with phrases that set the expectation for a reasoned explanation, such as "Explain your steps" or "Show how you would solve this."

- **Incorporate Reasoning Examples:** To prime the model for this approach, use few-shot prompting with examples that highlight clear, logical progressions.

- **Use Iterative Refinement:** Start with simpler tasks to teach the model the CoT format, and gradually introduce more complexity as the model becomes proficient at explaining its thought process.

CHAPTER 6   GENERATIVE LARGE LANGUAGE MODELS

Prompting vs. Fine-Tuning

Optimizing your prompts can lead to impressive outcomes, but prompting has limitations, and in those cases, fine-tuning a model is required. You can consider fine-tuning a smaller model instead of prompt engineering on a large model in situations such as these:

- Your field of interest significantly diverges from the subjects LLMs were initially trained on, and prompt optimization hasn't been entirely successful.

- Your application requires proficiency in a language that isn't widely supported.

- Your work involves handling sensitive information subject to stringent data protection laws.

- Cost, privacy, or infrastructure constraints necessitate using a compact model.

Take a look at the self-explanatory decision flowchart in Figure 6-9 for when to use fine-tuning.

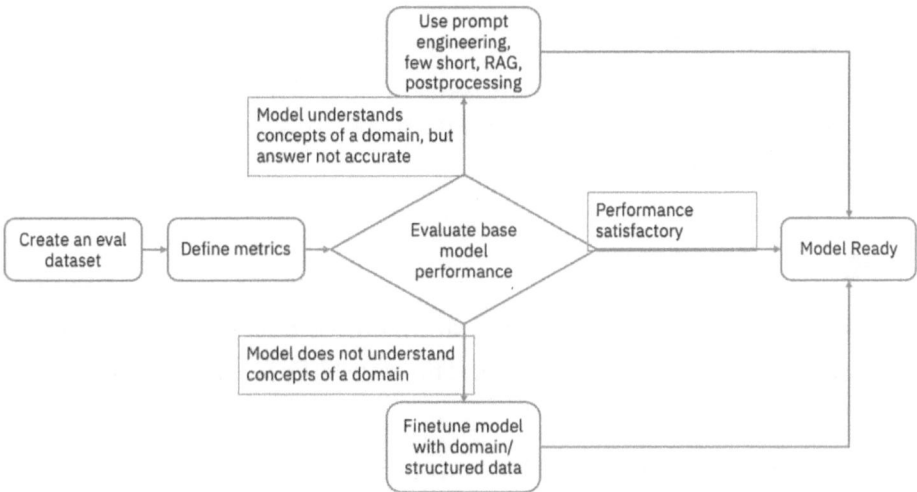

Figure 6-9. *Fine-tuning decision flowchart*

In these scenarios, ensuring access to, or the possibility of gathering, a sufficiently large and relevant dataset without incurring prohibitive expenses is crucial. Additionally, the process of fine-tuning a model demands adequate time and resources. Let's move on to our next section of fine-tuning an LLM.

Weighing the Trade-Offs: Prompting vs. Fine-Tuning

When enhancing LLMs for specific tasks, you need to weigh the options of prompt engineering against model fine-tuning. Each approach has its trade-offs in computational resources and the extent of customization they afford.

Prompt Engineering

Prompting is a technique that utilizes pre-existing knowledge of a model to perform a new task efficiently without additional training. However, it comes with the trade-offs.

- **Computational Efficiency:** Prompt engineering requires no additional training, thus conserving computational resources.

- **Quick Deployment:** It allows for faster experimentation and deployment since it relies on adjusting inputs rather than modifying the model.

- **Limited Customization:** The efficacy of prompt engineering is constrained by the model's existing knowledge and capabilities. It may not adequately capture domain-specific nuances or support languages with less online presence.

Fine-Tuning

On the other hand, fine-tuning involves re-training the model on a targeted dataset to adapt its understanding to specific contexts. The trade-offs for the fine-tuning are as follows:

- **Increased Resource Usage:** Fine-tuning demands significant computational power and time, especially for large models, and may incur higher costs.

- **Customization:** This approach allows the model to internalize the specifics of a domain or language, potentially leading to better performance on specialized tasks.

- **Data Requirements:** Fine-tuning requires a curated dataset representative of the task at hand, which can be challenging to assemble.

- **Privacy and Security:** When dealing with sensitive information, it is often necessary to fine-tune on-premises or in a secure environment, which adds to infrastructure considerations.

Decision-Making Framework

To decide between the two, consider the following framework:

- **Task Alignment:** Prompting is likely sufficient if the task aligns well with the LLM's pre-training. If not, fine-tuning may be necessary.

- **Data Availability:** Do you have access to a high-quality, task-specific dataset? If so, fine-tuning might be more beneficial.

- **Performance Requirements:** If the highest possible accuracy is essential and the model's pre-trained knowledge isn't cutting it, fine-tuning is often the better route.

- **Resource Constraints:** Evaluate the available computational resources and budget. Limited resources may favor prompting, while more abundant resources may allow for fine-tuning.

- **Time to Market:** If rapid deployment is crucial, prompting is quicker. Fine-tuning, while slower, may offer long-term performance benefits.

Example Application

For instance, a company operating in a niche market with unique jargon might find that LLMs do not perform adequately using prompt engineering alone. Here, fine-tuning on specialized corpus can imbue the model with the knowledge it otherwise lacks. Conversely, prompt engineering could suffice for a startup wanting to implement a chatbot quickly, especially when the conversations fall within the scope of the LLM's training.

By considering these trade-offs, you can make informed decisions tailored to their specific needs, balancing between the efficiency of prompting and the customization potential of fine-tuning.

CHAPTER 6 GENERATIVE LARGE LANGUAGE MODELS

Fine-Tuning LLMs

In our journey of understanding large language models, we have come across the concept of fine-tuning. You might wonder why fine-tuning is essential and how it benefits us in specific tasks or projects.

Fine-tuning a language model allows us to tailor it to our needs and tasks. While pre-trained language models are trained on vast amounts of internet text, they might not fully grasp the intricacies of our particular job or domain. This is where fine-tuning comes in.

By exposing the pre-trained model to our task-specific data during fine-tuning, we enhance its performance and enable it to better to understand the nuances relevant to our target task. Whether it's text classification, named entity recognition, machine translation, or question-answering, fine-tuning helps us achieve task-specific excellence.

One significant benefit of fine-tuning is its data efficiency. Rather than training a model from scratch, fine-tuning requires less labeled data. The pre-trained model already possesses a deep understanding of general language, and fine-tuning allows it to transfer this knowledge to our task with fewer labeled examples. This saves us time and effort and makes our training process more efficient.

Furthermore, fine-tuning enables the model to adapt to our specific domain or dataset. If our task revolves around a particular industry, like healthcare or finance, fine-tuning the model on domain-specific data equips it with the necessary vocabulary, jargon, and relevant language patterns. This adaptation ensures that the model speaks our language and understands our unique requirements.

By leveraging fine-tuning, we can take advantage of transfer learning by keeping most of the layers of the pre-trained transformer frozen and only updating a few layers, as shown in Figure 6-10. We can benefit from the pre-trained model's general language understanding and knowledge and transfer it to our specific task. The transfer of knowledge empowers our model to perform better and converge faster on our target task, giving us a head start in achieving our goals.

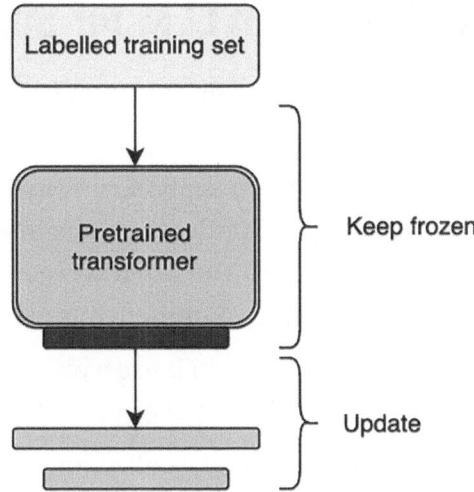

Figure 6-10. *Fine-tuning LLM*

Another advantage of fine-tuning is customization and adaptability. We have the flexibility to tailor the pre-trained model to our specific needs. Whether incorporating task-specific features, accommodating constraints, or making modifications, fine-tuning allows us to align the model more closely with our requirements. This customization empowers us to create a model that suits our unique context and objectives.

As we venture into the world of fine-tuning, it's essential to approach it carefully. We should consider factors such as the availability of task-specific data, the similarity between the pre-training and fine-tuning tasks, and the potential risks of bias or overfitting. Proper validation and evaluation of the fine-tuned model are crucial to ensure its effectiveness and performance on our target task.

We need a large dataset of labeled data and a lot of computational resources to do this. The dataset of labeled data must be large enough to capture the nuances of the task for which we are trying to fine-tune the LLM.

The computational resources needed to fine-tune an LLM can be significant, mainly if we use a large one.

Case Study: Fine-Tuning an LLM for Sentiment Analysis

Let's review a case study exploring how to fine-tune an LLM for sentiment analysis in customer reviews.

> **Objective:** Imagine you want to analyze customer reviews for your e-commerce platform to understand customer sentiment (positive, negative, or neutral). You can leverage an LLM fine-tuned for sentiment analysis to automate this process.
>
> **Model selection:** Let's assume you'll use a pre-trained LLM that has been trained on a massive text dataset.
>
> **Fine-Tuning Dataset:** You'll need a collection of customer reviews labeled with their corresponding sentiment (positive, negative, or neutral). This dataset should represent the reviews you typically receive on your platform. Aim to balance positive, negative, and neutral reviews to avoid biasing the model toward a specific sentiment.

Fine-Tuning Steps

You need to follow these steps (generally) to fine-tune an LLM:

1. **Data Pre-processing:** Clean and pre-process your customer review dataset. This might involve removing irrelevant information such as punctuation, stop words, and HTML tags. You might also need to handle sarcasm and slang appropriately.

2. **Model Selection:** Choose an appropriate pre-trained LLM, like BERT or GPT-3, trained on diverse languages and domains.

3. **Fine-Tuning/Transfer Learning:** To preserve the general language knowledge, it is recommended to freeze most of the pre-trained model's layers. Afterward, customize the last few layers of the LLM with the prepared dataset using an appropriate learning rate and epoch count. Then, fine-tune the model on your labeled customer review dataset. During the training process, it is essential to monitor it carefully to avoid overfitting and ensure that the model learns effectively. To prevent overfitting, techniques like early stopping can be implemented.

4. **Evaluation:** Evaluate the model's sentiment classification accuracy on a separate hold-out validation set of labeled reviews using precision, recall, F1 score, and accuracy metrics.

Common Pitfalls and How to Overcome Them

Here are some common issues and how to solve them:

- **Data Scarcity:** If labeled customer reviews are limited, consider data augmentation techniques like synonym replacement or back-translation to expand your dataset artificially.

- **Domain Mismatch:** The pre-trained model's training data might not perfectly align with the domain of your customer reviews. To bridge the gap, consider techniques like transfer learning with a pre-trained model on a similar domain (e.g., product descriptions).

- **Overfitting:** Fine-tuning for too long or using a complex model with limited data can lead to overfitting. To mitigate this risk, employ techniques like early stopping, regularization (e.g., dropout), and data augmentation.

By following these steps and addressing potential pitfalls, you can fine-tune an LLM for sentiment analysis or any other NLP task tailored to your needs. Remember, fine-tuning is an iterative process. Analyze the evaluation results, refine your approach, and retrain the model to achieve optimal performance.

Parameter Efficient Fine-Tuning

Full fine-tuning of an LLM starts with selecting a base large language model. Then, fine-tune the entire neural network weights by propagating gradients on a downstream task, with loss computed only on the output/response portion of labeled sample data. This process requires a lot of GPU memory to store model parameters and optimizer states, gradients, and activations. This is an expensive process, and with full fine-tuning, we need to create a new copy of the model for each task. So, we could end up with a large number of fine-tuned models that become difficult to store and manage. See Figure 6-11.

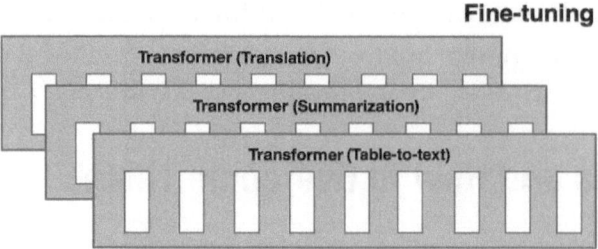

Figure 6-11. *Full fine-tuning of LLM (source: adapted from "COS597G: Understanding Large Language Models, Lecture 6, Princeton course archive (https://www.cs.princeton.edu/courses/archive/fall22/cos597G/lectures/lec06.pdf))*

However, if we tuned a subset of the parameters for each task, we could alleviate storage and compute costs. This is *parameter efficiency*. There are many different types of parameter efficient fine-tuning (PEFT), such as adapter-based, prompt tuning, and reparametrization-based. With PEFT, we can fine-tune LLMs by only adjusting a small number of additional parameters while keeping most pre-trained model parameters fixed. This approach dramatically reduces the computational and storage costs associated with fine-tuning.

One advantage of PEFT is its ability to overcome the challenge of *catastrophic forgetting*, a phenomenon observed when fully fine-tuning LLMs. By selectively updating only a fraction of the model's parameters, PEFT minimizes the risk of forgetting previously learned knowledge. This makes PEFT approaches particularly effective in scenarios with limited data availability and exhibits better generalization to out-of-domain situations.

Another compelling aspect of PEFT is its impact on portability. With PEFT methods, you can fine-tune your models and obtain compact checkpoints that occupy only a few megabytes of storage. This starkly contrasts the large checkpoints generated through full fine-tuning, which can occupy gigabytes of space. For example, using traditional fine-tuning on the bigscience/mt0-xxl model would result in 40GB checkpoints for each downstream dataset. However, with PEFT methods, the checkpoints for each downstream dataset would only be a few megabytes. Remarkably, this reduction in size does not compromise performance, as PEFT achieves comparable results to full fine-tuning.

The beauty of PEFT lies in its simplicity. You can effortlessly utilize the same pre-trained LLM for multiple tasks by adding only a small number of trained weights on top of the existing model. This means you don't have to replace the entire model for each new task, saving time and computational resources.

In summary, PEFT approaches empower you to achieve comparable performance to full fine-tuning while working with a much smaller number of trainable parameters. With PEFT, we can balance computational efficiency, model portability, and performance, opening new possibilities in leveraging large language models for diverse tasks.

Fine-Tuning LLM for Question Answering

In this section, we will go through an exercise of fine-tuning an LLM, Flan-T5-XXL, using a PEFT technique called Low-Rank Adaptation (LORA). The objective of fine-tuning the LLM is to extract long answers for medical research datasets that are accurate and not hallucinating. LoRA is an innovative approach that aims to make the fine-tuning process more efficient and effective. With LoRA[25], you can fine-tune your models while significantly reducing the number of trainable parameters, resulting in improved computational efficiency and reduced storage costs.

The main idea behind LoRA is to represent weight updates using smaller matrices called update matrices, which are obtained through low-rank decomposition (Figure 6-12). By decomposing the weight updates into these smaller matrices, we can adapt the model to new data while keeping the overall number of changes low. The original pre-trained weights remain frozen and unchanged, preserving the knowledge captured during pre-training. The output activations of original (frozen) pre-trained weights on the left are augmented by low-rank matrices (weight matrices A and B) on the right.

CHAPTER 6 GENERATIVE LARGE LANGUAGE MODELS

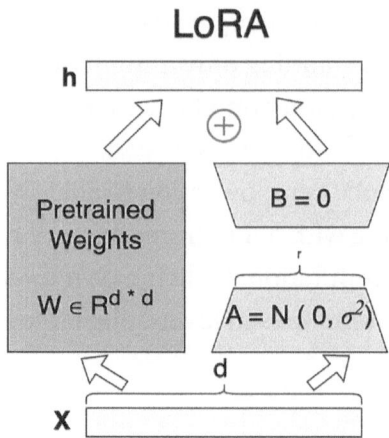

Figure 6-12. *LoRA fine-tuning of LLM*[26]

Using LoRA, we overcome the challenges associated with catastrophic forgetting, a phenomenon observed when fine-tuning large language models. With LoRA, we only fine-tune a small number of model parameters while keeping the majority of the pre-trained parameters frozen. This approach prevents catastrophic forgetting and allows the model to retain its ability to generate high-quality responses.

In addition to its computational benefits, LoRA has been shown to perform well in low-data regimes and generalize better to out-of-domain scenarios. This means despite limited training data, LoRA can produce reliable and accurate results. Furthermore, LoRA can be applied to various modalities, as we saw in Chapters 1–4, such as image classification and stable diffusion dream booths, making it a versatile technique for multiple applications.

One of the key advantages of using LoRA is its impact on model portability. With LoRA, you can create lightweight and portable models by adding small weights to the pre-trained language model rather than replacing the entire model. This allows you to use the same pre-trained language model for multiple tasks, saving storage space, and facilitating model deployment.

The small trained weights from LoRA can be easily combined with the pre-trained language model, enabling you to achieve comparable performance to full fine-tuning while using only a fraction of the storage space. For example, while full fine-tuning of a model like big science/mt0-XXL may require 40GB of storage, LoRA methods can reduce the storage requirements to just a few megabytes for each downstream dataset.

QLoRA builds on top of LoRA by quantizing the adapters, as shown in Figure 6-13. *Quantization* is a process in which data representation is reduced from a larger set of possible values to a smaller set of discrete values. In other words, it involves mapping continuous or high-resolution data to a limited number of discrete levels. In machine learning and deep learning, quantization is commonly used to reduce models' memory and computational requirements. For example, in neural network models, weights and activations are typically represented as 32-bit floating-point numbers, which can consume significant memory and computational resources. Quantization techniques can be applied to convert these 32-bit floating-point numbers into lower-bit representations, such as 16-bit or even 8-bit fixed-point numbers. By quantizing the adapters, QLoRA makes it possible to fine-tune LLMs on devices with limited memory and compute resources.

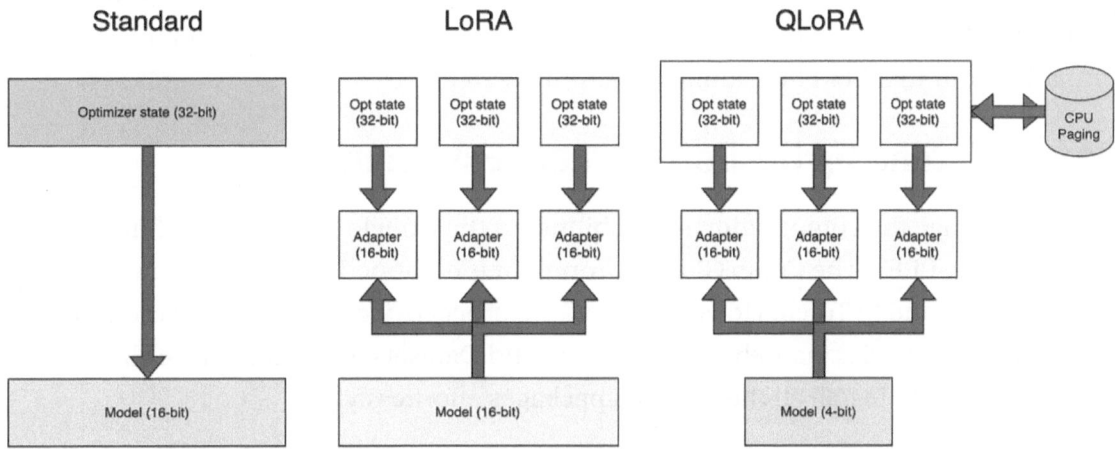

Figure 6-13. *QLoRA fine-tuning LLM*

LoRA and QLoRA introduce innovative techniques to enhance the fine-tuning process for language models. LoRA integrates new trainable low-rank matrices to each LLM layer while keeping the original parameters frozen, significantly reducing memory usage during fine-tuning. QLoRa takes it a step further by introducing three key steps.

1. **Four-bit NormalFloat quantization:** This method improves quantile quantization by ensuring an equal number of values in each quantization bin. This prevents computational issues and errors for outlier values.

2. **Double quantization:** QLoRa employs quantizing the quantization constants, resulting in additional memory savings.

3. **Paging with unified memory:** QLoRa automatically manages page-to-page transfers between the CPU and GPU using NVIDIA Unified Memory. This ensures seamless GPU processing, even in scenarios where the GPU may have limited memory.

These steps significantly reduce memory requirements for fine-tuning while maintaining performance close to standard fine-tuning. Embracing LoRa and QLoRa can significantly optimize the fine-tuning process and empower language models to perform at their best.

Strap in! We'll delve into each step with meticulous detail for this initial PEFT exercise focusing on fine-tuning an LLM. This comprehensive explanation sets the stage for future exercises that adopt a more concise approach. So, without further ado, let's embark on our first LLM fine-tuning adventure with PEFT!

Step 1: Setting Up the Development Environment

We will leverage the computational capabilities of the Google Colab GPU to fine-tune an LLM using PEFT. The Colab GPU environment is equipped with preconfigured CUDA drivers and PyTorch. However, before we proceed, we need to install the essential Hugging Face libraries, namely, Transformers and Datasets. Executing the following code cell, you can install all the required packages effortlessly.

```
# install Hugging Face Libraries
!pip install "peft==0.2.0"
!pip install "transformers==4.27.1" "datasets==2.9.0" "accelerate==0.17.1" "evaluate==0.4.0" "bitsandbytes==0.37.1" loralib --upgrade --quiet
# installing additional dependencies for logging training
!pip install rouge-score tensorboard py7zr
```

Step 2: Data Preprocessing

We will use the PubMedQA[27] dataset, a unique biomedical question-answering (QA) dataset collected from PubMed abstracts. The dataset is designed explicitly to answer research questions with answers based on the corresponding abstracts.

PubMedQA comprises three types of instances.

- 1,000 expert-annotated instances
- 61,200 unlabeled instances
- 211,300 artificially generated instances

Each instance in PubMedQA consists of the following components:

- **Question:** The question is either an existing research article title or derived from one.
- **Context:** The context refers to the corresponding abstract without its conclusion.
- **Long Answer:** The long answer represents the abstract's conclusion and is expected to answer the research question.
- **Yes/No/Maybe Answer:** This answer summarizes the conclusion and indicates whether the answer is affirmative, negative, or uncertain.

PubMedQA is the first QA dataset that requires reasoning over biomedical research texts, particularly their quantitative contents, to answer the questions.

Here is an example of a data entry from the Pubmedqa dataset:

```
{'pubid': 21645374,
 'question': 'Do mitochondria play a role in remodelling lace plant leaves
 during programmed cell death?',
 'context': {'contexts': ['Programmed cell death (PCD) is the regulated
 death.....','...', '...'],
  'labels': ['BACKGROUND', 'RESULTS'],
  'meshes': ['Alismataceae', '...', '...', '...', '...'],
  'reasoning_required_pred': ['y', 'e', 's'],
  'reasoning_free_pred': ['y', 'e', 's']},
 'long_answer': 'Results depicted mitochondrial dynamics in vivo......',
 'final_decision': 'yes'}
```

Step 2.1: Load the Dataset

Let's begin by loading the dataset from the Hugging Face Hub, as shown here:

```
from datasets import load_dataset

# Download and Load dataset from HuggingFace Hub
dataset = load_dataset('pubmed_qa', 'pqa_labeled')

print(f"Train dataset size: {len(dataset['train'])}")
```

In this code block, we loaded a dataset using the Hugging Face library's `Datasets` module. We will specifically load the `pubmed_qa` dataset with the variant `pqa_labeled`.

Using the `load_dataset()` function, we can easily fetch the desired dataset from the Hugging Face Hub. In this case, you will fetch the `pubmed_qa` dataset.

Once the dataset is loaded, we will print the size of the training dataset using the `len()` function. This will provide information about the number of samples in the training dataset.

Step 2.2: Train-Test Split

We will split the dataset into train and test to fine-tune our model on the training dataset and then evaluate the fine-tuned model on the test dataset.

```
# Extract 1000 examples for the train dataset
train_dataset = dataset['train'].shuffle().select(range(800))

# Extract 200 examples for the test dataset
test_dataset = dataset['train'].shuffle().select(range(800, 1000))
```

First, we select 800 examples from the `train` dataset by using the `shuffle()` function to randomize the order of the examples and the `select()` function to choose a specific range of examples.

Next, we extract 200 examples for the test dataset. Similar to the previous step, we again use the `shuffle()` function to randomize the order of examples, but this time we select samples from the range 800 to 1000. These 200 examples will be used to evaluate the model's performance.

By performing this extraction process, we create a smaller subset of examples from the original dataset specifically for training and testing purposes. This allows us to work with a manageable amount of data while maintaining sample diversity.

Step 2.3: Transform the Dataset

The context column of the Pubmed_QA dataset is a dictionary with a key as context and values as a list of sentences provided to answer a particular research question.

{'contexts': ["Quality of Life (QoL) assessment remains integral in the investigation of women with lower urinary tract dysfunction. Previous work suggests that physicians tend to underestimate patients' symptoms and the bother that they cause. The aim of this study was to assess the relationship between physician and patient assessed QoL using the Kings Health Questionnaire (KHQ).",
  'Patients complaining of troublesome lower urinary tract symptoms (LUTS) were recruited from a tertiary referral urodynamic clinic. Prior to their clinic appointment they were sent a KHQ, which was completed before attending. After taking a detailed urogynecological history, a second KHQ was filled in by the physician, blinded to the patient responses, on the basis of their impression of the symptoms elicited during the interview. These data were analyzed by an independent statistician. Concordance between patient and physician assessment for individual questions was assessed using weighted kappa analysis. QoL scores were compared using Wilcoxons signed rank test.',
  'Seventy-five patients were recruited over a period of 5 months. Overall, the weighted kappa showed relatively poor concordance between the patient and physician responses; mean kappa: 0.33 (range 0.18-0.57). The physician underestimated QoL score in 4/9 domains by a mean of 5.5% and overestimated QoL score in 5/9 domains by a mean of 6.9%. In particular, physicians underestimated the impact of LUTS on social limitations and emotions (P<0.05).'],
 'labels': ['AIMS', 'METHODS', 'RESULTS'],
 'meshes': ['Adult',
  'Aged',
  'Data Interpretation, Statistical',
  'Female',
  'Humans',
  'Middle Aged',
  'Patients',

```
    'Physicians',
    'Prospective Studies',
    'Quality of Life',
    'Sample Size',
    'Surveys and Questionnaires',
    'Urologic Diseases',
    'Young Adult'],
 'reasoning_required_pred': ['y', 'e', 's'],
 'reasoning_free_pred': ['y', 'e', 's']}
```

Therefore, we need to transform this column to create a passage that can be tokenized and used for training our model. Here is the code snippet for transforming the dataset:

```
# Create a new dataset with the transformed context column
transformed_train_dataset = train_dataset.from_dict({
    'context': [". ".join(example['context']['contexts']) for example in
    train_dataset],
    'question': train_dataset['question'],
    'answer': train_dataset['long_answer'],
    # 'labels': train_dataset['labels'],
    # 'meshes': train_dataset['meshes'],
    # 'reasoning_required_pred': train_dataset['reasoning_required_pred'],
    # 'reasoning_free_pred': train_dataset['reasoning_free_pred']
})

# Create a new dataset with the transformed context column
transformed_test_dataset = test_dataset.from_dict({
    'context': [". ".join(example['context']['contexts']) for example in
    test_dataset],
    'question': test_dataset['question'],
    'answer': test_dataset['long_answer'],
})

# Print the transformed context for the first example
print("Transformed Context:", transformed_train_dataset['context'][0])
```

CHAPTER 6   GENERATIVE LARGE LANGUAGE MODELS

In this code block, we will create new datasets with a transformed column called context based on the original datasets.

First, we iterate through each example in the training dataset and concatenate the individual context sentences into a single string using the join() function. This transformed context is then assigned to the context column of the new transformed_ train_dataset. We also copy the question and long_answer columns from the original training dataset to the corresponding columns in the new dataset.

Similarly, we perform the same transformation on the context column for the test dataset and create a new transformed_test_dataset with the transformed context, question, and long_answer columns.

Finally, we print out the transformed context of the first example in the transformed training dataset using transformed_train_dataset['context'][0].

Here is the output of the transformed context:

```
Transformed Context: Obesity may be associated with lower prostate specific
antigen through hemodilution. We examined the relationship between body
mass index and prostate specific antigen by age in men without prostate
cancer in a longitudinal aging study to determine whether prostate specific
antigen must be adjusted for body mass index.. The study population
included 994 men (4,937 observations) without prostate cancer in the
Baltimore Longitudinal Study of Aging. Mixed effects models were used to
examine the relationship between prostate specific antigen and body mass
index in kg/m(2) by age. Separate models were explored in men with prostate
cancer censored at diagnosis, for percent body fat measurements, for weight
changes with time and adjusting for initial prostate size in 483 men (2,523
observations) with pelvic magnetic resonance imaging measurements.. In men
without prostate cancer body mass index was not significantly associated
with prostate specific antigen after adjusting for age (p = 0.06). A
10-point body mass index increase was associated with a prostate specific
antigen difference of -0.03 ng/ml (95% CI -0.40-0.49). Results were similar
when men with prostate cancer were included, when percent body fat was
substituted for body mass index, and after adjusting for prostate volume.
Longitudinal weight changes also had no significant association with
prostate specific antigen.
```

By transforming the context column this way, we are consolidating the multiple sentences into a single coherent context, which can be more convenient for processing and analysis.

Step 2.4: Tokenize the Dataset

To train our model, we need to convert the textual inputs into a format the model can understand and process. This conversion is accomplished using the transformers tokenizer.

A tokenizer is responsible for breaking down the text into smaller units called *tokens*. Depending on the tokenizer's configuration, these tokens can be individual words, subwords, or characters. Each token is assigned a unique token ID, a numerical representation of that token.

The tokenizer takes care of several important tasks during this conversion process. It handles tokenization, which involves splitting the text into tokens based on specific rules or algorithms. It also performs additional tasks such as adding special tokens to mark the beginning and end of the input, handling out-of-vocabulary words, and applying various text normalization techniques.

Once the text is tokenized and assigned token IDs by the tokenizer, the model can efficiently process it during training. The model learns to associate these token IDs with meaningful representations and uses them to make predictions or perform language-understanding tasks.

Therefore, the tokenizer is a crucial component that converts our input text into token IDs, enabling the model to effectively understand and work with the textual data.

Let's begin by loading a tokenizer as shown here:

```
from transformers import AutoTokenizer, AutoModelForSeq2SeqLM

model_id="google/flan-t5-xxl"

# Load tokenizer of FLAN-t5-XL
tokenizer = AutoTokenizer.from_pretrained(model_id)
```

We are importing the necessary components from the Hugging Face library's Transformers module in the previous code block. We specifically import the `AutoTokenizer` and `AutoModelForSeq2SeqLM` classes.

Next, we define a variable called model_id and set it to the identifier of the model we want to use. In this case, the model is google/flan-t5-xxl. This model is a pre-trained language model called FLAN-t5-XXL[28], which was developed by Google.

To use this model, we need to load its tokenizer. The tokenizer is responsible for converting our input text into a format the model can understand. We achieve this by calling the AutoTokenizer.from_pretrained() method and passing in the model_id as the argument.

The from_pretrained() method automatically retrieves the tokenizer for the specified model from the Hugging Face model Hub. It downloads the necessary files if not already cached on your system. Once the tokenizer is loaded, it is assigned to the variable tokenizer, and we can use it to tokenize our text inputs.

By utilizing the AutoTokenizer and loading the appropriate tokenizer for the FLAN-t5-XXL model, we ensure that our text inputs are properly processed and converted into a tokenized format suitable for the model's input requirements.

Before training begins, data preprocessing is necessary. Our model receives a question and context as input and generates an answer as output. To ensure efficient data batching, we must determine the lengths of our input and output sequences for optimal organization and utilization of computational resources.

```
from datasets import concatenate_datasets
import numpy as np

# Tokenize the combined inputs
tokenizedQueries = concatenate_datasets([transformed_train_dataset]).map(
    lambda x: tokenizer(x["context"], x["question"], truncation=True),
    batched=True,
    remove_columns=["context", "question", "answer"]
)

queryLengths = [len(x) for x in tokenizedQueries["input_ids"]]

# Take the 100th percentile of max length for better utilization. You can
reduce this if you are running out of memory.
maxQueryLength = int(np.percentile(queryLengths, 100))
print(f"Max query length: {maxQueryLength}")

# This is the total sequence length for the response text after
tokenization.
```

CHAPTER 6   GENERATIVE LARGE LANGUAGE MODELS

```
# Sequences longer than responseLengths will be truncated; sequences
shorter will be padded."
tokenizedResponses = concatenate_datasets([transformed_train_dataset]).map(
    lambda x: tokenizer(x["answer"], truncation=True),
    batched=True,
    remove_columns=["context", "question", "answer"]
)
responseLengths = [len(x) for x in tokenizedResponses["input_ids"]]
# Take the 100th percentile of max length for better utilization. You can
reduce this if you are running out of memory.
maxResponseLength = int(np.percentile(responseLengths, 100))
print(f"Max response length: {maxResponseLength}")
```

We combine our preprocessed training dataset into a single dataset using the `concatenate_datasets` function from the `Datasets` library.

Next, we tokenize the combined inputs using the tokenizer. The `map` function is used to apply the tokenizer to each input in batch mode, where the inputs are provided as a list of dictionaries. We remove the `context`, `question`, and `answer` columns from the tokenized inputs.

To determine the maximum length of the source (input) sequences, we calculate the lengths of the tokenized inputs and store them in the `queryLengths` list. We obtain a value representing the maximum source length by taking the 100th percentile of these lengths using `np.percentile`. You can reduce this value if you run out of memory in your GPU environment, but there is a risk of losing the right context for getting the answer. The result is printed as `maxQueryLength`.

Similarly, we tokenize the target (output) sequences using the same approach. We calculate the lengths of the tokenized targets and store them in the `responseLengths` list. We determine the maximum target length by taking the 100th percentile of these lengths. The result is printed as `maxResponseLength`.

The output of this step is shown here:

```
Max query length: 512
Max response length: 171
```

Step 2.5: Preprocess the Dataset

Now, we can use the input and output length information from the last step to apply the preprocessing function to the whole dataset. So, we preprocess our dataset before training and save it to disk as shown here:

```
def preprocess_dataset(sample, padding="max_length"):
    tokenizer.pad_token = tokenizer.eos_token

    # Combine 'context' and 'question' columns
    inputs = ['Answer the question based on the context below. ' + context
    + ' ' + question
             for context, question in zip(sample["context"],
             sample["question"])]

    # Tokenize inputs
    model_inputs = tokenizer(inputs, max_length=maxQueryLength,
    padding=padding, truncation=True)

    # Tokenize labels with the text_target keyword argument
    labels = tokenizer(text_target=sample["answer"], max_
    length=maxResponseLength, padding=padding, truncation=True)

    # Replace all tokenizer.pad_token_id in the labels by -100 when we want
    to ignore padding in the loss.
    if padding == "max_length":
        labels["input_ids"] = [
            [(l if l != tokenizer.pad_token_id else -100) for l in label]
            for label in labels["input_ids"]
        ]
    model_inputs["labels"] = labels["input_ids"]
    # Only take the first token in the target sequence as the label
#     model_inputs["labels"] = labels["input_ids"][:min(len(labels["input_
    ids"]), 800)]
    return model_inputs

tokenized_train_dataset = transformed_train_dataset.map(preprocess_dataset,
batched=True, remove_columns=['context', 'question', 'answer'])
```

CHAPTER 6   GENERATIVE LARGE LANGUAGE MODELS

```
tokenized_test_dataset = transformed_test_dataset.map(preprocess_dataset,
batched=True, remove_columns=['context', 'question', 'answer'])
print(f"Keys of tokenized dataset: {list(tokenized_train_dataset.
features)}")

# save datasets to disk for loading it later
tokenized_train_dataset.save_to_disk("data/train")
tokenized_test_dataset.save_to_disk("data/eval")
```

The previous code block involves the preprocessing of datasets by combining the `context` and `question` columns, tokenizing the inputs and targets, and saving the tokenized datasets to disk.

The `preprocess_dataset` function inputs a sample and a padding option. It starts by setting the tokenizer's pad token to the end-of-sentence token. Then, it combines the `context` and `question` columns by creating a new list called `inputs`. Each element of `inputs` is formed by concatenating the string "Answer the question based on the context below. " with the corresponding `context` and `question` values from the sample. The prefixed line "Answer the question based on the context below. " acts as the instruction while fine-tuning the LLM. You must add the same instruction prompt during the inference/prediction time.

Next, the `inputs` are tokenized using the tokenizer. The `tokenizer` function is called with the `inputs` list and several arguments such as `max_length` (maximum allowed length of the tokens), `padding` (whether to pad the inputs), and `truncation` (whether to truncate the inputs if they exceed the maximum length). The result is stored in the `model_inputs` variable.

The targets are also tokenized using the tokenizer, with the `answer` column provided as the `text_target` argument. Similar to the inputs, the targets are tokenized with options for the maximum length, padding, and truncation. The resulting tokens are stored in the `labels` variable.

If padding is set to `max_length`, the function replaces any tokenizer padding tokens (identified by `tokenizer.pad_token_id`) in the labels with -100. This replacement indicates that these tokens should be ignored during the loss calculation.

The `model_inputs` dictionary is updated to include the labels, which are the input IDs of the tokenized labels. This allows the labels to be used during training.

The `preprocess_dataset` function returns the `model_inputs` dictionary.

CHAPTER 6 GENERATIVE LARGE LANGUAGE MODELS

The tokenized_train_dataset and tokenized_test_dataset functions are created by applying the preprocess_dataset function to the respective datasets (transformed_train_dataset and transformed_test_dataset). The datasets are processed in batches (batched=True), and the context, question, and answer columns are removed from the tokenized datasets.

The keys of the tokenized dataset features are printed to display the available keys.

Finally, the tokenized datasets are saved to disk for easy loading in the future. The train dataset is saved as data/train, and the test dataset is saved as data/eval.

Step 3: Model Training/Fine-Tuning

We will use the bitsandbytes LLM.int8() technique to quantize out the frozen LLM to int8, as shown in Figure 6-14. Model quantization to int8 converts a model's weights and activations from floating-point values (e.g., 32-bit floating-point) to 8-bit integer values. This will allow us to reduce the needed memory for FLAN-T5 XXL by about four times.

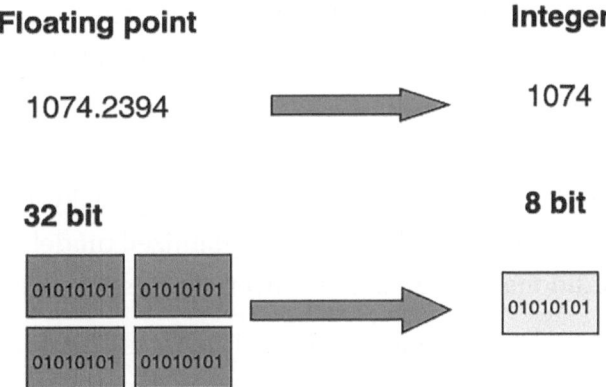

Figure 6-14. Quantization

Here are the steps involved in quantizing the LLM to int8:

1. Install the bitsandbytes library.
2. Load the frozen LLM model.
3. Quantize the model to int8 using the LLM.int8() function.
4. Save the quantized model.

Once the model is quantized to int8, you can load it and use it as usual. The quantized model will take up much less memory than the original model so that you can deploy it on devices with limited memory.

Here are the benefits of quantizing the LLM to int8:

- **Reduced Memory Footprint:** The quantized model will take up much less than the original model.

- **Increased Speed:** The quantized model will be faster than the original model.

- **Improved Accuracy:** The quantized model can be as accurate as the original model.

On the other hand, some of the disadvantages of quantizing LLMs to int8 are as follows:

- **Loss of Accuracy:** Quantization can introduce some loss of accuracy, especially for models trained on large datasets. This is because the quantized model cannot represent the same level of precision as the original model.

- **Increased Computational Complexity:** Quantization can increase the computational complexity of the model, especially for models used for inference. This is because the quantized model has to perform additional operations to convert the weights to int8 format.

- **Reduced Portability:** Quantized models are less portable than the original models. This is because the quantized models are specific to a particular hardware platform.

Overall, quantization is a trade-off between accuracy, computational complexity, and portability.

Step 3.1: Load the Model

The first step is to load the model to begin our training process. In this case, we will use the model `philschmid/flan-t5-xxl-sharded-fp16`, a sharded version of the model `google/flan-t5-xxl`. The sharding technique allows us to avoid memory constraints when loading the model.

```
from transformers import AutoModelForSeq2SeqLM

model_id = "philschmid/flan-t5-xxl-sharded-fp16"

model = AutoModelForSeq2SeqLM.from_pretrained(model_id, load_in_8bit=True,
device_map="auto")
```

In the previous code block, we use the `AutoModelForSeq2SeqLM` class from the Transformers library. The `model_id` variable is set to `philschmid/flan-t5-xxl-sharded-fp16`, representing the specific model we want to load from the Hugging Face model hub.

Next, we load the model using the `from_pretrained` method of the `AutoModelForSeq2SeqLM` class. We pass `model_id` as the argument to specify which model to load. Additionally, we set `load_in_8bit=True` to indicate that we want to load the model in 8-bit integer format (int8) after quantization. This helps reduce memory usage and improve computational efficiency during inference.

The `device_map="auto"` argument enables automatic mapping of the model to the available devices, ensuring efficient utilization of resources.

Therefore, the previous code block allows us to load the sharded model `philschmid/flan-t5-xxl-sharded-fp16` from the Hugging Face model hub, taking advantage of sharding and model quantization in int8 format to optimize memory usage and computational efficiency.

Step 3.2: Prepare the Model

You are ready to prepare your model for LoRA int-8 training using peft.

First, you need to define a LoRA config. This config specifies the parameters of the low-rank matrices, such as the r value, the `lora_alpha` value, the target modules, the `lora_dropout` value, and the bias.

```
lora_config = LoraConfig(
 r=16,
 lora_alpha=32,
 target_modules=["q", "v"],
 lora_dropout=0.05,
 bias="none",
 task_type=TaskType.SEQ_2_SEQ_LM
)
```

CHAPTER 6   GENERATIVE LARGE LANGUAGE MODELS

Here is an explanation of all the parameters in the code block:

- `r`: This is the inner dimension of the low-rank matrices to train. A higher rank means more trainable parameters, resulting in a more accurate model but also requiring more memory. Other values for `r` could be 4, 8, 16, 32, 128, or 256.

- `lora_alpha`: Alpha is a scaling factor, so this hyperparameter controls the strength of the LoRA layers. A higher value means LoRA layers act more strongly on the base model.

- `target_modules`: This is a list of modules for applying the LoRA update matrices. The default value is `["q", "v"]`, which means that the weight updates will be applied to the query and value heads of the model. The idea of using key/value/query formulation in attention mechanisms is inspired by the paper "Attention Is All You Need." Understanding the concept of queries, keys, and values is akin to how retrieval systems work. Let's consider an example to grasp this concept better.

  Imagine you are using a search engine, like Google, to find a relevant document. In this scenario, the input phrase is the query. The indices of documents/websites serve as the set of keys. The actual documents/website contents are the values. In essence, the key/value/query formulation allows attention mechanisms to focus on specific information (values) in the input data by using the queries and comparing them with the corresponding keys, enabling the model to generate more accurate and contextually relevant outputs. In the context of LoRA:

  - **q (Query Projection):** In the LoRA model, q represents the transformation applied to the input data (queries) before passing it through the LoRA layer. This transformation helps reduce the input data's dimensionality and capture relevant features for the specific task. The query projection generates a compact representation of the input data that is more suitable for the LoRA layer's learning process.

- **v (Value Projection):** Similarly, v represents the transformation applied to the output of the LoRA layer before combining it with the original pre-trained model's output. The value projection helps map the LoRA's output to the same dimension as the original model's output. This ensures that the LoRA's contribution to the final prediction properly aligns with the original model's output.
- **lora_dropout:** This is the dropout probability of LoRA layers. Dropout helps to prevent overfitting, so it is generally a good idea to use a dropout rate of at least 0.05.
- **bias:** This specifies if the bias parameters should be trained. The values can be `none, all,` or `lora_only`.
- **task_type:** This specifies the type of task the model will use. The default value is `TaskType.SEQ_2_SEQ_LM`, meaning the model will be used for sequence-to-sequence language modeling.

Next, you need to prepare your model for int8 training. This involves converting the model's weights to int8 format.

```
model = prepare_model_for_int8_training(model)
```

The `prepare_model_for_int8_training()` function takes a model as input and returns a new model with the weights converted to int8 format. Once you have called the `prepare_model_for_int8_training()` function, your model will be ready for LoRA int8 training.

Finally, you need to add the LoRA layers to your model. This is done using the `get_peft_model()` function.

```
model = get_peft_model(model, lora_config)
```

Once you have added the LoRA layers, you can print the number of trainable parameters in your model.

```
model.print_trainable_parameters()
```

The output of the code block should be similar to the following:

```
trainable params: 18874368 || all params: 11154206720 || trainable%: 0.16921300163961817
```

This means your model now has 18,874,368 trainable parameters, about 0.17% of your model's total number of parameters.

As you can see, we are only training 0.17% of the model's parameters! This considerable memory gain is possible because the LoRA layers apply only to a small subset of the model's parameters. This means we can fine-tune the model without running out of memory.

The QLoRA layers work by quantizing the model's weights to int8 format. This reduces the memory footprint of the model by a factor of 8. However, the QLoRA layer also introduces some errors in the model's predictions. This error can be reduced by increasing the resolution of the LoRA adapter.

In this case, we are using a resolution of 16, which means that the error introduced by the LoRA adapter is relatively tiny. This allows us to fine-tune the model without sacrificing too much accuracy.

The memory gain that we achieve by using the LoRA adapter is significant. This means we can fine-tune large language models on devices with limited memory. This capability is valuable for many applications, such as mobile devices and embedded systems.

Step 3.3: Create a Data Collator

The next step in our process is to create a `DataCollator` that will handle the padding of our inputs and labels. We will use the `DataCollatorForSeq2Seq` class from the Transformers library to achieve this.

```
from transformers import DataCollatorForSeq2Seq

# Ignoring tokenizer pad token in the loss
label_pad_token_id = -100
# Defining data collator
data_collator = DataCollatorForSeq2Seq(
    tokenizer,
    model=model,
    label_pad_token_id=label_pad_token_id,
    pad_to_multiple_of=8
)
```

In the provided code block, we import the `DataCollatorForSeq2Seq` class. This class allows us to define a data collator designed for sequence-to-sequence tasks.

To initialize the `DataCollatorForSeq2Seq`, we provide the following arguments:

- `tokenizer`: The tokenizer we are using for our model. It is used to tokenize the inputs and labels.

- `model`: The model object we have loaded. It is used to determine the necessary model configurations.

- `label_pad_token_id`: The parameter `label_pad_token_id` is used to specify the token ID that will be used for padding the target (label) sequences during training. In the context of sequence-to-sequence tasks, the target sequences are typically represented as labels for the model to learn from. By setting `label_pad_token_id = -100`, we choose a specific token ID (in this case, -100) that the model will ignore during the training process. By ignoring the padding token in the loss computation, the model will not be penalized for generating the padding tokens during training. It ensures that the padding tokens do not contribute to the loss calculation and, hence, do not affect the learning process. The model will only be updated based on the tokens in the target sequences that are not padding tokens, thus focusing on the relevant parts of the sequence and improving the overall training efficiency and performance.

- `pad_to_multiple_of`: We set `pad_to_multiple_of` to 8 to ensure that the input sequences and labels are padded to a length that is a multiple of 8. This is often done for optimization, as some hardware accelerators work more efficiently with data aligned to specific memory boundaries.

By utilizing the `DataCollatorForSeq2Seq` in our training process, we can handle the padding of inputs and labels consistently and optimally, considering the requirements of our model and task.

Step 3.4: Define Training Hyperparameters

Now, we will define the hyperparameters, such as learning rate, number of epochs, batch size, etc., to use for our training as shown here:

CHAPTER 6 GENERATIVE LARGE LANGUAGE MODELS

```
from transformers import Seq2SeqTrainer, Seq2SeqTrainingArguments

output_dir="lora-flan-t5-xxl"

# Define training args
training_args = Seq2SeqTrainingArguments(
    output_dir=output_dir,
    auto_find_batch_size=True,
    learning_rate=1e-3, # higher learning rate
#     num_train_epochs=5,
    max_steps=1, #10,
    logging_dir=f"{output_dir}/logs",
    logging_strategy="steps",
    logging_steps=1,
    save_strategy="no",
    report_to="tensorboard",
)
```

The code block defines the following hyperparameters:

- **output_dir:** This is the directory where the model and training logs will be saved.
- **auto_find_batch_size:** This flag tells the trainer to find the optimal batch size automatically.
- **learning_rate:** This is the learning rate that will be used for training. A higher learning rate will result in faster training, which may also lead to overfitting.
- **num_train_epochs:** The model will be trained for this number of epochs. An epoch is a complete pass through the training dataset.
- **max_steps:** This is the maximum number of steps that the model will be trained for. The training will be stopped if the model reaches this number of steps before completing all epochs.
- **logging_dir:** This is the directory where the training logs will be saved.

- **logging_strategy:** This specifies how often the training logs will be saved. The default value is steps, meaning the logs will be saved every logging_steps steps.

- **logging_steps:** This specifies the number of steps between each logging event.

- **save_strategy:** This determines how often the model will be saved. The default value is no, meaning the model will not be saved during training.

- **report_to:** This specifies where the training metrics will be reported. The default value is tensorboard, meaning the metrics will be reported to TensorBoard.

Once we have defined the hyperparameters, we can create a Seq2SeqTrainer instance. This instance will be used to train the model.

```
# Create a Trainer instance
trainer = Seq2SeqTrainer(
    model=model,
    args=training_args,
    data_collator=data_collator,
    train_dataset=tokenized_train_dataset,
)
model.config.use_cache = False  # silence the warnings. Please re-enable for inference!
```

The Seq2SeqTrainer instance takes the following arguments:

- **model:** This is the model that will be trained.

- **args:** This is the dictionary of hyperparameters that we defined earlier.

- **data_collator:** This function will be used to collate the training data.

- **train_dataset:** This is the training dataset.

Finally, we set the model.config.use_cache flag to False. This silences the warnings generated by the model when used for training. We will re-enable the caching for inference.

Step 3.5: Training the Model

Finally, we are ready to train our model. The code block `trainer.train()` will train the model using the hyperparameters that we defined earlier. The training process will be logged to TensorBoard, and the model will be saved for every `logging_steps` step.

In the case of T5, we keep some layers in `float32` for stability purposes. This is because T5 is a large model, and quantizing all of the layers to int8 can introduce too much error. By keeping some of the layers in float32, we can improve the stability of the model during training.

The training process may take some time, depending on the size of the model and the computational resources you have available. Once the training is complete, you can evaluate the model on the validation dataset to see how well it performs. Figure 6-15 shows some sample output of the training cell block.

Step	Training Loss
1	1.946800
2	1.537800
3	1.939600
4	0.958500
5	0.948300
6	1.421500
7	1.910300
8	1.118200
9	2.289000

[1000/1000 45:02, Epoch 1/2]

Figure 6-15. *Training loss output*

The training took a couple of dollars (for GPU compute resource) and 45 minutes. This is a significant cost saving compared to a full fine-tuning on FLAN-T5-XXL, which requires 8x A100 40GBs and costs hundreds of dollars for the same duration.

Step 3.6: Saving the Model

We can now save our model to use for inference and evaluation. We will save it locally, but you can also upload it to the Hugging Face Hub utilizing the `model.push_to_hub` method.

```
# Save fine-tuned LoRA model & tokenized results
peft_model_id="flan-t5-pubmed"
trainer.model.save_pretrained(peft_model_id)
tokenizer.save_pretrained(peft_model_id)
# Uncomment below to save the base model for inferencing
# trainer.model.base_model.save_pretrained(peft_model_id)
```

In the previous code block, we save our LoRA model and tokenizer results for future use. First, we specify the `peft_model_id` as "flan-t5-pubmed". This will serve as the identifier or name for our saved model and tokenizer.

Next, we use the `save_pretrained()` method of the `model` object to save our LoRA model. By calling `trainer.model.save_pretrained(peft_model_id)`, we save the entire model, including any modifications or adaptations made during the training process.

Similarly, we use the `save_pretrained()` method of the `tokenizer` object to save our tokenizer. By calling `tokenizer.save_pretrained(peft_model_id)`, we save the tokenizer and any special tokens, vocabulary, and settings used during training.

Optionally, if you want to save only the base model without any modifications made during training, you can use the `base_model` attribute of the `model` object. Only the base model will be saved by calling `trainer.model.base_model.save_pretrained(peft_model_id)`.

By saving our LoRA model and tokenizer, we can quickly reload them in the future for inference, fine-tuning, or any further analysis.

The output of the previous code block will be similar to the following:

```
('flan-t5-pubmed/tokenizer_config.json',
 'flan-t5-pubmed/special_tokens_map.json',
 'flan-t5-pubmed/tokenizer.json')
```

Our LoRA checkpoint is only 84 MB, yet it includes all the knowledge the model has learned for PubmedQA. This is a significant reduction in size compared to the full FLAN-T5-XXL model, which is 111 GB. The LoRA checkpoint is also much faster to load and use, making it ideal for deployment on devices with limited memory.

Table 6-1 summarizes the size of the two models.

CHAPTER 6 GENERATIVE LARGE LANGUAGE MODELS

Table 6-1. Model vs. LoRA Checkpoint Size

Model	Size
FLAN-T5-XXL	111 GB
LoRA checkpoint	84 MB

Step 4: Model Evaluation

After the training is done, we want to evaluate and test it. The most commonly used metrics to evaluate question-answering tasks are F1-score, BLEU score, ROGUE score, and Sentence Similarity scores.

Note For our FLAN-T5 XXL model, we need at least 18GB of GPU memory.

Table 6-2 summarizes the most commonly used metrics for question-answering tasks.

Table 6-2. Most Frequently Used Metrics For Question-Answering Tasks

Metric	Description
F1-score	This is the most commonly used metric for question-answering tasks. It measures the accuracy and fluency of the model's answers.
BLEU score	This metric measures the fluency of the model's answers. It is calculated by comparing the model's answers to a reference answer.
ROUGE score	This metric measures the overlap between the model's answers and a reference answer. It is calculated by comparing the n-grams of the two answers.
Sentence Similarity score	This metric measures the similarity between the model's answers and a reference answer. It is calculated by using a similarity metric, such as cosine similarity.

Step 4.1: Load the Fine-Tuned Model

We will use the PEFT and Transformers libraries to load the fine-tuned model. First, you need to load the LoRA config. The LoRA config contains the parameters of the LoRA adapter, such as the r value, the `lora_alpha` value, the target modules, the `lora_dropout` value, and the bias.

```
peft_model_id = "flan-t5-pubmed"
config = PeftConfig.from_pretrained(peft_model_id)
```

The previous code defines the `peft_model_id` directory and loads the LoRA config from the directory.

Next, you need to load the base LLM model and tokenizer. The base LLM model is the model that was used to initialize the LoRA model. The tokenizer is used to preprocess the input text.

```
model = AutoModelForSeq2SeqLM.from_pretrained(config.base_model_name_or_path, load_in_8bit=True, device_map={"":0})
tokenizer = AutoTokenizer.from_pretrained(config.base_model_name_or_path)
```

The previous code loads the base LLM model and tokenizer from the `config.base_model_name_or_path` directory. The `load_in_8bit=True` flag tells the model to load the model in int8 format. The `device_map={"":0}` flag tells the model to load the model on the CPU.

Finally, you need to load the LoRA model. The LoRA model is loaded from the `peft_model_id` directory. The model is then set to evaluation mode.

```
model = PeftModel.from_pretrained(model, peft_model_id, device_map={"":0})
model.eval()
```

This line loads the LoRA model from the `peft_model_id` directory and sets the model to evaluation mode.

Step 4.2: Test the Fine-Tuned Model

Before running the fine-tuned model on the entire test dataset, try it on one example and manually evaluate the results.

First, you need to define the context and the question. The context provides information about the medical research and asks a question based on the context.

```
prompt = f"""
Answer the question based on the context below.
 Context: To study whether nontriploid partial hydatidiform moles truly
exist. We conducted a reevaluation of pathology and ploidy in 19 putative
nontriploid partial hydatidiform moles using standardized histologic
diagnostic criteria and repeat flow cytometric testing by the Hedley
technique. On review of the 19 moles, 53% (10/19) were diploid nonpartial
```

moles (initially pathologically misclassified), and 37% (7/19) were triploid
partial moles (initial ploidy misclassifications). One additional case (5%)
was a diploid early complete mole (initially pathologically misclassified).
 Question: Do nontriploid partial hydatidiform moles exist?
""".strip()
```

Next, you need to tokenize the context and the question. This will convert the text into a format that the model can understand.

```
encoding = tokenizer(prompt, return_tensors="pt")
```

Finally, you need to generate the answer. The model will generate a sequence of tokens that represent the answer to the question.

```
with torch.inference_mode():
    outputs = model.generate(
        input_ids=encoding.input_ids,
        attention_mask=encoding.attention_mask,
        max_new_tokens=maxResponseLength,
    )

generated_output = tokenizer.decode(outputs[0], skip_special_tokens=True)

print("Generated Output: ", generated_output)
```

The output of the code block is the generated answer to the question. The answer is a text string containing the model's response to the question.

```
Generated Output:  Nontriploid partial hydatidiform moles do not exist.
The initial pathologic diagnosis of these moles is often incorrect. Flow
cytometric testing by the Hedley technique is the most reliable method for
determining ploidy. The initial pathologic diagnosis of these moles should
be reevaluated. Flow cytometric testing by the Hedley technique is the most
reliable method for determining ploidy. The initial pathologic diagnosis of
these moles should be reevaluated.
```

We have fine-tuned the same model to provide a more descriptive answer than the pre-trained model generating "Yes/No" responses. Figure 6-16 shows this analysis.
```

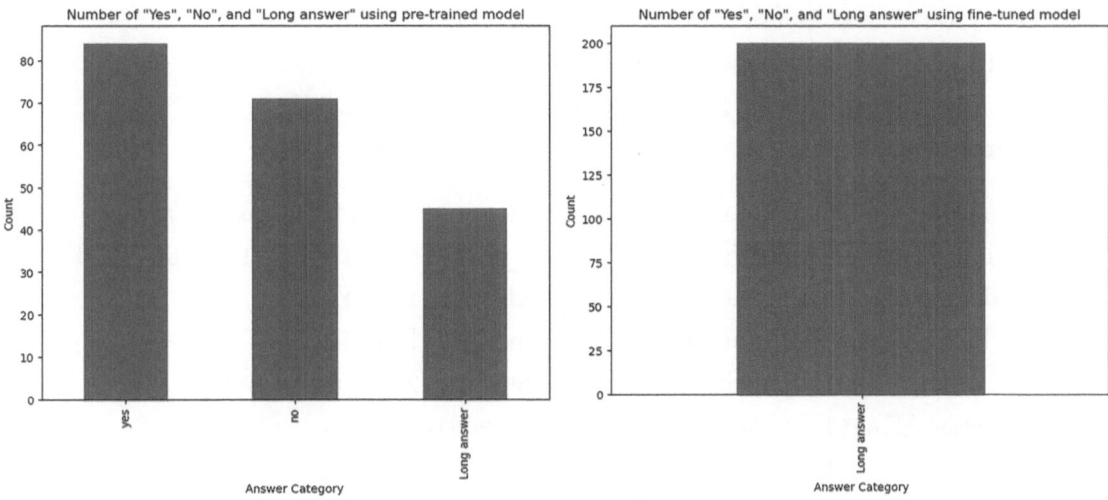

Figure 6-16. *Before and after fine-tuning output comparison*

Step 4.3: Evaluate the Fine-Tuned Model on the Test Dataset

We have manually compared an answer quality with the pre-trained model. Let's dive into the statistical comparison of the results before and after fine-tuning the model.First, you need to load the evaluation dataset. The evaluation dataset contains a set of questions and answers that the model has not seen before.test_dataset = load_from_disk('data/eval/').with_format('torch')Next, you need to define the metrics that will be used to evaluate the model. In this example, the metrics are ROUGE-1, ROUGE-2, ROUGE-L, ROUGE-Lsum, F1 score, BLEU score, and SentenceSim score.rouge_metric = evaluate.load('rouge')bleu_metric = corpus_bleu

```
def compute_f1(a_gold, a_pred):
    gold_toks = get_tokens(a_gold)
    pred_toks = get_tokens(a_pred)
    common = collections.Counter(gold_toks) & collections.
    Counter(pred_toks)
    num_same = sum(common.values())
    if len(gold_toks) == 0 or len(pred_toks) == 0:
        return int(gold_toks == pred_toks)
    if num_same == 0:
        return 0
```

```
    precision = 1.0 * num_same / len(pred_toks)
    recall = 1.0 * num_same / len(gold_toks)
    f1 = (2 * precision * recall) / (precision + recall)
    return f1

def calculate_sentencesim_score(generated_answer, actual_answer):
    model = SentenceTransformer('average_word_embeddings_glove.6B.300d')
    embeddings1 = model.encode([generated_answer])[0]
    embeddings2 = model.encode([actual_answer])[0]
    similarity = np.dot(embeddings1, embeddings2) / (np.linalg.
    norm(embeddings1) * np.linalg.norm(embeddings2))
    return similarity
```

Finally, it would be best if you evaluated the model on the evaluation dataset. The code block will print out the scores for each metric.

```
predictions, references = [], []
for sample in tqdm(test_dataset):
    p, l = evaluate_peft_model(sample)
    predictions.append(p)
    references.append(l)

# rouge = rouge_metric.compute(predictions=predictions,
references=references, use_stemmer=True)
bleu = bleu_metric([[ref] for ref in references], predictions, auto_
reweigh=True)
f1_scores = [compute_f1(ref, pred) for ref, pred in zip(references,
predictions)]
f1_avg = sum(f1_scores) / len(f1_scores)

sentencesim_scores = [calculate_sentencesim_score(pred, ref) for pred, ref
in zip(predictions, references)]
sentencesim_avg = sum(sentencesim_scores) / len(sentencesim_scores)

print(f"Rouge1: {rouge['rouge1'].mid.fmeasure* 100:.2f}%")
print(f"Rouge2: {rouge['rouge2'].mid.fmeasure* 100:.2f}%")
print(f"RougeL: {rouge['rougeL'].mid.fmeasure* 100:.2f}%")
print(f"RougeLsum: {rouge['rougeL'].mid.fmeasure* 100:.2f}%")
```

```
print(f"Avg F1: {f1_avg* 100:.2f}%")
print(f"BLEU: {bleu* 100:.2f}%")
print(f"Avg SentenceSim: {sentencesim_avg * 100:.2f}%")
```

The results are shown below:

```
Rouge1: 23.13%
Rouge2: 4.76%
RougeL: 14.72%
RougeLsum: 14.72%
Avg F1: 19.81%
BLEU: 23.38%
Avg SentenceSim: 78.57%
```

Step 4.4: Create an Evaluation DataFrame

Let's create a DataFrame to store the question, actual answers, and generated answers so that we can reuse it to compare the results later rather than train the model again to get the output. The DataFrame will contain the following columns:

- **Question:** The question that was asked.
- **Prediction:** The answer that the model generated.
- **Answer:** The ground truth answer.

First, you need to load the evaluation dataset. The evaluation dataset contains questions and answers that the model has never seen.

```
test_dataset = load_from_disk('data/eval/').with_format('torch')
```

Next, you need to iterate over the evaluation dataset and extract the questions, predictions, and references.

```
questions, predictions, references = [], [], []

for sample in tqdm(test_dataset):
    p, l = evaluate_peft_model(sample)

    # Assuming that the questions are in the input_ids in the sample
    question = tokenizer.decode(sample['input_ids'], skip_special_
    tokens=True)
```

CHAPTER 6 GENERATIVE LARGE LANGUAGE MODELS

```
        predictions.append(p)
        references.append(l)
        questions.append(question)
```

Finally, you can create the DataFrame and print out the first few rows (Figure 6-17).

```
df_flan_t5_eval = pd.DataFrame(
    {'Question': questions,
     'Prediction': predictions,
     'Answer': references,
    })
df_flan_t5_eval.head()
```

	Question	Prediction	Answer
0	Answer the question based on the context below...	When compared with the control group, patients...	Left ventricular dimensions are not influenced...
1	Answer the question based on the context below...	The high prevalence of complex coronary lesion...	Complex coronary lesions such as bifurcation a...
2	Answer the question based on the context below...	There was no consensus about the degree of mar...	Results of this survey highlight the wide vari...
3	Answer the question based on the context below...	The prevalence of chronic conditions increased...	Analyzing the prevalence of 11 chronic conditi...
4	Answer the question based on the context below...	Compared with the control group, women with a ...	NT thickness does not show a significative inc...

Figure 6-17. Evaluation dataframe

You can also write the DataFrame to a CSV file, as shown in Figure 6-18.

```
df_flan_t5_eval.to_csv('df_flan_t5_eval.csv')
```

	F1 Score	BLEU Score	ROUGE Score	SentenceSim Score
count	200.000000	200.000000	200.000000	200.000000
mean	0.232534	0.037779	0.264530	0.728594
std	0.079427	0.034505	0.087622	0.120750
min	0.061224	0.004558	0.071429	0.325754
25%	0.175651	0.012999	0.204246	0.660566
50%	0.227458	0.026530	0.253695	0.742767
75%	0.282490	0.050444	0.322581	0.815835
max	0.496894	0.192402	0.511628	0.975518

Figure 6-18. Evaluation metrics dataframe

CHAPTER 6   GENERATIVE LARGE LANGUAGE MODELS

Let's compare the result with the pre-trained FLAN-T5-XXL model (Figure 6-19). You can execute the Evaluating-Pre-trained-model-closed-QnA.ipynb notebook from our GitHub repo to get the results from the pre-trained model.

	F1 Score	BLEU Score	ROUGE Score	SentenceSim Score
count	200.000000	200.000000	200.000000	200.000000
mean	0.063637	0.006748	0.067779	0.152722
std	0.131048	0.022264	0.139672	0.288905
min	0.000000	0.000000	0.000000	-0.114598
25%	0.000000	0.000000	0.000000	-0.016939
50%	0.000000	0.000000	0.000000	0.023571
75%	0.000000	0.000000	0.000000	0.093360
max	0.545455	0.183504	0.636364	0.930428

Figure 6-19. *Evaluation metrics pre-trained model*

The F1 score of 22.69% significantly improved over the F1 score of 6% for the pre-trained model. So, there is a 283% lift in the F1 score. This means our LoRA checkpoint is more likely to make correct predictions than the pre-trained model.

The fact that our LoRA checkpoint is only 80 MB is also impressive. This means it can be used on devices with limited memory.

Overall, these results are very promising. They suggest that our LoRA checkpoint can be used to improve the performance of large language models on various tasks.

Summary

In this chapter, we explored generative large language models and their transformative impact on natural language processing. We began with practical NLP tasks where these models excel, such as sentiment analysis, entity extraction, and topic modeling.

We then navigated the creative potential of LLMs, showcasing their prowess in tasks such as creative writing, text summarization, and dialog generation. These sections highlighted the models' capabilities and provided you with the insights to harness them effectively in your own NLP projects.

Diving deeper, we examined advanced prompting techniques that optimize LLM performance. We discussed the nuanced differences between few-shot prompting, chain-of-thought prompting, and when to choose fine-tuning over prompting. Emphasizing the importance of model customization, we introduced the concept of parameter efficient fine-tuning. We walked you through a detailed, step-by-step process of fine-tuning an LLM for a question-answering task, from setting up the development environment to data pre-processing, model training, and evaluation.

This chapter serves as a springboard for further exploring LLMs and NLP. Here are some resources to delve deeper into specific areas:

For Specific NLP Tasks: Consider exploring online tutorials, workshops, or courses focusing on tasks that particularly interest you. Platforms like Hugging Face offer a wealth of resources and pre-trained models for various NLP applications: `https://huggingface.co/learn/nlp-course/`.

Advanced LLM Techniques: If you're interested in pushing the boundaries of LLM capabilities, you can explore research papers or blog posts on topics like prompt engineering for complex reasoning tasks, leveraging large language models for code generation or exploring the ethical considerations surrounding bias in LLMs. The Allen Institute for Artificial Intelligence publishes cutting-edge research on LLMs and NLP: `https://allenai.org/papers`.

Fine-Tuning for Different Tasks: The chapter covered fine-tuning for question answering. If you're interested in fine-tuning for other tasks, there are numerous online resources and tutorials available. Explore libraries like PEFT (`https://huggingface.co/docs/peft/en/index`), which provides efficient functionalities for fine-tuning on various NLP tasks.

Remember, the field of NLP and LLMs is rapidly evolving. Staying updated with the latest research, advancements, and best practices will ensure you can leverage these powerful tools to their full potential in your NLP endeavors.

CHAPTER 7

Advanced Techniques for Large Language Models

Introduction

In this chapter, we dive into advanced tuning techniques for large language models (LLMs). We explore powerful methods that push the boundaries of what LLMs can achieve, enabling us to shape their outputs and enhance their performance in various domains.

LLMs have revolutionized natural language processing, demonstrating their prowess in generating coherent and contextually relevant text. However, as with any technology, there is always room for improvement and fine-tuning to meet specific requirements and achieve desired outcomes.

This chapter focuses on advanced fine-tuning techniques that allow us to take LLMs to new heights, building upon what we learned in the previous chapter.

In this chapter, we will continue with the instruction fine-tuning using Parameter Efficient Fine-Tuning (PEFT), such as Low-Rank Adaptation (LORA), for different task types–abstractive summarization. Finally, we explore the exciting world of Reinforcement Learning from Human Feedback (RLHF). Through techniques like Supervised Fine-tuning (SFT), reward modeling, and reinforcement learning, we can train LLMs to generate more refined and sophisticated outputs by incorporating human feedback.

Throughout this chapter, we present real-world examples, discuss challenges, and provide practical implementation insights. We aim to equip you with the knowledge and tools necessary to leverage these advanced techniques effectively, empowering you to push the boundaries of what LLMs can accomplish. We will explore the following topics and implementations using advanced LLM techniques in this chapter.

The "Fine-Tuning LLMs for Abstractive Summarization" section delves into two approaches for fine-tuning LLMs to generate summaries of factual text documents. Here's a breakdown of what you can expect:

- **Fine-Tuning an Encoder-Decoder Model:** We'll explore how to fine-tune a standard encoder-decoder architecture commonly used for machine translation tasks. This approach can be practical for creating concise and informative summaries that accurately capture the key points of the source document.

- **Fine-Tuning a Decoder-Only Model for a More Creative and Abstract Summary:** We'll look at an alternative approach using a decoder-only model. This method can be beneficial when you desire a more creative or abstract summary that goes beyond simply relaying factual information.

The Reinforcement Learning from Human Feedback" section will showcase the power of RLHF in shaping LLM outputs based on human feedback. We'll explore a specific case study.

- **Controlled Review Generation: Tuning the Model for Positive Review Generation:** Here, we'll demonstrate how to fine-tune an LLM using RLHF to generate positive reviews. This example highlights how RLHF can guide the LLM toward a specific sentiment or outcome, but we'll also discuss ethical considerations and potential biases when using such techniques.

Whether you are an AI researcher, developer, or enthusiast, this chapter comprehensively explores advanced techniques for LLMs, enabling you to unlock their full potential and create unique, generative AI systems.

So, let's embark on this journey together and explore the exciting realm of advanced techniques for large language models!

Fine-Tuning LLMs for Abstractive Summarization

Abstractive summarization is a technique used in natural language processing to create a short, coherent version of a longer text document. This method involves understanding the main ideas and concepts within the original text and then expressing those ideas in new, concise terms. Unlike extractive summarization, which simply extracts key sentences or phrases directly from the text without altering them, abstractive summarization generates new sentences, often synthesizing information or rephrasing content to produce a fluent summary representative of the original text's meaning. This process can be likened to how a human might read an article and then explain it in their own words, focusing on the essential points and possibly using different expressions and vocabulary.

Fine-Tuning an Encoder-Decoder Model

In this section, we will first learn how to fine-tune an encoder-decoder LLM for an abstractive summarization task, and then we will apply the same learning for a decoder-only LLM in the following section.

Step 1: Installing Libraries and Data Loading

Let's start by setting up the environment for fine-tuning the model.

```
!pip install datasets evaluate transformers[sentencepiece]
!pip install accelerate
```

Here, we're preparing our Python environment for the task at hand. We install the `datasets` and `evaluate` libraries to handle and assess our data and the `transformers` library to leverage pre-trained models. The `accelerate` package is included for efficient computation, which is especially useful when dealing with large models and datasets.

Next, we will load and explore the dataset.

```
from datasets import load_dataset
xsum_dataset = load_dataset("EdinburghNLP/xsum")
xsum_dataset
```

In this snippet, we load the dataset we'll use for summarization. The `xsum_dataset`[29] is a popular choice for summarization tasks, known for its quality and diversity. We're using the `load_dataset` function from the `datasets` library to fetch this dataset.

Let's explore the dataset before we dive into the data preparation step.

```
def show_samples(dataset, num_samples=3, seed=42):
    sample = dataset["train"].shuffle(seed=seed).select(range
    (num_samples))
    for example in sample:
        print(f"\n'>> Summary: {example['summary']}'")
        print(f"'>> Document: {example['document']}'")
show_samples(xsum_dataset)
```

Here, we define a function called show_samples to inspect a few samples from our dataset. This function shuffles the dataset and selects a specified number of samples, allowing us to glimpse what our data looks like, both the documents and their summaries.

```
'>> Summary: As Chancellor George Osborne announced all English state
schools will become academies, the Welsh Government continues to reject the
model here.'
'>> Document: In Wales, councils are responsible for funding and overseeing
schools.
But in England, Mr Osborne's plan will mean local authorities will cease to
have a role in providing education.
Academies are directly funded by central government and head teachers
have more freedom over admissions and to change the way the school
works...........................
```

Step 2: Data Pre-processing

Let's prepare our dataset for training by initializing a tokenizer as shown here:

```
from transformers import AutoTokenizer
model_checkpoint = "t5-small"
tokenizer = AutoTokenizer.from_pretrained(model_checkpoint)
```

In this cell, we're initializing a tokenizer from the transformers library. We're using the t5-small[30] model checkpoint, a smaller variant of the T5 model, suitable for tasks like summarization. The tokenizer is crucial for converting text into a format our model can understand.

Let's see an example of the tokenized input.

CHAPTER 7　ADVANCED TECHNIQUES FOR LARGE LANGUAGE MODELS

```
inputs = tokenizer("Clean-up operations are continuing across the
Scottish Borders and Dumfries and Galloway after flooding caused by
Storm Frank.")
inputs
tokenizer.convert_ids_to_tokens(inputs.input_ids)
```

Here, we demonstrate how to use the tokenizer. We take a sample sentence and pass it through the tokenizer. This process converts the text into input IDs, essentially numerical representations of the words. We also showcase how to convert these IDs into tokens, helping us understand the tokenization process.

We need to set the limit for tokenizing the input and target length. So, we need to analyze the word count in the dataset using the following code:

```
import numpy as np

def word_count_analysis(dataset):
    document_word_counts = []
    summary_word_counts = []

    for example in dataset:
        # Count words in the document and summary
        document_words = example["document"].split()
        # document_words = example['document'].apply(lambda x: x.split())

        summary_words = example["summary"].split()
        # summary_words = example['summary'].apply(lambda x: x.split())

        # Append word counts to respective lists
        document_word_counts.append(len(document_words))
        summary_word_counts.append(len(summary_words))

    return document_word_counts, summary_word_counts

# Perform the word count analysis on the dataset subset
document_word_counts, summary_word_counts = word_count_analysis(xsum_
dataset['train'])

# Calculating the mean and median word counts
mean_document_word_count = np.mean(document_word_counts)
median_document_word_count = np.median(document_word_counts)
```

```
mean_summary_word_count = np.mean(summary_word_counts)
median_summary_word_count = np.median(summary_word_counts)

print("Mean Document Word Count:", mean_document_word_count)
print("Median Document Word Count:", median_document_word_count)
print("Mean Summary Word Count:", mean_summary_word_count)
print("Median Summary Word Count:", median_summary_word_count)
```

This function calculates the word counts for documents and summaries in the dataset. It provides valuable insights into the length distribution of texts and summaries, which is critical for setting the appropriate tokenization parameters. The output of the previous function is shown here:

```
Mean Document Word Count: 373.8646328015879
Median Document Word Count: 295.0
Mean Summary Word Count: 21.09764512730035
Median Summary Word Count: 21.0
```

We will set the maximum input length and maximum target length for tokenization as shown here:

```
max_input_length = 512
max_target_length = 30

def preprocess_function(examples):
    model_inputs = tokenizer(
        examples["document"],
        max_length=max_input_length,
        truncation=True,
    )
    labels = tokenizer(
        examples["summary"], max_length=max_target_length, truncation=True
    )
    model_inputs["labels"] = labels["input_ids"]
    return model_inputs
```

Here, we define a preprocessing function to tokenize the dataset. The function truncates or pads the inputs and labels (summaries) to fixed lengths, ensuring they are in a consistent format suitable for model training.

We can now apply this preprocessing function to our dataset.

```
tokenized_datasets = xsum_dataset.map(preprocess_function,
batched=True)
```

This line applies the preprocessing function to the entire dataset. Using the map method, we efficiently process all data entries, preparing them for the subsequent modeling steps.

Step 2.1: Evaluation and Benchmark Setup

Let's take an example from the dataset to explore the evaluation criteria.

```
generated_summary = "Hurricane Patricia is a category 5 storm."
reference_summary = "Hurricane Patricia has been rated as a category
5 storm."
```

These lines set up an example of a generated summary and its corresponding reference summary. This setup is often used for evaluating the model's summarization quality.

We will use the Recall-Oriented Understudy for Gisting Evaluation (ROUGE) score to evaluate the generated output. The ROUGE score is a standard metric for evaluating text summarization in NLP. It compares a generated summary with reference summaries, focusing on recall. Key variants include ROUGE-N for N-gram overlap, ROUGE-L for longest common subsequence, and ROUGE-S for skip-bigram co-occurrence. These measures assess how well the generated summary captures key points from the reference, providing a quantitative evaluation of summarization quality.

Let's begin by installing the library for the ROUGE score.

```
!pip install rouge_score
import evaluate
rouge_score = evaluate.load("rouge")
```

Here, we install the `rouge_score` package and load the ROUGE metric from the evaluate library.

Now, we can compute the ROUGE score on the example we selected earlier.

```
scores = rouge_score.compute(predictions=[generated_summary],
references=[reference_summary])
```

This code computes the ROUGE scores for our example summaries. It quantitatively measures how well the generated summary matches the reference, considering various aspects like overlap in words and phrases. The output is shown here:

```
{'rouge1': 0.7058823529411764,
 'rouge2': 0.5333333333333333,
 'rougeL': 0.7058823529411764,
 'rougeLsum': 0.7058823529411764}
```

We need to set a baseline to compare our evaluation results. A basic approach in text summarization is the lead-3 baseline, where the first three sentences of a text are used as a summary. To accurately identify sentence boundaries, especially in cases with acronyms like U.S. or U.N., we utilize the nltk library, which offers a sophisticated sentence boundary detection algorithm. Install nltk using this:

```
!pip install nltk
```

Then, acquire the necessary punctuation rules with the following:

```
import nltk
nltk.download("punkt")
```

Afterward, we use nltk's sentence tokenizer to devise a function that retrieves the first three sentences of a text. In line with summarization practices, we separate these sentences with newlines. Let's apply this method to a sample from our training dataset.

```
from nltk.tokenize import sent_tokenize

def three_sentence_summary(text):
    return "\n".join(sent_tokenize(text)[:3])

print(three_sentence_summary(xsum_dataset["train"][1]["document"]))
```

In the previous code block, we use NLTK's sentence tokenizer. The function three_sentence_summary is defined to extract the first three sentences from a text, providing a simple baseline summarization method. This is useful for creating a basic summary, which can serve as a comparison point for more advanced models.

Now, let's define a baseline summarization method.

```
def evaluate_baseline(dataset, metric):
    summaries = [three_sentence_summary(text) for text in
    dataset["document"]]
    return metric.compute(predictions=summaries,
    references=dataset["summary"])
```

This function evaluates the baseline summarization method (the three-sentence summary) using the specified dataset and evaluation metric (like ROUGE). It's an important step to understand the performance of a basic summary before proceeding with more complex models.

We will run the baseline summarization on the validation dataset as shown here:

```
import pandas as pd
score = evaluate_baseline(xsum_dataset["validation"], rouge_score)
rouge_names = ["rouge1", "rouge2", "rougeL", "rougeLsum"]
rouge_dict = dict((rn, round(score[rn] * 100, 2)) for rn in
rouge_names)
rouge_dict
```

Here, we evaluate our baseline summarizer on the validation set and calculate various ROUGE scores. The results are stored in a dictionary and displayed. This gives us a clear understanding of the summarizer's performance regarding different ROUGE metrics. The output is shown here:

{'rouge1': 18.46, 'rouge2': 2.52, 'rougeL': 11.98, 'rougeLsum': 14.51}

Step 3: Model Training

Let's initialize the model.

```
from transformers import AutoModelForSeq2SeqLM
model = AutoModelForSeq2SeqLM.from_pretrained(model_checkpoint, device_
map="auto")
```

In this step, we initialize our sequence-to-sequence model using the AutoModelForSeq2SeqLM class from the Transformers library. The model is loaded with pre-trained weights from the specified model_checkpoint. The device_map="auto" argument optimizes the model's device allocation, automatically choosing between CPU and GPU based on availability and compatibility.

CHAPTER 7   ADVANCED TECHNIQUES FOR LARGE LANGUAGE MODELS

Next, we are going to set training arguments.

```
from transformers import Seq2SeqTrainingArguments
batch_size = 8
num_train_epochs = 8
logging_steps = len(tokenized_datasets["train"]) // batch_size
model_name = model_checkpoint.split("/")[-1]
args = Seq2SeqTrainingArguments(
    output_dir=f"{model_name}-finetuned-xsum",
    evaluation_strategy="epoch",
    learning_rate=5.6e-5,
    per_device_train_batch_size=batch_size,
    per_device_eval_batch_size=batch_size,
    weight_decay=0.01,
    save_total_limit=3,
    num_train_epochs=num_train_epochs,
    predict_with_generate=True,
    logging_steps=logging_steps,
)
```

Here, we define the training arguments for our model using `Seq2SeqTrainingArguments`. Parameters like batch size, number of training epochs, learning rate, and weight decay are set here. These arguments configure essential aspects of the training process, such as how often to log progress, how to handle evaluation, and where to save the model checkpoints.

We will set up the evaluation method before starting the training process.

```
import numpy as np
def calculate_evaluation_metrics(prediction_data):
    predicted, true_labels = prediction_data
    translated_predictions = tokenizer.batch_decode(predicted, skip_
    special_tokens=True)
    true_labels = np.where(true_labels != -100, true_labels, tokenizer.
    pad_token_id)
    translated_labels = tokenizer.batch_decode(true_labels, skip_special_
    tokens=True)
```

```
    formatted_predictions = ["\n".join(sent_tokenize(pred_text.strip()))
    for pred_text in translated_predictions]
    formatted_labels = ["\n".join(sent_tokenize(label_text.strip())) for
    label_text in translated_labels]
    rouge_results = rouge_score.compute(
        predictions=formatted_predictions, references=formatted_labels,
        use_stemmer=True
    )
    scaled_results = {metric: score * 100 for metric, score in rouge_
    results.items()}
    return {metric_key: round(score_value, 4) for metric_key, score_value
    in scaled_results.items()}
```

We define a function in the previous code block to calculate evaluation metrics for the model's predictions. It involves decoding the predicted and true labels, formatting them, and computing ROUGE scores. The ROUGE scores are used to evaluate the quality of the generated summaries by comparing them with the true summaries.

Next, we are going to initialize the data collator.

```
from transformers import DataCollatorForSeq2Seq
data_collator = DataCollatorForSeq2Seq(tokenizer, model=model)
```

The data collator is initialized to handle the dynamic padding of input data during training. It ensures that all inputs and labels in a batch are padded to the same length, a necessary step for efficient and error-free processing by the model during training.

We will remove the column names from the tokenized dataset as shown here:

```
tokenized_datasets = tokenized_datasets.remove_columns(
    xsum_dataset["train"].column_names
)
```

We clean up the tokenized datasets in this step by removing unnecessary columns. This helps streamline the data and reduce memory usage, focusing only on the essential columns required for model training.

Next, we are going to prepare for training.

CHAPTER 7   ADVANCED TECHNIQUES FOR LARGE LANGUAGE MODELS

```
features = [tokenized_datasets["train"][i] for i in range(2)]
data_collator(features)
```

Here, we prepare for training by selecting a few features (data points) from our tokenized training dataset. These features are then passed through the data collator to demonstrate how it processes and prepares data for the training. This step is crucial for understanding how the data is transformed before being fed into the model.

Finally, let's initialize the trainer.

```
from transformers import Seq2SeqTrainer
trainer = Seq2SeqTrainer(
    model,
    args,
    train_dataset=tokenized_datasets["train"],
    eval_dataset=tokenized_datasets["validation"],
    data_collator=data_collator,
    tokenizer=tokenizer,
    compute_metrics=calculate_evaluation_metrics,
)
```

In this previous code block, we set up the `Seq2SeqTrainer`, a utility from the Transformers library designed to simplify the training process. The trainer is configured with the model, training arguments, datasets for training and evaluation, data collator, tokenizer, and function to compute evaluation metrics. This setup is central to training the model effectively, as it encapsulates all the components needed for the training loop.

Finally, it's time to start the training process.

```
trainer.train()
```

This line of code is where the actual training of the model begins. Using our trainer object's `train()` method, the model undergoes training with the data, arguments, and configurations we have set. This process involves feeding the training data to the model in batches, adjusting the model's weights based on the loss, and iterating over the entire dataset for the specified number of epochs.

CHAPTER 7   ADVANCED TECHNIQUES FOR LARGE LANGUAGE MODELS

You can observe (Figure 7-1) that the training and validation loss has been decreasing and ROUGE scores increasing with each training epoch. The ROUGE scores are much higher compared to the baseline scores of {'rouge1': 18.46, 'rouge2': 2.52, 'rougeL': 11.98, 'rougeLsum': 14.51}.

[204048/204048 9:04:09, Epoch 8/8]

Epoch	Training Loss	Validation Loss	Rouge1	Rouge2	Rougel	Rougelsum
1	2.727200	2.380859	30.112200	9.072700	24.192000	24.189500
2	2.537900	2.303072	30.897300	9.653600	24.913500	24.910300
3	2.458600	2.262487	31.340100	10.078000	25.317800	25.321800
4	2.407500	2.238034	31.435700	10.262800	25.480000	25.476800
5	2.370500	2.219620	31.691000	10.498600	25.717400	25.708400
6	2.343100	2.209808	32.084100	10.714300	25.995200	25.991500
7	2.324100	2.201328	32.067600	10.723100	25.965300	25.963900
8	2.311700	2.198619	32.104700	10.783600	26.031600	26.020500

Figure 7-1. T5 fine-tuning

Step 4: Model Evaluation

Now that we have trained the model, let's evaluate the model.

This step involves evaluating the trained model on the validation dataset using the evaluate() method. This process assesses the model's performance on data not seen during training, providing metrics that help gauge its effectiveness in summarizing text. The evaluation score (Figure 7-2) shows that our model has outperformed the baseline set for evaluation in the previous step, which is also displayed in Table 7-1.

```
trainer.evaluate()
```

```
{'eval_loss': 2.1986191272735596,
 'eval_rouge1': 32.1047,
 'eval_rouge2': 10.7836,
 'eval_rougeL': 26.0316,
 'eval_rougeLsum': 26.0205,
 'eval_runtime': 873.7662,
 'eval_samples_per_second': 12.969,
 'eval_steps_per_second': 1.622,
 'epoch': 8.0}
```

Figure 7-2. T5 evaluation results

CHAPTER 7   ADVANCED TECHNIQUES FOR LARGE LANGUAGE MODELS

Table 7-1. Summarization Evaluation of Finetuned Model Over Baseline

ROUGE	Baseline	Finetuned evaluation score
ROUGE 1	18.46	32.10
ROUGE 2	2.52	10.78
ROUGE L	11.98	26.03
ROUGE LSum	14.51	26.02

Let's initialize the summarization pipeline for inference.

```
from transformers import pipeline
model_id = "t5-small-finetuned-xsum/checkpoint-204000/"
summarizer = pipeline("summarization", model=model_id)
```

Note Please update the checkpoint-20400 based on the latest checkpoint that you see in the t5-small-finetuned-xsum folder.

In this step, we initialize the summarization pipeline using the Transformers library. The pipeline function creates a pipeline for the summarization task, specifying the model (t5-small-finetuned-xsum). This pipeline will take input text, feed it to the model, and generate a summary.

Let's define a function for printing summaries.

```
def print_summary(idx):
    document = xsum_dataset["test"][idx]["document"]
    summary = summarizer(xsum_dataset["test"][idx]["document"])[0]
    ["summary_text"]
    print(f"'>>> Document: {document}'")
    print(f"\n'>>> Summary: {summary}'")
```

Here, we define a function `print_summary` that takes an index (`idx`) and prints the document and its summary from the test dataset. The document at the given index is passed through the summarizer pipeline to generate a summary, which is printed alongside the original document. This function is helpful for quickly visualizing the model's summarization capabilities.

Next, we are going to display a summary of a specific document.

```
print_summary(100)
```

We call this cell the `print_summary` function with the index 100. This action triggers the summarization of the 100th document in the test dataset and displays the original document and its generated summary (shown next). It's a practical demonstration of the model's summarization on a specific example.

```
'>>> Summary: Tasers will be used by police in Scotland in a bid to protect the public from terrorism, the police chief has said in '
```

Let's try to get another summary for a different document.

```
print_summary(0)
```

Similarly, this cell uses the `print_summary` function to display the summary for the first document (shown next) in the test dataset (index 0). It showcases another instance of the model's summarization ability, providing an additional example for evaluation.

```
'>>> Summary: Hundreds of one-bedroom flats have been built in Wales to help homeless people, a charity has said . the need for housing has been questioned'
```

Abstractive Summarization Using a Decoder-Only Model

We will continue to use the same xsum dataset from Hugging Face to fine-tune a decoder-only model for abstractive summarization. We will use the Llama-2-7B-chat model, a variant of Llama2-7B developed by Meta and available in the open-source domain.

Step 1: Installing Libraries and Data Loading

This step is the same as the previous section, so you can refer to the previous instructions to run the following code:

```
!pip install -U \
accelerate bitsandbytes datasets  \
peft safetensors transformers trl
from datasets import load_dataset

# load 1k samples from the train split
train_dataset = load_dataset("EdinburghNLP/xsum", split='train[:1000]')
train_dataset[0]
```

The only difference from the previous section is that we will use only a subset of the dataset, i.e., the first 1,000 rows, to avoid catastrophic forgetting, as explained in Chapter 6. The output of the previous code is shown here:

```
{'document': 'The full cost of damage in Newton Stewart, one of the areas worst affected, is still being assessed.\nRepair work is ongoing in Hawick and many roads in Peeblesshire remain badly affected by standing water.\nTrains on the west ..........................................',
 'summary': 'Clean-up operations are continuing across the Scottish Borders and Dumfries and Galloway after flooding caused by Storm Frank.',
 'id': '35232142'}
```

Step 2: Data Pre-processing

We will now prepare the data for fine-tuning the Llama-2-chat-7B model. So, we will start by defining a utility function to format prompts for instructing the LLM to behave in a certain manner, which in this case is to summarize the document.

```
def prompt_formatter(sample):
    return f"""<s>### Instruction:
You are a helpful, respectful, and honest assistant. \
Your task is to summarize the following dialogue. \
Your answer should be based on the provided dialogue only.

### Dialogue:
```

CHAPTER 7   ADVANCED TECHNIQUES FOR LARGE LANGUAGE MODELS

```
{sample['document']}

### Summary:
{sample['summary']} </s>"""

n = 0
print(prompt_formatter(train_dataset[n]))
```

In the previous code block, we have a function named `prompt_formatter`, which takes a single argument called `sample`. This function is designed to format a prompt that will be used for a language model, instructing it on what task to perform. Let me break down what each part of the function does for you.

The function returns a formatted string that includes special tokens and instructions for a language model. The `<s>` and `</s>` are likely special tokens that indicate the start and end of a text sequence for the model.

Within this string, there are three main sections.

- **Instruction:** This section provides clear instructions to the language model, characterizing it as helpful, respectful, and honest, and tasks it specifically with summarizing a dialogue. This helps set the context for the model's behavior and the nature of the task it needs to perform.

- **Dialogue:** Here, the function includes the actual text of the dialogue from the `sample` that needs to be summarized. The `sample['document']` part fetches the dialogue text from the provided `sample` dictionary.

- **Summary:** Finally, the function specifies where the summary should be placed, using `sample['summary']`. The expected output, or the summary of the provided dialogue, is included from the `sample`.

The `prompt_formatter` function is then called with `train_dataset[n]` as its argument, where n is initialized to 0. This means the function is formatting the first element (n=0) from a collection of data samples named `train_dataset`. The result of this formatted prompt is then printed out.

Using this function, we can ensure that each prompt given to the language model is consistent and structured. This is particularly important when working with LLMs for a specific task such as summarization. This consistency allows the model to understand and perform the summarization task as accurately as possible.

Now, we will set up the tokenizer for our dataset using the following code:

```
tokenizer = AutoTokenizer.from_pretrained(model_id)
tokenizer.pad_token = tokenizer.eos_token
tokenizer.padding_side = "right"
```

First, we create the tokenizer by loading it from a pre-trained model using the identifier `model_id`. This Llama-2-chat-7B pre-trained model has been trained on a large dataset and understands how to convert text into a numerical format that a machine learning model can process.

Next, we set the tokenizer's `pad_token` to be the same as the `eos_token`. The `eos_token` typically signifies the end of a sentence, and by equating the padding token to this, we are indicating that padding should be treated the same way as the end of a sentence by the model.

Finally, we specify that padding should be added to the "`right`" or end of the sequences. This is important for models sensitive to the sequence order, ensuring consistent processing of the text inputs.

In summary, we're preparing the tokenizer with the necessary configurations to process our text data correctly for the pre-trained model we are using.

Step 3: Model Training

We will begin by setting up the pre-trained model, which is Llama-2-chat-7B, for training using the quantization technique as shown here:

```
import torch
from transformers import AutoModelForCausalLM, BitsAndBytesConfig
from peft import prepare_model_for_kbit_training

# model_id = "meta-llama/Llama-2-7b-chat-hf"
model_id = "daryl149/llama-2-7b-chat-hf"

# load model in NF4 quantization with double quantization,
# set compute dtype to bfloat16

bnb_config = BitsAndBytesConfig(
    load_in_4bit=True,
    bnb_4bit_quant_type="nf4",
    bnb_4bit_use_double_quant=True,
```

```
    bnb_4bit_compute_dtype=torch.bfloat16
)
model = AutoModelForCausalLM.from_pretrained(
    model_id,
    quantization_config=bnb_config,
    use_cache=False,
    device_map="auto",
)
model = prepare_model_for_kbit_training(model)
```

Let's break down the previous code block step-by-step.

1. **Imports:** We import the necessary modules from PyTorch and the Hugging Face `transformers` library.

2. **Model Identifier:** We specify `model_id`, the identifier for the pre-trained model we want to use. This ID tells the `from_pretrained` method exactly which model to load.

3. **Quantization Configuration:** We set up a `BitsAndBytesConfig` object for the model quantization. Here, we choose 4-bit quantization (`load_in_4bit=True`) with a specific type (`nf4`) and enable double quantization (`bnb_4bit_use_double_quant=True`). We also set the compute data type to `bfloat16`, a lower precision than the standard floating point, allowing faster computation with less memory usage.

4. **Model Loading:** The model is loaded with the specified quantization configuration, with caching disabled (`use_cache=False`) and the device allocation set to `auto`, allowing the library to automatically choose the best device (CPU or GPU) available.

5. **Model Preparation:** Finally, we call `prepare_model_for_kbit_training` from the `peft` module, which likely further optimizes the model for training with quantization.

In essence, we are preparing a causal language model for training that is more memory-efficient, enabling us to train larger models on hardware with less memory.

CHAPTER 7   ADVANCED TECHNIQUES FOR LARGE LANGUAGE MODELS

Now let's set up the trainer for Parameter Efficient Fine Tuning (PEFT) as we have learned in the previous chapter using the following code:

```
from transformers import TrainingArguments, AutoTokenizer
from peft import LoraConfig, get_peft_model
from trl import SFTTrainer

#
# construct a Peft model.
# the QLoRA paper recommends LoRA dropout = 0.05 for small models (7B, 13B)
#
peft_config = LoraConfig(
    r=8,
    lora_alpha=16,
    lora_dropout=0.05,
    bias="none",
    task_type="CAUSAL_LM",
)
model = get_peft_model(model, peft_config)

#
# set up the trainer
#
tokenizer = AutoTokenizer.from_pretrained(model_id)
tokenizer.pad_token = tokenizer.eos_token
tokenizer.padding_side = "right"

args = TrainingArguments(
    output_dir="llama2-7b-chat-samsum",
    num_train_epochs=2,
    per_device_train_batch_size=4,
    gradient_accumulation_steps=2,
    logging_steps=4,
    save_strategy="epoch",
    learning_rate=2e-4,
    optim="paged_adamw_32bit",
    bf16=True,
```

```
    fp16=False,
    tf32=True,
    max_grad_norm=0.3,
    warmup_ratio=0.03,
    lr_scheduler_type="constant",
    disable_tqdm=False,
)
trainer = SFTTrainer(
    model=model,
    train_dataset=train_dataset,
    peft_config=peft_config,
    max_seq_length=1024,
    tokenizer=tokenizer,
    packing=True,
    formatting_func=prompt_formatter,
    args=args,
)
```

In the previous code block, we are setting up a specialized training environment for a pre-trained language model with the help of the `transformers` and `peft` libraries and initializing a trainer from the `trl` library.

- **LoRA Configuration:** We create a `LoraConfig` object, configuring the LoRA parameters such as rank (`r`), scale (`lora_alpha`), and dropout rate (`lora_dropout`). These settings are recommended for small models like the one we're using (7B parameters).

- **Model Adaptation:** We adapt the pre-trained model with Parameter Efficient Fine-Tuning (PEFT) using the Low-Rank Adaptation (LoRA) technique by calling `get_peft_model`. This allows us to fine-tune the model efficiently with less computational cost.

- **Tokenizer Setup:** We prepare the tokenizer corresponding to our model, setting the padding token to the end-of-sentence token and the padding side to the right, ensuring consistent text processing

CHAPTER 7 ADVANCED TECHNIQUES FOR LARGE LANGUAGE MODELS

- **Training Arguments:** We define `TrainingArguments` to specify the training parameters, such as the number of epochs, batch size, learning rate, optimizer type, etc. These parameters control how the model will be fine-tuned.

- **Trainer Initialization:** We initialize the Supervised Fine-Tuning Trainer (`SFTTrainer`), providing it with the model, dataset, PEFT configuration, tokenizer, and other settings necessary for training.

We are fine-tuning a language model using an efficient and tailored method for smaller models, focusing on causal language generation tasks. The `SFTTrainer` is then ready to train the model on the provided dataset.

It is time to start the fine-tuning of the Llama-2-chat-7B model by calling the train method on the trainer class as shown here:

```
trainer.train()
```

You can observe the training loss during the fine-tuning process. See Figure 7-3.

CHAPTER 7   ADVANCED TECHNIQUES FOR LARGE LANGUAGE MODELS

[154/154 10:04, Epoch 1/2]

Step	Training Loss
4	2.197500
8	2.078900
12	2.011400
16	1.920500
20	1.904100
24	1.853400
28	1.861300
32	1.806500
36	1.791800
40	1.771600
44	1.691900
48	1.696300
52	1.672100
56	1.733800
60	1.713300
64	1.736500
68	1.742500
72	1.669100
76	1.709100
80	1.657100
84	1.700500
88	1.774000
92	1.633500
96	1.703900
100	1.697800

Figure 7-3. *Llama2-7B fine-tuning*

Once the model is fine-tuned, you can save the LoRA layers using the `save_model` method, as shown here:

```
trainer.save_model()
```

Step 4: Model Inference and Evaluation

Since you have saved your LoRA layers, you can load them later to run inference on the fine-tuned model by loading the LoRA layers. Loading the fine-tuned model is similar to loading the pre-trained model, except you need to load both the LoRA layers and the base model, as shown here:

```
import torch
from peft import AutoPeftModelForCausalLM
from transformers import AutoTokenizer

model_folder = "llama2-7b-chat-samsum"

# load both the adapter and the base model
model = AutoPeftModelForCausalLM.from_pretrained(
    model_folder,
    low_cpu_mem_usage=True,
    torch_dtype=torch.float16,
    load_in_4bit=True,
    device_map='auto'
)
tokenizer = AutoTokenizer.from_pretrained(model_folder)
```

We are using the `AutoPeftModelForCausalLM` class to load both the LoRA layers and the base model, specifying the quantization of 4 bits in this case. You can also print the loaded model (`print(model)`) to ensure that your LoRA layers have the right architecture and that you are not loading a pre-trained model with an empty LoRA layer.

```
PeftModelForCausalLM(
  (base_model): LoraModel(
    (model): LlamaForCausalLM(
      (model): LlamaModel(
        (embed_tokens): Embedding(32000, 4096, padding_idx=0)
        (layers): ModuleList(
          (0-31): 32 x LlamaDecoderLayer(
            (self_attn): LlamaAttention(
              (q_proj): lora.Linear4bit(
```

```
          (base_layer): Linear4bit(in_features=4096, out_
          features=4096, bias=False)
          (lora_dropout): ModuleDict(
            (default): Dropout(p=0.05, inplace=False)
          )
          (lora_A): ModuleDict(
            (default): Linear(in_features=4096, out_features=8,
            bias=False)
          )
          (lora_B): ModuleDict(
            (default): Linear(in_features=8, out_features=4096,
            bias=False)
          )
          (lora_embedding_A): ParameterDict()
          (lora_embedding_B): ParameterDict()
        )
        (k_proj): Linear4bit(in_features=4096, out_features=4096,
        bias=False)
        (v_proj): lora.Linear4bit(
          (base_layer): Linear4bit(in_features=4096, out_
          features=4096, bias=False)
...
      (lm_head): Linear(in_features=4096, out_features=32000, bias=False)
    )
  )
)
```

Let's pick a sample, construct a prompt, and send it to the fine-tuned model.

```
sample = dataset[10]

prompt = f"""### Instruction:
You are a helpful, respectful and honest assistant. \
Your task is to summarize the following dialogue. \
Your answer should be based on the provided dialogue only.
```

```
### Dialogue:
{sample['document']}

### Summary:
"""
print(prompt)
```

Finally, it's time to generate the summary using our fine-tuned model as shown here:

```
input_ids = tokenizer(prompt, return_tensors="pt").input_ids.cuda()
outputs = model.generate(input_ids=input_ids, max_new_tokens=50,
temperature=0.7)

print('Output:\n',
      tokenizer.batch_decode(outputs.detach().cpu().numpy(), skip_special_
      tokens=True)[0][len(prompt):])
print('\nGround truth:\n', sample['summary'])
```

The previous code generates the following output:

```
Output:
 Head teachers are warning that schools are facing a "perfect storm" of
funding cuts and rising costs.
```

```
Ground truth:
 Head teachers say they are axing GCSE and A-level subjects, increasing
class sizes and cutting support services as they struggle with school
funding.
```

You can observe that the generated output differs from the ground truth but represents the same meaning. So, this is a classic example of *abstractive summarization* where you don't want to repeat the exact text in the summary, which is available in the text by adding a component of creativity using a generative LLM.

Guidelines on Fine-Tuning a Large Language Model

We have run many experiments on fine-tuning a large language model and believe that the experimentation details can help you find the most suitable solution. Table 7-2 shows some critical notes on SFT.

Table 7-2. Memory Requirements for SFT an LLM

Model	Memory Requirements
OPT-1.3B	Cannot be fine-tuned on a 32GB GPU without PEFT
FLAN-T5	Needs at least 18GB single GPU for inference and evaluation
Falcon 7B	Results are not good on question-answering task
OPT-350M	Can be fine-tuned without PEFT

Token length plays a significant role in memory consumption. Table 7-3 shows the memory requirements for different token lengths.

Table 7-3. Memory Requirements Based on Token Lengths

Token Length	Memory Requirements
128 tokens	1.3 GB
256 tokens	2.6 GB
512 tokens	5.2 GB

You can use smaller tokens to reduce memory consumption. However, this will also reduce the context window and the size of the generated text.

Types of SFT (Supervised Fine-Tuning)

There are two main types of supervised fine-tuning carried out for the LLMs.

- **Task Tuning:** Task tuning is fine-tuning a large language model (LLM) for a specific task, such as question-answering (QA) or summarization. Depending on the particular task, several different datasets can be used for task tuning. For example, the CoQA dataset is a good dataset for QA, while the Samsum dataset is a good dataset for summarization.

- **Domain Tuning:** This is fine-tuning a large language model for a specific domain, such as the legal or healthcare domain. Depending on the particular domain, several different datasets can be used for domain tuning. For example, the Legal dataset is a good dataset for the legal domain, while the Healthcare dataset is a good dataset for the healthcare domain.

Memory Consumption During SFT

Table 7-4 shows the main factors that affect memory consumption during fine-tuning.

Table 7-4. *Memory Consumption During Fine-tuning*

Factor	Description
Loading the model to CUDA	This is the most memory-intensive step. You can use a sharded model to reduce memory usage.
Data size	The size of the dataset also affects memory consumption. You can use a smaller dataset or smaller tokens to reduce memory usage.
Batch size	The batch size is the most critical factor for memory consumption. You can reduce the batch size to fit the memory limitation.
SFT step	The GPU should have sufficient memory left to fine-tune the model after loading the dataset and the model.
Final inferencing	You can load the quantized model, which will use less than about 4.5 GB GPU memory for a 7B model.

Reinforcement Learning from Human Feedback

Natural language processing (NLP) has witnessed tremendous progress in recent years, with language models (LMs) like GPT-3 demonstrating astonishing capabilities in generating human-like text. However, these models often lack fine-grained control and may produce biased, offensive, or factually incorrect outputs. To address these concerns, researchers have been exploring new training approaches, one of which is RLHF.

What Is RLHF?

RLHF is a training technique that leverages human feedback to fine-tune pre-trained language models. It aims to improve model behavior and control the output generated by the language model while maintaining its fluency and natural language understanding capabilities.

How Does RLHF Work?

Figure 7-4 shows the stages of RLHF.

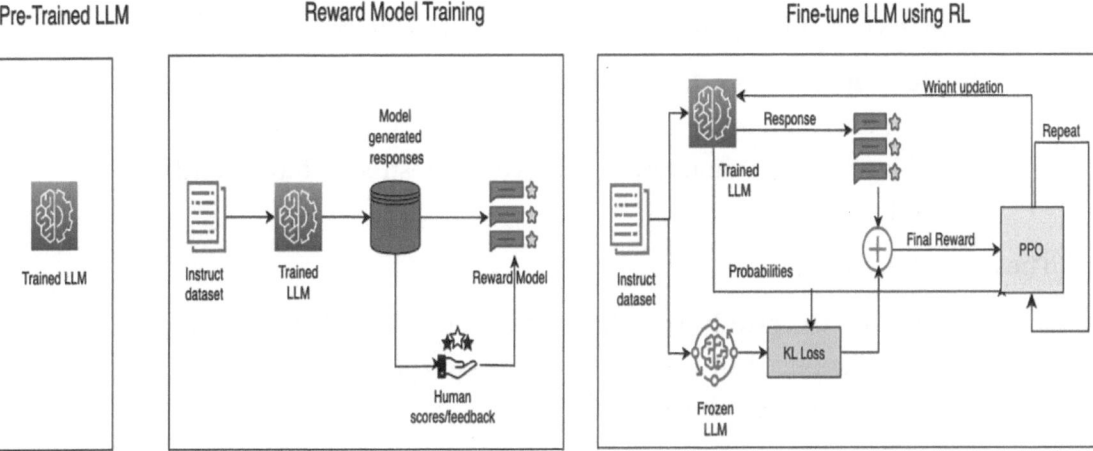

Figure 7-4. *RLHF stages*

The stages are as follows:

1. **Pre-training the Language Model:** RLHF starts with a pre-trained language model, which has been trained on vast amounts of text data to learn language patterns and representations.

2. **Interactive Prompts:** Human AI trainers create interactive prompts that guide the language model's responses. These prompts can include demonstrations of desired behavior or comparisons of different model outputs.

3. **Model Response Ranking:** The AI trainers rank model responses based on various criteria, such as fluency, relevance, and appropriateness. This ranking serves as the reward signal for reinforcement learning.

4. **Reinforcement Learning:** Using the ranked responses as rewards, the language model undergoes reinforcement learning. The model's parameters are updated to maximize the probability of generating desired outputs according to human feedback.

5. **Fine-Tuning:** After reinforcement learning, the language model is fine-tuned using the newly generated training data, which consists of human-generated prompts and model responses.

6. **Evaluation and Iteration:** The fine-tuned model is evaluated for performance and appropriateness. If necessary, the process of RLHF is repeated iteratively to achieve the desired behavior and control.

These are the benefits:

- **Controlled Output:** RLHF allows developers to guide the language model's responses and ensure it adheres to specific guidelines, minimizing undesirable outputs.

- **Ethical Considerations:** By incorporating human feedback, RLHF helps reduce bias, harmful content, and offensive language, making language models more ethically responsible.

- **Customization:** Developers can tailor the language model for specific applications and domains with interactive prompts, leading to more valuable and specialized outputs.

These are the challenges and considerations:

1. **Training Data Quality:** The quality and diversity of human feedback play a crucial role in shaping the language model's behavior. Ensuring high-quality feedback is essential for effective RLHF.

2. **Bias Mitigation:** While RLHF aims to reduce bias, it's essential to be mindful of potential bias in human feedback. Efforts should be made to ensure diverse perspectives are considered during training. You can follow these guidelines for mitigating bias in human feedback:

    - **Diverse Feedback Panels:** To minimize bias in human feedback, ensure your evaluation team comprises individuals from various backgrounds and perspectives. This helps reduce the influence of any single viewpoint on the model's training.

    - **Blind Evaluation:** Consider implementing blind evaluation techniques in which evaluators are unaware of the source of the generated text (human or LLM). This helps them focus on the content and reduces the risk of bias based on the source.

    - **Active Learning:** Employ active learning techniques where the model identifies ambiguous cases and requests clarification from human experts. This focuses human effort on areas where the model is most uncertain, reducing bias amplification from already clear-cut examples.

3. **Optimizing Reward Functions:** Designing appropriate reward functions that accurately capture desired model behavior is challenging and requires careful consideration. Here are a few good practices for optimizing reward functions:

    - **Shaping Rewards:** Design reward functions that go beyond simple binary classifications (good/bad) and incorporate intermediate levels of quality or relevance. This provides the model with more nuanced feedback for improved learning. For example, use a three or five-point scoring mechanism for the feedback score.

    - **Calibrated Rewards:** Carefully calibrate rewards to avoid unintended consequences. For instance, overemphasizing factual accuracy might discourage creativity, while solely focusing on fluency could lead to nonsensical outputs. So, capture different dimensions of the generated output such as factual accuracy, cohesiveness, completeness, etc.

- **Human-in-the-Loop Training:** Incorporate a human-in-the-loop training approach where experts can review the model's outputs and adjust the reward function dynamically. This allows for continuous improvement and adaptation of the reward system based on real-world performance.

Conclusion

RLHF represents a promising approach in the ongoing quest to develop more controlled and ethical language models. By integrating human feedback and reinforcement learning, RLHF enables us to fine-tune language models for safer, more reliable, and contextually relevant language generation, opening up exciting possibilities in various applications across industries and domains. Nonetheless, continuous research and thoughtful implementation are essential to overcome challenges and unlock the full potential of RLHF for advancing the field of NLP.

Reward Model Implementation

RLHF is a widely used technique to fine-tune LLMs based on feedback from human users. In RLHF, the weight updates of the LLM are driven by the reward or feedback a user provides for the completions generated by the model. Assigning an appropriate reward is a complex task, and one approach is to have a human evaluator assess all model completions against a specific alignment metric, such as usefulness. This feedback is then quantified as a scalar value, and the LLM weights are iteratively updated to maximize this reward obtained from the human evaluator.

However, obtaining human feedback can be time-consuming and costly. To address this, we can train another model known as the Reward Model, which acts as a proxy for human feedback. The primary goal of the reward model is to evaluate how closely a model's response aligns with human preferences. The reward model inputs a (prompt response) pair and outputs a reward score. This can be formulated as a regression or classification task. Building an effective reward model requires high-quality data, as determining what is considered good or bad may vary among individuals, making it challenging to map to a scalar value.

CHAPTER 7 ADVANCED TECHNIQUES FOR LARGE LANGUAGE MODELS

To create a dataset for training the reward model, one approach is asking labelers to compare two responses and decide which is better, as shown in Figure 7-5. This kind of dataset is referred to as a *comparison dataset*, where each record consists of a prompt, chosen response, and rejected response.

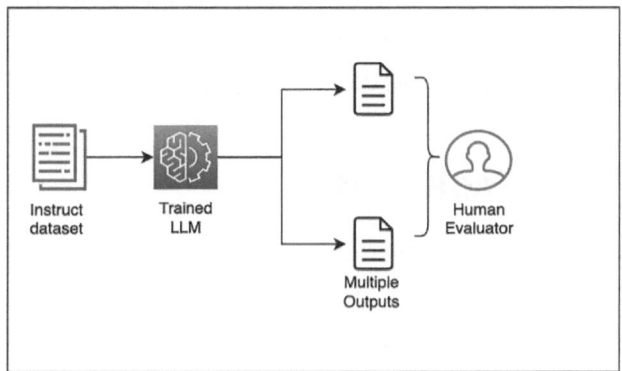

Figure 7-5. *Reward model data collection*

During the training of the reward model, the comparison dataset should be in the format of (prompt, chosen response, rejected response), where the better option comes first. The order is crucial for designing the loss function. Any model capable of processing variable-length text input and outputting a scalar value can be used. Commonly, a supervised (SFT) model is employed. The last de-embedding layer is removed, and a single neuron is added to the last layer for the scalar output. Figure 7-6 shows the process of training a reward model.

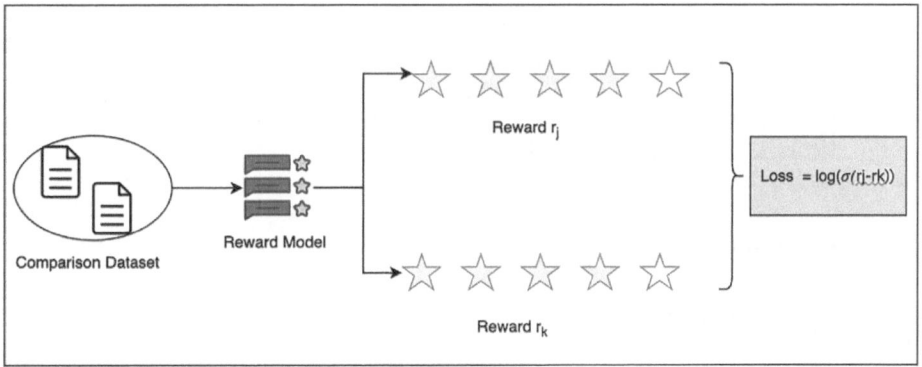

Figure 7-6. *Reward model training*

The training of the reward model involves two passes in each epoch. In the first pass, the prompt and chosen response are fed to the reward model, producing Rchosen as the output. In the second pass, the same prompt is used along with the rejected response, generating Rrejected as the output. The loss function is designed to maximize the gap between the chosen response score and the rejected response score. A high reward score for the selected response and a low reward score for the rejected response result in a loss of 0, aligning with the intuition behind the loss function.

Controlled Review Generation

In this section, we conduct fine-tuning on GPT2 (small) using the Yelp polarity dataset to generate positive reviews. Instead of training the model from scratch, we provide the beginning of real reviews and task the model with producing positive continuations. We utilize a BERT classifier to evaluate the positivity of these generated sentences. The classifier's outputs act as reward signals for proximal policy optimization (PPO) training, guiding the model to generate more positive and coherent continuations, as shown in Figure 7-7.

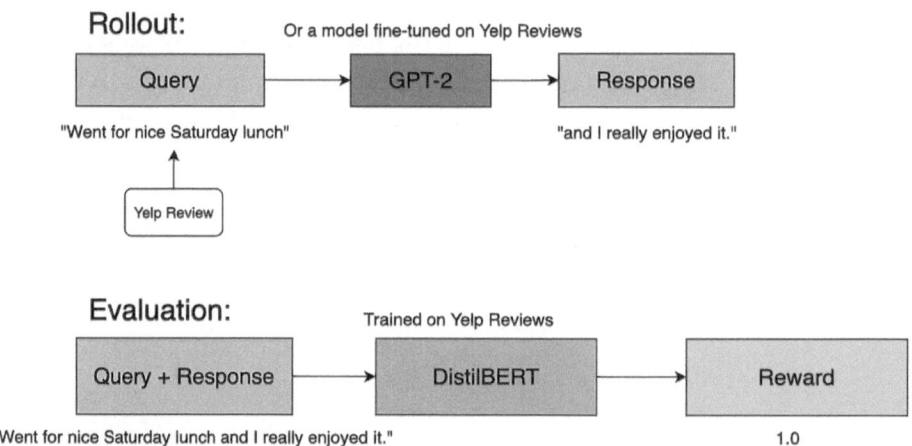

Figure 7-7. *Reward tuning for LLMs*

Setting Up the Environment

We will start with setting up the environment and installing the necessary packages for training and evaluating a PPO model with Transformers.

Import Dependencies

Let us begin by installing and importing the dependencies.

```
%load_ext autoreload
%autoreload 2
%pip install transformers trl wandb
import torch
from tqdm import tqdm
import pandas as pd

tqdm.pandas()

from transformers import pipeline, AutoTokenizer
from datasets import load_dataset

from trl import PPOTrainer, PPOConfig, AutoModelForCausalLMWithValueHead
from trl.core import LengthSampler
```

First, we load the `autoreload` extension to reload modules when changes are made automatically. Then, we install the required packages: `transformers`, `trl` library, and `wandb` (Weights & Biases) for logging and monitoring.

We also import essential libraries, including `torch` for PyTorch, `tqdm` for progress bar visualization, and `pandas` for data manipulation.

Next, we load the `datasets` module to access the Yelp polarity dataset and the `pipeline` and `AutoTokenizer` from the Transformers library for text generation tasks.

In addition, we import specific components from the TRL library. The TRL library provides tools for training and evaluating text-adaptive models. We import `PPOTrainer` for PPO training, `PPOConfig` to configure the PPO training process, `AutoModelForCausalLMWithValueHead` for initializing the model with a value head, and `LengthSampler` to sample sequences of different lengths during training.

Configure Environment

We will now configure the PPO training process and specify additional parameters for sentiment analysis.

CHAPTER 7   ADVANCED TECHNIQUES FOR LARGE LANGUAGE MODELS

```
ppo_config = PPOConfig(
    model_name="gpt2",
    learning_rate=1.41e-5,
    log_with="wandb",
)
sentiment_kwargs = {"return_all_scores": True, "function_to_apply": "none", "batch_size": 16}
```

In the previous code block, we set up the configuration and parameters for training a PPO algorithm and performing sentiment analysis on generated text.

1. **PPO Configuration (ppo_config)**:
   - model_name: Specifies the name of the base language model to be used for training. We are using the base gpt2 model. However, you can also start with a fine-tuned model in this step. We have skipped the supervised fine-tuning step in this section, as it was already covered in the previous section.
   - learning_rate: Sets the learning rate to 1.41e-5 for the PPO training process.
   - log_with: Determines the logging method for training progress. Here, we set it to wandb, meaning that the training progress will be logged using the Weights & Biases platform.

2. **Sentiment Analysis Parameters (sentiment_kwargs)**:
   - return_all_scores: A boolean flag that indicates whether the sentiment analysis should return all sentiment scores or just the top sentiment score. We set it to True, which will return all sentiment scores.
   - function_to_apply: Specifies a function to apply to the sentiment scores. In this case, we set it to none, which means no specific function will be applied to the sentiment scores.
   - batch_size: Sets the batch size for sentiment analysis. We chose a batch size of 16, indicating that the sentiment analysis will be performed on batches of 16 generated sentences at a time.

Overall, these configurations and parameters are crucial for setting up the PPO training process and sentiment analysis, which will be used to generate positive reviews based on the GPT-2 model.

The next step is to set up the (Weights & Biases) W&B for monitoring.

```
import wandb

wandb.init()
```

After executing this code, you will be prompted to log in to the Weights & Biases platform.

Loading the Dataset

The yelp_polarity[31] dataset is an extensive collection of Yelp reviews used for binary sentiment classification. It consists of 560,000 highly polarized Yelp reviews for training and 38,000 reviews for testing. The dataset was originally part of the Yelp Dataset Challenge 2015.

The dataset was created by Xiang Zhang and used as a text classification benchmark in the research paper "Character-level Convolutional Networks for Text Classification" presented at Advances in Neural Information Processing Systems 28 (NIPS 2015).

The reviews in the dataset are categorized into two classes: negative and positive. Stars 1 and 2 are negative, while 3 and 4 are positive. Each class has 280,000 training samples and 19,000 testing samples, resulting in 560,000 and 38,000 testing samples. We will use a small sample of this dataset to implement a reward model for controlled (positive) review generation.

We will define a utility function to load the yelp_reviews dataset into a DataFrame and filter for reviews of at least 200 characters.

```
def build_dataset(config, dataset_name="yelp_polarity", min_text_length=2,
max_text_length=8):
    """
    Build dataset for training. This function builds the dataset from
    'load_dataset'. You should
    customize this function to train the model on your dataset.

    Args:
        dataset_name ('str'):
            The name of the dataset to be loaded.
```

CHAPTER 7   ADVANCED TECHNIQUES FOR LARGE LANGUAGE MODELS

```
    Returns:
        dataloader ('torch.utils.data.DataLoader'):
            The dataloader for the dataset.
    """
    tokenizer = AutoTokenizer.from_pretrained(config.model_name)
    tokenizer.pad_token = tokenizer.eos_token
    # Load yelp_polarity dataset using datasets library
    dataset = load_dataset(dataset_name, split="train")
    dataset = dataset.shuffle().select(range(25000))
    dataset = dataset.rename_columns({"text": "review"})
    dataset = dataset.filter(lambda x: len(x["review"]) > 200,
    batched=False)

    input_sampler = LengthSampler(min_text_length, max_text_length)

    def tokenize(sample):
        sample["input_ids"] = tokenizer.encode(sample["review"])[:input_
        sampler()]
        sample["query"] = tokenizer.decode(sample["input_ids"])
        return sample

    dataset = dataset.map(tokenize, batched=False)
    dataset.set_format(type="torch")
    return dataset
```

The previous code defines a function called build_dataset used to create a dataset for training a model. The function takes several arguments, including a configuration (config), the name of the dataset to be loaded (dataset_name), and the minimum and maximum text lengths for the input data (min_text_length and max_text_length, respectively).

Inside the function, the AutoTokenizer class from the transformers library creates a tokenizer based on the provided configuration. The tokenizer's pad token is then set to the end-of-sequence token.

Next, the Yelp polarity dataset is loaded using the load_dataset function from the datasets library. The dataset is then shuffled, and a subset of 25,000 samples is selected for training. The column name text is renamed to review for better readability.

After that, the dataset is filtered to keep only those samples with review texts longer than 200 tokens. The LengthSampler class is used to create a sampler that will be used later to ensure each sample has a token length within the specified range (min_text_length and max_text_length).

A tokenize function is defined to tokenize the reviews using the previously created tokenizer. The function truncates or pads the review text to match the specified token length range. The tokenized input IDs are stored in the input_ids field, and the corresponding decoded text is stored in each sample's query field.

The map function is then used to apply the tokenize function to each sample in the dataset. Finally, the dataset is formatted as a torch dataset using set_format, and the resulting dataset is returned. This dataset can be used for training a language model on the Yelp polarity dataset.

Let us now implement the previous function and create a data collator.

```
data_loader = build_dataset(ppo_config)

def custom_collator(batch_data):
  return dict((key, [data[key] for data in batch_data]) for key in batch_data[0])
```

In the previous code, we create a data loader for training using the build_dataset function. The build_dataset function loads the Yelp Polarity dataset and prepares it for training with the provided configuration ppo_config. The resulting data_loader will be used to efficiently feed batches of data into the model during training.

Next, we create a custom collator function, which is used to process data batches during training. A collator takes a batch of data samples and organizes them into a suitable format before inputting them into the model. In this case, the custom_collator function takes a batch of data and returns a dictionary. The keys in the dictionary correspond to the data keys in each sample (e.g., input_ids, attention_mask, etc.), and the values are lists containing the corresponding data for each sample in the batch.

These utility functions will handle the Yelp Polarity dataset and organize the data for efficient model training.

Loading the Models

Let's discuss loading the models.

Load the Pre-trained GPT-2 Models

We utilize the GPT2 model with a value head and its tokenizer. The model is loaded twice: one model is optimized during training, while the other model serves as a reference to compute the KL-divergence from the initial point. This KL-divergence acts as an extra reward signal in the PPO training process, ensuring that the optimized model does not deviate significantly from the original language model. This approach helps enhance the training stability and maintain the essential properties of the language model throughout the optimization process.

```
updated_model = AutoModelForCausalLMWithValueHead.from_pretrained(ppo_
config.model_name)
reference_model = AutoModelForCausalLMWithValueHead.from_pretrained(ppo_
config.model_name)
updated_tokenizer = AutoTokenizer.from_pretrained(ppo_config.model_name)

updated_tokenizer.pad_token = updated_tokenizer.eos_token
```

In the previous code, we are loading two versions of the GPT2[32] model with a value head and a tokenizer. The first model, updated_model, will be optimized during training. The second model, called reference_model, calculates the KL-divergence from the starting point. This helps provide an additional reward signal during the PPO training to ensure that the updated model does not deviate too much from the original language model.

Initialize PPOTrainer

We will initialize the PPOTrainer, which handles device placement and optimization.

```
reinforcement_trainer = PPOTrainer(ppo_config, updated_model, reference_
model, tokenizer=updated_tokenizer, dataset=data_loader, data_
collator=custom_collator)
```

The code creates a PPOTrainer object named reinforcement_trainer with the provided configurations, models, tokenizer, dataset, and data collator. The PPOTrainer is used to train the model using the PPO algorithm with reinforcement learning.

Load BERT Classifier

We will load a BERT classifier fine-tuned on the Yelp Polarity dataset.

```
device = reinforcement_trainer.accelerator.device
if reinforcement_trainer.accelerator.num_processes == 1:
    device = 0 if torch.cuda.is_available() else "cpu"  # to avoid a 'pipeline' bug
sentiment_pipeline = pipeline("sentiment-analysis", model="VictorSanh/roberta-base-finetuned-yelp-polarity", device=device)
```

We set up the device for training the reinforcement learning model in the previous code. If running on a single GPU or CPU, we set the device to 0 (GPU) if available or `cpu` otherwise. This is to avoid a bug related to the `pipeline` function.

Next, we create a sentiment analysis pipeline using the Hugging Face `pipeline` function. We use the pre-trained model `VictorSanh/roberta-base-finetuned-yelp-polarity` for sentiment analysis. This pipeline will evaluate the sentiment of the generated sentences and provide reward signals during the PPO training.

The output of the fine-tuned BERT model consists of logits for both the negative and positive classes. During the language model training, we will utilize the logits corresponding to the positive type as a reward signal.

Let's try on a sample review.

```
sample_text = "this restaurant was really good!!"
sentiment_pipeline(sample_text, **sentiment_kwargs)

[[{'label': 'LABEL_0', 'score': -4.56001091003418}, {'label': 'LABEL_1', 'score': 4.22826623916626}]]
```

You can observe that the logit for the positive class is 4.22, and the negative class is -4.56, as the review is positive.

Text Generation Settings

We employ simple sampling techniques without top-k and nucleus sampling during response generation to ensure minimal length. This approach allows us to produce responses effectively.

```
generation_kwargs = {"min_length": -1, "top_k": 0.0, "top_p": 1.0, "do_sample": True, "pad_token_id": updated_tokenizer.eos_token_id}
```

Optimizing the Model for Positive Review Generation

The next step is to optimize the LLM to generate positive reviews. We will start with the setup of the training loop.

Training

The training loop involves the following key steps:

1. Obtain query responses from the policy network (GPT-2).
2. Extract sentiments for the query and responses using BERT.
3. Optimize the policy using PPO based on the (query, response, and reward) triplet.

In the next code block, we train a large language model using the PPO algorithm to generate improved responses.

```
min_output_length = 4
max_output_length = 16
output_length_sampler = LengthSampler(min_output_length, max_output_length)

generation_params = {
    "min_length": -1,
    "top_k": 0.0,
    "top_p": 1.0,
    "do_sample": True,
    "pad_token_id": updated_tokenizer.eos_token_id,
}

for epoch, data_batch in tqdm(enumerate(reinforcement_trainer.dataloader)):
    input_query_tensors = data_batch["input_ids"]

    # Get a response from GPT-2
    response_tensors = []
    for input_query in input_query_tensors:
        gen_len = output_length_sampler()
        generation_params["max_new_tokens"] = gen_len
        response = reinforcement_trainer.generate(input_query,
        **generation_params)
```

```
        response_tensors.append(response.squeeze()[-gen_len:])
data_batch["response"] = [updated_tokenizer.decode(r.squeeze()) for r
in response_tensors]

# Compute sentiment score
texts = [q + r for q, r in zip(data_batch["query"], data_
batch["response"])]
pipeline_outputs = sentiment_pipeline(texts, **sentiment_kwargs)
rewards = [torch.tensor(output[1]["score"]) for output in pipeline_
outputs]

# Run PPO step
statistics = reinforcement_trainer.step(input_query_tensors, response_
tensors, rewards)
reinforcement_trainer.log_stats(statistics, data_batch, rewards)
```

Here are the steps broken down:

1. `min_output_length` and `max_output_length` define the minimum and maximum lengths of the generated responses by the language model.

2. `output_length_sampler` is a utility function that randomly samples lengths between `min_output_length` and `max_output_length` for the generated responses.

3. `generation_params` is a dictionary containing various parameters for response generation using the language model. These parameters include `min_length`, `top_k`, `top_p`, `do_sample`, and `pad_token_id`.

4. The code runs a loop over the data batches from the `reinforcement_trainer.dataloader`. Each batch contains input queries represented by `input_ids`.

5. The language model generates a response using each input query's specified generation parameters (`generation_params`). The length of the generated response is randomly sampled using `output_length_sampler`.

CHAPTER 7 ADVANCED TECHNIQUES FOR LARGE LANGUAGE MODELS

6. The generated responses are then converted from tensor format to text format using `updated_tokenizer.decode`.

7. Sentiment analysis is performed on the text pairs formed by concatenating the original input queries with the generated responses. The sentiment analysis uses the fine-tuned BERT model, which gives sentiment scores for each text pair.

8. The sentiment scores obtained from BERT are used as reward signals for the PPO training. These rewards represent how positive the generated responses are.

9. The PPO training step (`reinforcement_trainer.step`) is performed with the input queries, generated responses, and computed rewards.

10. The training statistics, including rewards and other relevant information, are logged using `reinforcement_trainer.log_stats` for analysis and monitoring during training, as shown in Figure 7-8.

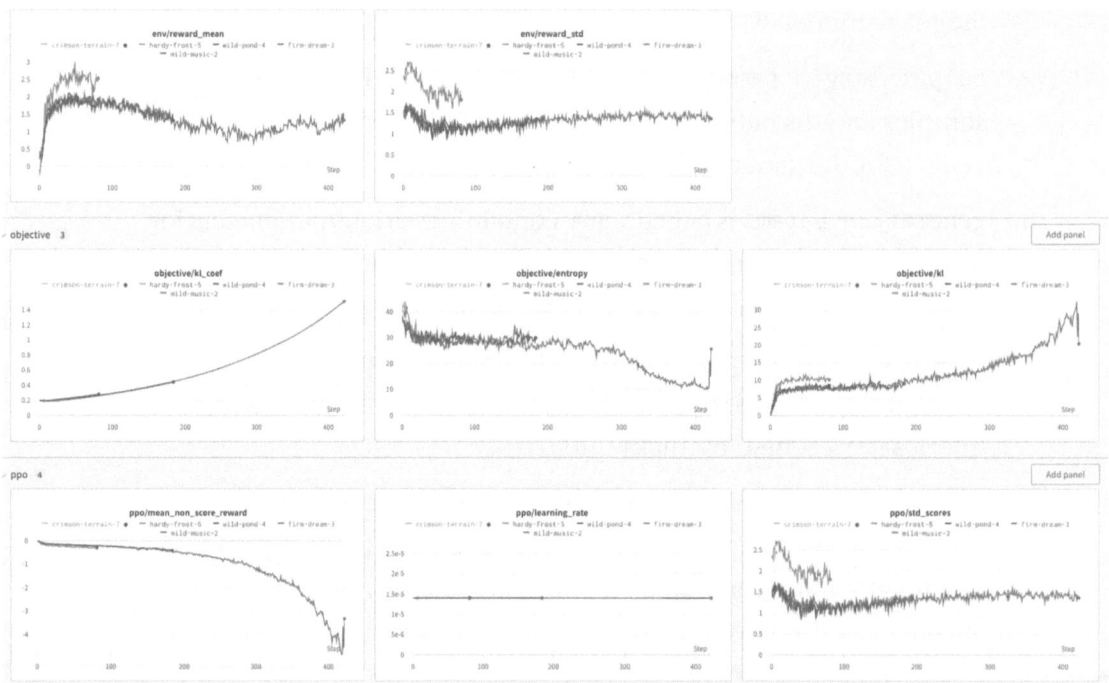

Figure 7-8. PPO monitoring

CHAPTER 7 ADVANCED TECHNIQUES FOR LARGE LANGUAGE MODELS

Overall, this code block implements the PPO training loop, where the language model is optimized to generate improved (positive) responses based on sentiment rewards obtained from the BERT sentiment analysis model. The goal is to encourage the model to produce more positive and contextually appropriate responses.

Model Evaluation

Let's review some examples from the Yelp polarity dataset and see how the trained model performs on them. We will compare the tuned model against the model before optimization.

```
#### Get a batch from the dataset
bs = 16
game_data = dict()
data_loader.set_format("pandas")
df_batch = data_loader[:].sample(bs)
game_data["query"] = df_batch["query"].tolist()
query_tensors = df_batch["input_ids"].tolist()

response_tensors_ref, response_tensors = [], []

#### get response from gpt2 and gpt2_ref
for i in range(bs):
    gen_len = output_length_sampler()
    output = reference_model.generate(
        torch.tensor(query_tensors[i]).unsqueeze(dim=0).to(device), max_
        new_tokens=gen_len, **generation_kwargs
    ).squeeze()[-gen_len:]
    response_tensors_ref.append(output)
    output = updated_model.generate(
        torch.tensor(query_tensors[i]).unsqueeze(dim=0).to(device), max_
        new_tokens=gen_len, **generation_kwargs
    ).squeeze()[-gen_len:]
    response_tensors.append(output)

#### decode responses
game_data["response (before)"] = [updated_tokenizer.decode(response_
tensors_ref[i]) for i in range(bs)]
```

CHAPTER 7 ADVANCED TECHNIQUES FOR LARGE LANGUAGE MODELS

```
game_data["response (after)"] = [updated_tokenizer.decode(response_
tensors[i]) for i in range(bs)]

#### sentiment analysis of query/response pairs before/after
sample_texts = [q + r for q, r in zip(game_data["query"], game_
data["response (before)"])]
game_data["rewards (before)"] = [output[1]["score"] for output in
sentiment_pipeline(sample_texts, **sentiment_kwargs)]

sample_texts = [q + r for q, r in zip(game_data["query"], game_
data["response (after)"])]
game_data["rewards (after)"] = [output[1]["score"] for output in sentiment_
pipeline(sample_texts, **sentiment_kwargs)]

# Store results in a dataframe
df_rl_results = pd.DataFrame(game_data)
df_rl_results
```

In this code block, we are evaluating the performance of the language model using reinforcement learning and comparing the responses generated by two versions of the model: the original model (`reference_model`) and the updated model (`updated_model`) after applying PPO training.

1. `bs = 16`: The batch size (`bs`) is 16, meaning we will process 16 samples in each iteration.

2. `game_data` is an empty dictionary that will store information about the query, responses, and rewards for the samples in the batch.

3. The dataset loader is configured to return data in Pandas format (`data_loader.set_format("pandas")`) for easy manipulation.

4. A batch of data (`df_batch`) is randomly sampled from the dataset using the data loader, and the queries and their corresponding input tensors (`query_tensors`) are extracted from the batch.

5. Two empty lists, `response_tensors_ref` and `response_tensors`, are created to store the generated responses from the `reference_model` and `updated_model`, respectively.

6. For each sample in the batch, the language models (reference_model and updated_model) generate responses using the specified generation parameters (generation_kwargs) and the randomly sampled output length (output_length_sampler()). The generated responses are stored in response_tensors_ref and response_tensors.

7. The generated responses are decoded using the tokenizer to obtain human-readable text. The original responses are stored in game_data["response (before)"], and the updated responses are stored in game_data["response (after)"].

8. Sentiment analysis is performed on the pairs of query/response before and after the PPO training. The sentiment analysis is done using the fine-tuned BERT model, which gives sentiment scores for each text pair. The sentiment scores for the original responses are stored in game_data["rewards (before)"], and the scores for the updated responses are stored in game_data["rewards (after)"].

9. The results are stored in a Pandas DataFrame called df_rl_results, which contains the query, original response, updated response, sentiment rewards before PPO training, and sentiment rewards after PPO training for each sample in the batch.

You can view the results in Figure 7-9. Reward after PPO optimization is positive in most cases.

CHAPTER 7　ADVANCED TECHNIQUES FOR LARGE LANGUAGE MODELS

	query	response (before)	response (after)	rewards (before)	rewards (after)
0	This place was an	effort to develop relief materials, I visit	amazing trip. Here are my top three	-2.352868	4.272993
1	Ace	Lock focus to ensure that Stunner sends a Fals...	— on a note they wrote about. I love our way,	-1.268909	3.298889
2	Came here for a Sunday	supporting the Boyle family after 2 days away...	Day task. I loved sharing my work with you. I...	2.186583	4.155334
3	Went for nice Saturday lunch	.\n\nBUF reports that the	and I really enjoyed it. I always	1.612728	4.005006
4	The first thing is the name.	Then you look at a label like "terms for	I love it. I love that I love that	-1.635414	3.198051
5	Well it's now 11:49	AM and I am playing a game I never	p.m. and I've been happy	-1.263709	1.778251
6	Client dinner	and any other spur for remorse	! I liked it. it	-2.832374	2.841812
7	If my insurance didn't	cover you for a year	get covers, I still	-2.671310	-1.811401
8	I had an opportunity to hit this	island to see their hero and not just some st...	team, and it was simply amazing. I always	2.986287	4.170051
9	I came	up unassigned and in awe and without	back and I like it. It's an	-1.148255	3.338899
10	They are just OK, not impressed	already," he said	, but I love	-3.965868	-2.013681
11	Massage	group members who've complained about Girardi...	is incredible (!). It's really named after me...	-1.864852	3.820748
12	Loved Happy Hour!! Great Food	. RATING on my buddy CAS with a Munchies multiple	! Lots of drinks with a great number of mix of...	3.529850	4.281379
13	My first clue	when I learned that I had Parkinson's was whe...	is that this is a wonderful tool that can giv...	-1.070085	2.025400
14	Okay..	now that my attention	Nice! Nice to	-0.951111	1.960830
15	I am no expert when it	comes to specialties such as medication	comes to things, but I thoroughly	-1.034884	2.280433

Figure 7-9. PPO results

Overall, this code block evaluates the performance of the language model by comparing the sentiment rewards obtained before and after applying the PPO training to assess the improvement in generating positive and contextually appropriate responses. The results are stored in a DataFrame for further analysis.

We can also compare the overall reward before and after for quantitative evaluation as shown here:

```
print("mean:")
display(df_rl_results[["rewards (before)", "rewards (after)"]].mean())
print()
print("median:")
display(df_rl_results[["rewards (before)", "rewards (after)"]].median())
```

The result is shown here:

```
mean:
rewards (before)   -0.734012
rewards (after)     2.600187
dtype: float64

median:
rewards (before)   -1.205982
rewards (after)     3.248470
```

It is evident that PPO optimization led to a significant improvement in the reward, so we can say that our tuned model is much better at generating positive reviews.

Save Model

Save the model locally or push it to Hugging Face for later use.

```
updated_model.save_pretrained("gpt2-yelp-pos-v2")
updated_tokenizer.save_pretrained("gpt2-yelp-pos-v2")
```

Evaluating and Ensuring the Fairness of the Reward Model in RLHF

In the context of RLHF, the reward model plays a pivotal role in shaping the behavior of an LLM. It is crucial to scrutinize the reward model to prevent it from perpetuating or introducing biases based on the training data it receives. This section covers the comprehensive steps to ensure the objectivity and fairness of the reward model.

Balanced Data Curation:

To create a training dataset for the reward model that can handle diverse scenarios and users, it's important to collect feedback from a wide range of people to capture different perspectives. Here are some steps you can follow for this:

- Collect datasets from different geographical locations, cultures, and user backgrounds.
- Include scenarios that cover a broad range of inputs with varying levels of complexity and subject matter.

Multidimensional Reward Metrics:

It is recommended not to solely depend on one feedback aspect, such as usefulness or relevance. Instead, it is better to incorporate multiple criteria into the reward system to consider the multidimensional nature of quality responses.

- To achieve this, a reward metric should be defined that includes dimensions such as factual accuracy, ethical soundness, and user satisfaction.
- Multilabel classification techniques can be used where each dimension is scored independently and then integrated into an overall reward metric.

Fairness and Bias Auditing:

Conduct audits on the reward model using specialized tools and techniques to detect and measure biases. These audits should be integral to the reward model's development life cycle.

- Employ fairness metrics, such as demographic parity and equalized odds, to evaluate the reward model.
- Utilize bias and fairness assessment tools like AI Fairness 360 or Fairlearn to uncover biases systematically.

Adversarial Testing:

Regularly test the reward model with challenging cases designed to reveal biases or failures in the model's decision-making process.

- Develop adversarial examples that mimic potential edge cases or points of failure.
- Incorporate these adversarial examples into regular stress tests of the reward model.

Iterative Improvement and Retraining:

Adopt an iterative approach in which the reward model is continuously updated based on new data, especially those that highlight previous biases or errors.

- Collect new human feedback at regular intervals to provide fresh training data for the reward model.
- Retrain the reward model incrementally to adapt to evolving standards and societal norms.

Transparency in Reward Modeling:

Ensure that the reward model's decision-making process is transparent, allowing for easy interpretation and review of its behavior.

- Implement model interpretability tools, like SHAP or LIME, to understand the factors influencing the reward model's outputs.
- Document all decisions related to data curation and reward metrics, making the process auditable and transparent.

The reliability of the RLHF pipeline is inherently linked to the impartiality of the reward model. By instituting these measures, we can steer the reward model toward fairness, thus fostering more ethical and balanced LLMs. A rigorously vetted reward model not only uplifts the quality of the generated outputs but also earns the trust of users, which is indispensable in deploying AI solutions in real-world settings.

RLHF Summary

This chapter section explored controlled review generation using PPO training and fine-tuning techniques. The goal was to create a language model that could generate positive reviews based on the Yelp polarity dataset. We adopted a reinforcement learning approach with PPO as the optimization algorithm to achieve this.

We started by fine-tuning the GPT2 model with a value head using reinforcement learning. The model was optimized based on rewards obtained from a fine-tuned BERT classifier that analyzed the sentiment of the generated sentences. The BERT classifier acted as a proxy for human feedback, enabling us to train the language model efficiently without needing expensive and time-consuming manual feedback collection.

We used a comparison dataset to train the reward model (BERT classifier), where labelers were asked to choose the better of two responses for a given prompt. This comparison dataset was used to evaluate the alignment between model responses and human preferences, and it was formatted for a classification task.

The controlled review generation process involved the training loop, where we followed several main steps. First, we obtained query responses from the policy network (GPT2) and then computed sentiments for the query and responses using the fine-tuned BERT classifier. The sentiment scores acted as rewards in the PPO training, where we updated the policy network to optimize the response generation process based on the triplet (query, response, compensation).

We used sampling techniques during response generation and disabled top-k and nucleus sampling. This approach allowed us to obtain more diverse and contextually appropriate responses from the language model. By incorporating the PPO training and fine-tuning, we ensured that the optimized model did not deviate significantly from the original language model while generating positive and meaningful movie reviews.

Throughout the process, we carefully managed the output length of the responses to avoid excessively short or long reviews. We utilized the `PPOTrainer` and `PPOConfig` classes to facilitate the reinforcement learning process and effectively log the training statistics using `wandb`.

In summary, this chapter showcased how to implement controlled review generation using PPO training and fine-tuning techniques. By leveraging the strengths of different pre-trained models and reinforcement learning methods, we created a language model capable of generating positive and contextually relevant movie reviews.

Summary

Congratulations! You've reached the pinnacle of this chapter, exploring the cutting-edge world of advanced LLM techniques. We've tackled the intricacies of fine-tuning LLMs for abstractive summarization, demystifying both encoder-decoder and decoder-only models. Now, with your newfound knowledge, you can craft summaries that capture factual details and the essence and nuance of the original text.

But our journey was wider than summarization. We delved into the guidelines for fine-tuning LLMs, equipping you with a framework to optimize your models for diverse tasks. You've unlocked the potential of supervised fine-tuning (SFT), understanding its trade-offs and strategies for managing memory consumption.

Pushing the boundaries, we ventured into the realm of reinforcement learning from human feedback. You've grasped the concepts behind reward models and their implementation, paving the way for crafting LLMs that learn directly from human input. This opens doors to exciting possibilities like generating controlled review text tailored to your specific needs.

As you stand at this vantage point, remember this is just the beginning. Advanced LLM techniques are constantly evolving, offering an ever-expanding horizon of possibilities. Continue exploring, experimenting, and pushing the boundaries of what language models can achieve. With the knowledge and skills gained in this chapter, you're well-equipped to unlock the full potential of these powerful tools and shape the future of communication with language. So, we will use this knowledge to build applications using LLM in the following two chapters.

Happy learning and happy coding!

CHAPTER 8

Building Demo Applications Using LLMs

In the previous chapters on large language models (LLMs), you have learned about transformers, natural language generation tasks using LLMs, and advanced techniques for improving LLM performance. You will learn about the practical application of LLM to build demo applications in this chapter. Specifically, in this chapter, we will take a pragmatic approach and explore how to integrate LLMs into real-world applications to build powerful demo applications. By the end of this chapter, you will have the knowledge and skills to create impressive and interactive demo applications that showcase the capabilities of LLMs in solving various business problems. Let's dive in and unleash the potential of LLMs for building engaging and practical demo applications.

In the previous chapters, we learned how to download the LLM and use it locally. This requires computing resources and GPUs for inferencing, especially if the model is large and has billions of parameters. In this chapter, we will also use an API-based option to generate the output from LLM. This method is easier for demo applications as it doesn't require compute resources to load the model. However, this can incur pay-as-you-go costs if you are using a paid LLM, and you need to consider the data privacy issue since your data will be sent to the server for inferencing. One other factor to consider is the size of the model. The larger the size, the longer and better the output, but you might have a resource limitation to load a larger model with 176 billion parameters. However, a smaller model fine-tuned on a specific task can sometimes perform better. So, you must consider all these factors when planning to build an LLM-based app.

Making Sense of Website Content

When it comes to using an LLM on website content, it can do quite a few valuable things. Let me give you some examples.

- **Summarizing Content:** The LLM is pretty nifty at automatically generating short summaries of long articles, blog posts, or news stories. So, instead of reading the entire text, you can get a quick overview of the main points.

- **Answering Questions:** You can ask the LLM questions about the website content, and it'll dig into the text, find the relevant information, and give you accurate answers in natural language.

- **Finding Specific Information:** Need to find something specific on the website? The LLM can help with that, too! Just ask your query, and it'll sift through the content to retrieve the precise information you want. It's like having your own efficient and super-smart search assistant.

- **Personalized Recommendations:** The LLM can suggest personalized recommendations by understanding your website content and preferences. Whether it's related articles, blog posts, or even products, it aims to enhance your experience and keep you engaged.

- **Sentiment Analysis:** Have you ever wondered about the overall sentiment of articles, reviews, or discussions on a website? The LLM can analyze the sentiment expressed in the content, helping you understand the tone and emotions involved.

- **Language Translation:** If the website offers content in different languages, the LLM can help with on-the-fly translation. This means you can access the information in your preferred language without hassle.

- **Text Generation:** The LLM is quite clever at generating coherent and contextually relevant text based on the website content. This feature can be helpful when you need supplementary content, summaries, or descriptions.

These are just a few examples of how an LLM can be a game-changer for website content. The possibilities are vast, and you can tailor the specific use cases to meet your website's goals and requirements. It's all about enhancing user experiences, automating tasks, and extracting valuable insights.

CHAPTER 8 BUILDING DEMO APPLICATIONS USING LLMS

Let's create an app demonstrating the value of using LLM to extract insights from the website's content.

Data Scraping

The first step in deciphering the content of the website is to scrape the content. We will use the library BeautifulSoup to scrape the content and clean it. Let's begin by importing the required libraries.

```
import requests
import re
from bs4 import BeautifulSoup
import json
```

Let's use the same example from the previous use case. We will scrape the data from the manuals for Samsung.

```
url_to_scrape1 = 'http://downloadcenter.samsung.com/content/
PM/202001/20200128062847846/EB/ATT_G970U_G973U_G975U_EN_FINAL_200110/
camera_d1e10944.html'
```

Now we will make a GET request to get the content of the website.

```
# Make a GET request
response = requests.get(url_to_scrape1)
```

The GET request is made to the specified URL (url_to_scrape1) using the requests.get() method. The response from the request is stored in the response variable.

Now, we need to parse the HTML and extract the desired element. This step requires some analysis of the website content. You can right-click the web page and select Inspect Element. Then, as shown in Figure 8-1, you can expand on the HTML components and identify the highlighted component on the left as you hover over different HTML components on the right. In this case, guide_contents is the element we want to scrape.

CHAPTER 8   BUILDING DEMO APPLICATIONS USING LLMS

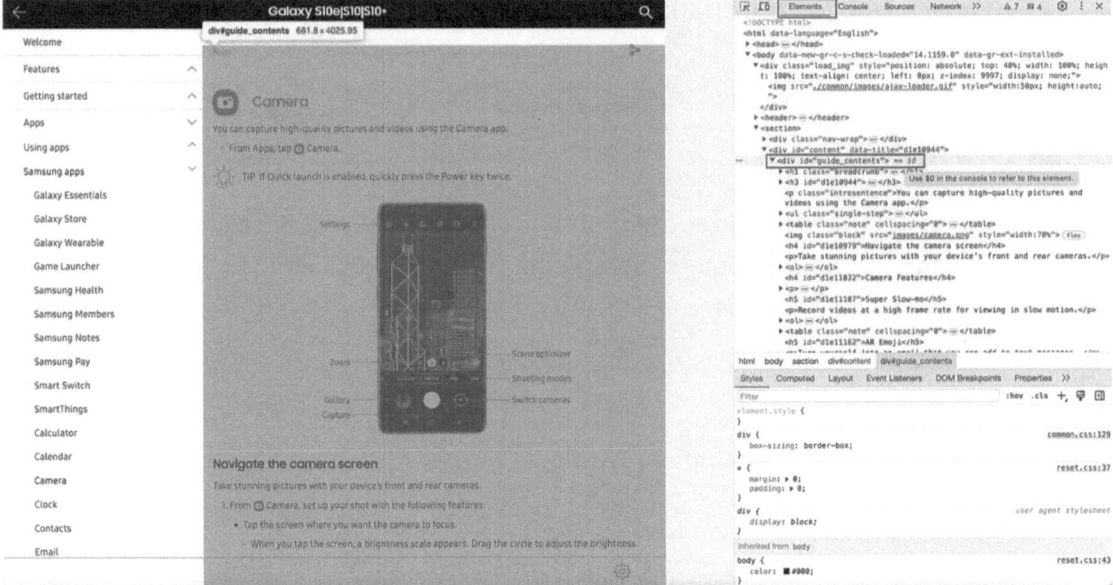

Figure 8-1. *Website scraping*

```
# Parse the HTML
soup = BeautifulSoup(response.content, 'html.parser')

# Find the desired element by ID
guide_contents_div = soup.find('div', id='guide_contents')
```

The previous code creates a BeautifulSoup object named soup by passing the response.content to the BeautifulSoup constructor with the specified parser type html.parser. It then uses the soup.find() method to locate the <div> element with the ID attribute of guide_contents and assigns it to the variable guide_contents_div.

Once we have the content, we need to get the text from the content.

```
# Extract the text content from the element
context_doc = guide_contents_div.get_text(strip=True)
```

```
# Print the extracted content
print(context_doc)
```

352

The previous code uses the get_text() method on the guide_contents_div element to extract the text content. The strip=True argument is used to remove any leading or trailing whitespace. The extracted content is then stored in the variable context_doc. You can also print the extracted content to the console using print(context_doc) for debugging.

The scraped text is not always clean, so we must remove unwanted characters to get the best results when feeding this text to an LLM. Therefore, you need to clean the extracted content as shown here:

```
# Clean the extracted content
clean_context = re.sub(r'[^a-zA-Z0-9.]', ' ', context_doc)
clean_context
```

In this part, the code utilizes the re.sub() function from the re module to perform a substitution. It replaces any characters that are not alphanumeric or a dot (.) with a space. The cleaned content is stored in the variable clean_context. The code then returns the result.

Question-answering

Now that you have scraped the content from the website, you can leverage it in many ways. Let's try out the question-answering pattern. We have already learned in Chapter 6 about fine-tuning an LLM for question-answering. Although fine-tuning can help you solve a specific problem/task with high accuracy, it requires significant time, computing resources, and golden datasets, which are often unavailable. So, in-context learning can be an easier alternative when you want to quickly prototype a solution or build a demo/proof-of-concept (PoC). We will use the principle of *in-context learning* for this exercise. In-context learning for LLMs is a process where the model uses immediate prior examples to understand and perform a new task without explicit retraining or fine-tuning. Essentially, you give the LLM a few examples of what you want it to do—these are known as *prompts*—and the model then "learns" the desired behavior based on these examples.

Think of it like showing someone a few photos of different birds sitting and then asking them to identify a sitting bird in a new image. They use the context from the previous photos to make an educated guess about the new one. Similarly, LLMs use the context from the provided examples to generate responses or complete tasks per the demonstrated pattern. This ability to quickly adapt to new information or tasks, just from

CHAPTER 8   BUILDING DEMO APPLICATIONS USING LLMS

a few examples, showcases LLMs' flexibility and power in understanding and generating human-like text. We will dive deep into the concept of in-context learning again in the next chapter.

Let's pick an example question for this website like this one:

```
question1 = 'What are the advanced recording options available in my device camera?'
```

Let's create a function to pass the website's content as a context to the LLM and pass the question to be answered based on the content.

```python
def generate_model_output(context, question):
    # Select the first 700 words from the context to be within the limit
    words = context.split()[:700]
    clean_words_joined = ' '.join(words)
    context = clean_words_joined

    model_input = f"Answer the question based on the context below. " + \
                  "Context: "  + context + \
                  " Question: " + question

    json_data = {
        "inputs": model_input,
        "parameters": {
            'temperature': 0.5,
            'max_new_tokens': 100,
            'return_full_text': False
        }
    }

    API_URL = "https://api-inference.huggingface.co/models/bigscience/bloom"
    headers = {"Authorization": "Bearer hf_XXXXXXX"}

    # Send the request to Hugging Face API
    response = requests.post(API_URL, headers=headers, json=json_data)
    json_response = json.loads(response.content.decode("utf-8"))
    model_output = json_response[0]['generated_text']
    return model_output
```

```
# Context and question
context = clean_context
question = question1

# Generate model output
model_output = generate_model_output(context, question)
```

The previous function generate_model_output(context, question) takes two parameters: context and question. This function aims to generate a model output by sending a request to the Hugging Face API with the given context and question.

1. First, the context is split into individual words using the split() method, and the first 700 words are selected. This is done to limit the context to a certain length. Then, the selected words are combined with spaces in between using the join() method, creating a cleaned version of the context. The cleaned context is then assigned back to the context variable.

   ```
   words = context.split()[:700]
   clean_words_joined = ' '.join(words)
   context = clean_words_joined
   ```

2. The model input is constructed as a string. It includes a prompt instructing the model to answer the question based on the given context. The model input and other parameters are organized into a dictionary called json_data. It includes the inputs field containing the model input string and the parameters field specifying the temperature, maximum new tokens, and whether to return the full text.

   ```
   model_input = f"Answer the question based on the context below. " + \
   json_data = {...}
   ```

CHAPTER 8   BUILDING DEMO APPLICATIONS USING LLMS

3. The URL for the Hugging Face API endpoint is specified. This determines which model will be used to generate the output. The authorization header is created with the required authorization token. A POST request is sent to the Hugging Face API endpoint with the specified API URL, headers, and JSON data containing the model input and parameters. The response from the API is stored in the response variable.

   `API_URL="https://api-inference.huggingface.co/models/bigscience/bloom"`
   `headers = {...}`
   `response = requests.post(API_URL, headers=headers, json=json_data)`

4. The response content is decoded from bytes to a JSON string and parsed into a Python dictionary using the json.loads() function. The parsed JSON response is stored in the json_response variable.

   `json_response = json.loads(response.content.decode("utf-8"))`

5. The generated text output from the model is extracted from the JSON response. It is assumed that the output is stored in the first item of the response, and the generated text can be accessed using the key generated_text. The extracted model output is assigned to the model_output variable. The function returns the generated model output.

   `model_output = json_response[0]['generated_text']`
   `return model_output`

After defining the generate_model_output() function, the code uses it by providing the clean_context and question1 variables as arguments. The function generates the model output based on the provided context and question, and the result is stored in the model_output variable.

Overall, this code segment uses the generate_model_output() function to generate model outputs based on a given context and question by interacting with the Hugging Face API.

We get the following output using this function for the given question:

Answer: 1. Super Slow mo 2. Live focus 3. Super steady 4. Super slow mo 5. Slow motion 6. Hyperlapse 7. Panoram 8. Night 9. Live focus video 10. Food 11. Super steady 12. Super slow mo 13. Slow motion 14. Hyperlapse 15. Panoram 16. Night 17. Live focus video 18. Food 19. Super steady 20. Super slow mo 21. Slow motion 22. Hyperlapse 23. Panorama

This is just a simple example, and you can continue building upon this by tweaking/updating the following:

1. **Input Prompt:** Changing the instruction can lead to significant variations in the generated answers.

2. **Model:** Using a pre-trained model has some limitations. If you are looking for task-specific or domain-specific results, you should fine-tune the model to get more accurate results, as shown in Chapter 7.

3. **Model Type and Parameters:** You can use a T5-based model for short answers while a larger model like BLOOM for longer answers.

4. **Context Window:** Increase the context window when using a locally deployed LLM or a paid version of the remotely hosted LLM.

These use cases have many applications, such as assisting customer support agents. They can search for answers quickly on a website and provide valuable information to customers. Alternatively, when the customers visit the support page/link, they can also open the chatbot pop-up and post their queries directly, which can be answered by the LLM-powered chatbot reading the website's content. Let's extend the application by adding another natural language generation (NLG) use case in the same application.

Summarization

Now, we will add the summarization feature for any website. So, let's scrape an article from the Internet. I will use my blog here and scrape the passages to summarize them using the following code:

```python
import requests
from bs4 import BeautifulSoup

# URL of the article/blog you want to scrape
url = "https://shivamsolanki.net/posts/Tuning-LLMs/"

# Send a GET request to fetch the HTML content
response = requests.get(url)
html_content = response.text

# Create a BeautifulSoup object to parse the HTML
soup = BeautifulSoup(html_content, "html.parser")

# Find all the paragraph tags within the article
paragraphs = soup.find_all("p")

# Extract the text content from each paragraph and filter based on length
passages = [p.get_text(strip=True) for p in paragraphs if len(p.get_text(strip=True)) >= 100]

# Create a combined passage until its length is less than 800, considering full passages
combined_passage = ""
for passage in passages:
    if len(combined_passage) + len(passage) <= 2000:
        combined_passage += passage + " "
    else:
        break

# Print the extracted passages
print("Combined Passage:", combined_passage)
```

The previous scraping code can be divided into the following steps:

1. **Importing the Necessary Libraries:** First, we import the `requests` library for sending HTTP requests and the `BeautifulSoup` class from the `bs4` module for parsing HTML content.

2. **URL and GET Request:** Then we set the URL variable to the desired article or blog URL you want to scrape. Then, it sends a GET request to that URL using `requests.get(url)` to fetch the HTML content of the page.

   ```
   url = "https://shivamsolanki.net/posts/Tuning-LLMs/"
   response = requests.get(url)
   html_content = response.text
   ```

3. **Creating BeautifulSoup Object and Finding Paragraph Tags:** The next step is to create a `BeautifulSoup` object named `soup` by passing the HTML content (`html_content`) and the parser ("html.parser") to the `BeautifulSoup` class. This object allows us to navigate and search through the HTML structure. Then, we use the `find_all` method on the `soup` object to find all the `<p>` tags within the HTML content. It stores the result in the `paragraphs` variable, a list of `Tag` objects representing each paragraph.

   ```
   soup = BeautifulSoup(html_content, "html.parser")

   paragraphs = soup.find_all("p")
   ```

4. **Extracting Text Content and Filtering:** The code iterates over the `paragraphs` list and uses the `get_text` method on each `Tag` object (p) to extract the paragraph's text content. It filters out paragraphs with less than 100 characters using list comprehension. The filtered paragraphs are stored in the `passages` list.

   ```
   passages = [p.get_text(strip=True) for p in paragraphs if
   len(p.get_text(strip=True)) >= 100]
   ```

5. **Creating Combined Passage:** The code initializes an empty string variable named `combined_passage`. It then iterates over the `passages` list and checks if adding the current passage to `combined_passage` would keep the length to less than 2,000 characters (as specified by `len(combined_passage) + len(passage) <= 2000`). If the condition is satisfied, it appends the passage to `combined_passage` with a space character. If the condition is unsatisfied, it breaks out of the loop using break.

```
combined_passage = ""
for passage in passages:
    if len(combined_passage) + len(passage) <= 2000:
        combined_passage += passage + " "
    else:
        break
```

So, the previous code allows us to scrape the HTML content of an article or blog, extract the paragraphs, filter them based on length, and create a combined passage that satisfies the length constraint. We have truncated the length to 2,000 as we want to use a hosted Hugging Face model for summarization via API. A locally downloaded/fine-tuned model can increase the context window for better results.

User Interface/Application

The final step in prototyping this type of application to provide a sense of its feasibility and importance to stakeholders is to create a user interface (UI) that helps the user interact at a higher level of abstraction. Let's combine the steps from 2.1 to 2.3 to create a Streamlit app for creating app that can summarize and provide the ability to ask questions based on the content of the website.

```
import streamlit as st
import requests
import re
from bs4 import BeautifulSoup
import json

# HuggingFace Inference API endpoint
QA_ENDPOINT = "https://api-inference.huggingface.co/models/bigscience/bloom"
SUMMARIZATION_ENDPOINT = "https://api-inference.huggingface.co/models/tiiuae/falcon-7b-instruct"[33]

min_length = 100
max_length = 2000

def extract_combined_passage(url, min_paragraph_length, max_combined_length):
```

```python
    response = requests.get(url)
    html_content = response.text
    soup = BeautifulSoup(html_content, "html.parser")
    paragraphs = soup.find_all("p")
    passages = [p.get_text(strip=True) for p in paragraphs if len(p.get_text(strip=True)) >= min_paragraph_length]
    combined_passage = ""
    for passage in passages:
        if len(combined_passage) + len(passage) <= max_combined_length:
            combined_passage += passage + " "
        else:
            break

    return combined_passage

# Streamlit app
def main():
    st.title("Website Summarizer and Q&A")

    # Get user input
    url = st.text_input("Enter the URL of a website:")
    question = st.text_input("Ask a question based on the text:")

    if st.button("Summarize and Answer"):
        if url:
            # Scrape text from the website
            # text = scrape_text_from_website(url)
            text = extract_combined_passage(url, min_length, max_length)

            # Generate a summary using the HuggingFace model
            summary = generate_summary(url)

            # Answer the question based on the summary
            answer = answer_question(url, question)

            # Display the results
            st.subheader("Summary:")
            st.write(summary)
```

```python
            st.subheader("Answer:")
            st.write(answer)
        else:
            st.warning("Please enter a valid URL.")

# Function to generate summary using HuggingFace model
def generate_summary(url):
    # Set up the HuggingFace API request
    headers = {"Authorization": "Bearer XXXXXXXX"}
    combined_passage = extract_combined_passage(url, min_length,
    max_length)

    model_input = f"Summarize the following article. " + \
        "Article: "  + combined_passage

    json_data = {
            "inputs": model_input,
            "parameters": {'temperature': 0.5,
                           'max_new_tokens': 100,
                           'return_full_text': False,
                           },
        }

    # HuggingFace response
    response = requests.post(SUMMARIZATION_ENDPOINT, headers=headers,
    json=json_data)
    json_response = json.loads(response.content.decode("utf-8"))
    # print("-----------------------------------", json_response)
    summary = json_response[0]['generated_text']
    return summary

# Function to answer a question based on the context
def answer_question(url, question):
    # Set up the HuggingFace API request
    headers = {"Authorization": "Bearer XXXXXXXXX"}
    combined_passage = extract_combined_passage(url, min_length,
    max_length)
```

```
    model_input = f"Answer the question based on the context below. " + \
        "Context: "  + combined_passage + \
        " Question: " + question

    json_data = {
            "inputs": model_input,
            "parameters": {'temperature': 0.5,
                            'max_new_tokens': 100,
                            'return_full_text': False,
                            },
        }

    # Send a request to the HuggingFace Inference API
    response = requests.post(QA_ENDPOINT, headers=headers, json=json_data)
    print("===================", response.json())

    # Extract and return the answer from the API response
    answer = response.json()[0]["generated_text"]
    return answer

if __name__ == "__main__":
    main()
```

Let's look at the Streamlit app step-by-step.

```
import streamlit as st
import requests
import re
from bs4 import BeautifulSoup
import json
```

First, we import the necessary libraries, including `Streamlit`, `Requests`, `re`, `BeautifulSoup`, and `json`. These libraries will help us fetch the website's HTML content, parse it, and make API requests to Hugging Face models.

```
QA_ENDPOINT = "https://api-inference.huggingface.co/models/
bigscience/bloom"
```

```
SUMMARIZATION_ENDPOINT = "https://api-inference.huggingface.co/models/
tiiuae/falcon-7b-instruct"

min_length = 100
max_length = 2000
```

Next, we define some global constants. One constant is the Hugging Face API endpoint for question-answering, which will help us generate answers to your questions. Another constant is the API endpoint for summarization, which will generate summaries of the website's content. We also set the minimum and maximum lengths for the extracted passages.

```
def extract_combined_passage(url, min_paragraph_length, max_combined_
length):
    # Send a GET request to fetch the HTML content
    response = requests.get(url)
    html_content = response.text

    # Create a BeautifulSoup object to parse the HTML
    soup = BeautifulSoup(html_content, "html.parser")

    # Find all the paragraph tags within the article
    paragraphs = soup.find_all("p")

    # Extract the text content from each paragraph and filter based
    on length
    passages = [p.get_text(strip=True) for p in paragraphs if len(p.get_
    text(strip=True)) >= min_paragraph_length]

    # Create a combined passage until its length is less than the specified
    limit, considering full passages
    combined_passage = ""
    for passage in passages:
        if len(combined_passage) + len(passage) <= max_combined_length:
            combined_passage += passage + " "
        else:
            break

    return combined_passage
```

CHAPTER 8 BUILDING DEMO APPLICATIONS USING LLMS

To extract the combined passage from the website's content, we define the function `extract_combined_passage`. This function takes a URL, minimum paragraph length, and maximum combined length as inputs. We use `BeautifulSoup` to scrape the HTML content from the provided URL inside the function. Then, we find all the paragraph tags within the article and extract the text content from each paragraph. We filter the passages based on length, discarding those less than the minimum length. Finally, we concatenate the passages until the length reaches the specified maximum length or until no more full passages are left.

Now, let's move on to the main part of the app.

```python
def main():
    st.title("Website Summarizer and Q&A")

    # Get user input
    url = st.text_input("Enter the URL of a website:")
    question = st.text_input("Ask a question based on the text:")

    if st.button("Summarize and Answer"):
        if url:
            # Scrape text from the website
            # text = scrape_text_from_website(url)
            text = extract_combined_passage(url, min_length, max_length)

            # Generate a summary using the HuggingFace model
            summary = generate_summary(url)

            # Answer the question based on the summary
            answer = answer_question(url, question)

            # Display the results
            st.subheader("Summary:")
            st.write(summary)

            st.subheader("Answer:")
            st.write(answer)
        else:
            st.warning("Please enter a valid URL.")
```

We define the `main()` function, which sets up the Streamlit interface. It displays a title and provides text input fields for the URL and the question. There's also a button that you can click to trigger the summarization and question-answering process.

When you click the button, we perform the following steps:

1. If you have provided a URL, we scrape the website's content and extract the combined passage using the `extract_combined_passage` function.

2. We generate a summary of the combined passage by making an API request to the HuggingFace generative (decoder-only) model.

```python
def generate_summary(url):
    # Set up the HuggingFace API request
    headers = {"Authorization": "Bearer XXXXXXXX"}
    combined_passage = extract_combined_passage(url,
    min_length, max_length)

    model_input = f"Summarize the following article. " + \
        "Article: "  + combined_passage

    json_data = {
            "inputs": model_input,
            "parameters": {'temperature': 0.5,
                        'max_new_tokens': 100,
                        'return_full_text': False,
                        },
        }

    # HuggingFace response
    response = requests.post(SUMMARIZATION_ENDPOINT,
    headers=headers, json=json_data)
    json_response = json.loads(response.content.
    decode("utf-8"))
    # print("-----------------------------------",
    json_response)
    summary = json_response[0]['generated_text']
    return summary
```

3. We generate an answer to your question based on the combined passage by making an API request to the HuggingFace generative (decoder-only) model.

```python
def answer_question(url, question):
    # Set up the HuggingFace API request
    headers = {"Authorization": "Bearer XXXXXXXX"}
    combined_passage = extract_combined_passage(url, min_length, max_length)

    model_input = f"Answer the question based on the context below. " + \
        "Context: " + combined_passage + \
        " Question: " + question

    json_data = {
            "inputs": model_input,
            "parameters": {'temperature': 0.5,
                           'max_new_tokens': 100,
                           'return_full_text': False,
                          },
           }

    # Send a request to the HuggingFace Inference API
    response = requests.post(QA_ENDPOINT, headers=headers, json=json_data)
    print("====================", response.json())

    # Extract and return the answer from the API response
    answer = response.json()[0]["generated_text"]
    return answer
```

4. Finally, we display the generated summary and answer on the Streamlit interface.

You can run the Streamlit app locally using this command:

```
streamlit run Web-chatbot.py
```

You will be able to access the UI at http://localhost:8501/, as shown in Figure 8-2.

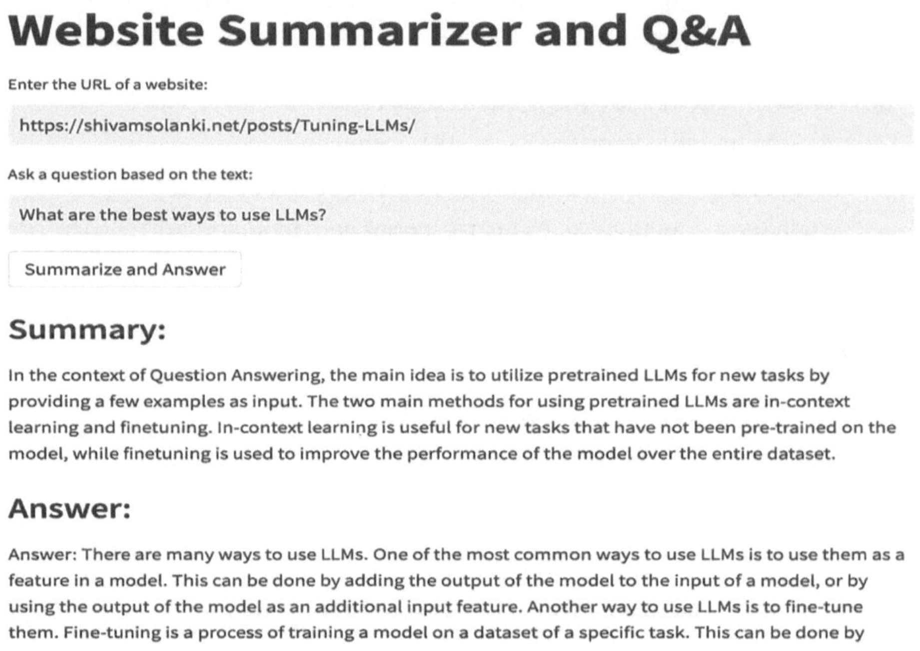

Website Summarizer and Q&A

Enter the URL of a website:

https://shivamsolanki.net/posts/Tuning-LLMs/

Ask a question based on the text:

What are the best ways to use LLMs?

[Summarize and Answer]

Summary:

In the context of Question Answering, the main idea is to utilize pretrained LLMs for new tasks by providing a few examples as input. The two main methods for using pretrained LLMs are in-context learning and finetuning. In-context learning is useful for new tasks that have not been pre-trained on the model, while finetuning is used to improve the performance of the model over the entire dataset.

Answer:

Answer: There are many ways to use LLMs. One of the most common ways to use LLMs is to use them as a feature in a model. This can be done by adding the output of the model to the input of a model, or by using the output of the model as an additional input feature. Another way to use LLMs is to fine-tune them. Fine-tuning is a process of training a model on a dataset of a specific task. This can be done by

Figure 8-2. Website summarizer and Q&A

That's how our Streamlit app works to leverage LLM and help you summarize website content and ask questions about the text. We have discussed many use cases for extracting insights from the scope of the websites by leveraging LLMs at the start of this section. Summarization and question-answering are just one of those examples, but you can keep expanding this application to meet your business requirements. Additionally, you can create applications by setting up the LLMs locally or in a virtual machine to increase the context window exponentially.

Uncovering Insights and Gaining a Quick Understanding of PDF Documents

In this section, we will explore extracting insights and gain a quick understanding of PDF documents. PDF files are widely used for sharing and distributing information, but extracting relevant knowledge and understanding their content can be time-consuming. However, with the advancements in LLMs and generative AI techniques, we can now automate the extraction of insights and generate summaries from PDF documents.

By leveraging LLMs, we can create an application to ask questions directly from the PDF and obtain answers based on its content. This application utilizes powerful language models to process the PDF text, extract critical information, and provide concise answers to our queries. Additionally, we can employ summarization techniques to generate a summary of the PDF document, enabling us to grasp its main points and key takeaways quickly.

With the ability to ask questions, obtain answers, and generate summaries from PDF documents, we can streamline our information retrieval process, save time, and gain valuable insights more efficiently. Whether for research purposes, studying, or extracting critical information from lengthy reports, this approach empowers us to unlock the knowledge contained within PDF documents swiftly and effectively.

In the upcoming sections, we will explore the implementation details and how to create an application that enables us to interact with PDF files, ask questions, and generate summaries, ultimately enhancing our understanding of these documents. Additionally, we will use different libraries like Langchain, OpenAPI, and the `pipeline` class of HuggingFace to equip you with various tools to build different types of applications depending on the business requirements.

Question-Answering for PDF

Let's build a simple demo application for question-answering based on PDF using the Streamlit framework.

```
from dotenv import load_dotenv
import streamlit as st
from PyPDF2 import PdfReader
from langchain.text_splitter import CharacterTextSplitter
from langchain.embeddings.openai import OpenAIEmbeddings
from langchain.vectorstores import FAISS
from langchain.chains.question_answering import load_qa_chain
from langchain.llms import OpenAI
from langchain.callbacks import get_openai_callback

import os

def main():
    # Load environment variables
```

```python
load_dotenv()
os.environ["OPENAI_API_KEY"] = "XXXXXX"

# Set up the Streamlit app
st.set_page_config(page_title="PDF-Chatbot")
st.header("Ask me anything about your PDF")

# Upload PDF file
pdf = st.file_uploader("Upload your PDF", type="pdf")

# Extract text from the PDF
if pdf is not None:
    pdf_reader = PdfReader(pdf)
    text = ""
    for page in pdf_reader.pages:
        text += page.extract_text()

    # Split the text into chunks
    text_splitter = CharacterTextSplitter(
        separator="\n",
        chunk_size=1000,
        chunk_overlap=200,
        length_function=len
    )
    chunks = text_splitter.split_text(text)

    # Create embeddings for text chunks
    embeddings = OpenAIEmbeddings()
    knowledge_base = FAISS.from_texts(chunks, embeddings)

    # Prompt the user to ask a question about the PDF
    user_question = st.text_input("Ask a question about your PDF:")
    if user_question:
        # Perform a similarity search to find relevant text chunks
        docs = knowledge_base.similarity_search(user_question)

        # Initialize LLM model and question-answering chain
        llm = OpenAI()
        chain = load_qa_chain(llm, chain_type="stuff")
```

```
        # Use the question-answering chain to get the answer
        with get_openai_callback() as cb:
            response = chain.run(input_documents=docs, question=user_
            question)
            print(cb)

        # Display the response to the user
        st.write(response)
if __name__ == '__main__':
    main()
```

Let's dive into the PDF-QA app now and understand the code step-by-step.

1. In the first block, we import the necessary libraries and modules, including dotenv for loading environment variables, streamlit for creating the app interface, and various modules from the langchain library for text processing and question-answering. We also import os to handle environment variables.

   ```
   from dotenv import load_dotenv
   import streamlit as st
   from PyPDF2 import PdfReader
   from langchain.text_splitter import CharacterTextSplitter
   from langchain.embeddings.openai import OpenAIEmbeddings
   from langchain.vectorstores import FAISS
   from langchain.chains.question_answering import load_qa_chain
   from langchain.llms import OpenAI
   from langchain.callbacks import get_openai_callback
   import os
   ```

2. In the main() function, we start by loading the environment variables using load_dotenv() and setting the OpenAI API key.

   ```
   load_dotenv()
   os.environ["OPENAI_API_KEY"] = "XXXXXX"
   ```

3. We configure the Streamlit app using set_page_config() to set the page title and display a header.

   ```
   st.set_page_config(page_title="PDF-Chatbot")
   st.header("Ask me anything about your PDF")
   ```

4. You are prompted to upload a PDF file using the file_uploader function from Streamlit. The uploaded file is stored in the pdf variable.

   ```
   pdf = st.file_uploader("Upload your PDF", type="pdf")
   ```

5. Then, we extract the text from the PDF.

   ```
   # Extract text from the PDF
   if pdf is not None:
       pdf_reader = PdfReader(pdf)
       text = ""
       for page in pdf_reader.pages:
           text += page.extract_text()
   ```

   In the previous code block, when you upload a PDF file, we read its contents using PdfReader from PyPDF2. We extract the text from each page and concatenate it into a single text string.

6. After extracting text from the PDF, we split the text into chunks as shown here:

   ```
   # Split the text into chunks
   text_splitter = CharacterTextSplitter(
       separator="\n",
       chunk_size=1000,
       chunk_overlap=200,
       length_function=len
   )
   chunks = text_splitter.split_text(text)
   ```

In the previous code block, we split it into smaller chunks using the `CharacterTextSplitter` from the `langchain` library. This helps in processing the text more efficiently. We specify the chunk size, overlap, and separator to control the splitting process.

7. The most crucial step is creating embedding for text chunks.

   ```
   embeddings = OpenAIEmbeddings()
   knowledge_base = FAISS.from_texts(chunks, embeddings)
   ```

   We create word embeddings for the text chunks using the `OpenAIEmbeddings` class from `langchain`. These embeddings capture the semantic meaning of the text and enable efficient similarity search.

8. You are prompted to ask a question about the PDF using `text_input` from Streamlit. The question is stored in the `user_question` variable.

   ```
   user_question = st.text_input("Ask a question about your PDF:")
   ```

9. When you enter a question, we perform a similarity search on the text chunks to find the most relevant chunks related to your question. This is done using the `similarity_search` method on the `knowledge_base` object.

   ```
   if user_question:
       # Perform similarity search to find relevant text chunks
       docs = knowledge_base.similarity_search(user_question)
   ```

10. Next, we initialize the LLM by creating an instance of the `OpenAI` class from `langchain`. We also load the question-answering chain using the `load_qa_chain` function from `langchain`. This chain is responsible for answering questions based on the given context.

    ```
    # Initialize LLM model and question-answering chain
    llm = OpenAI()
    chain = load_qa_chain(llm, chain_type="stuff")
    ```

11. Finally, we use the question-answering chain to find the answer to your question. This is done by calling the run method on the chain object and providing the relevant text chunks (docs) and your question. The answer is stored in the response variable and displayed using the write function from Streamlit.

```
with get_openai_callback() as cb:
    response = chain.run(input_documents=docs, question=user_
    question)
    print(cb)
# Display the response to the user
st.write(response)
```

12. In the __name__ == '__main__' block, we call the main() function to start the Streamlit app when the script is executed.

```
if __name__ == '__main__':
    main()
```

You can execute the previous application by running the following command:

```
streamlit run PDF-QA-Streamlit.py
```

You can view the UI in your favorite browser at http://localhost:8501/.

I have created a small PDF for this demo about the re-ranker, and you can ask any question about the Colbert algorithm, as shown in Figure 8-3.

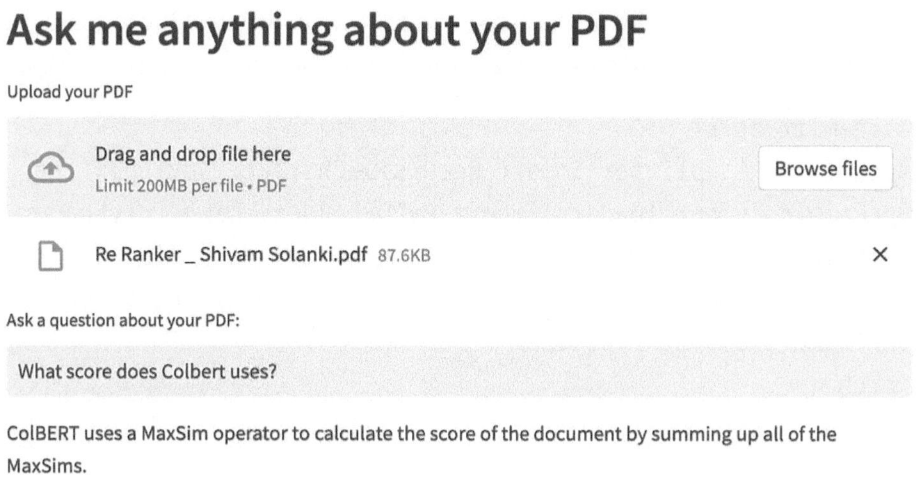

Ask me anything about your PDF

Upload your PDF

⬆ Drag and drop file here   Browse files
   Limit 200MB per file • PDF

📄 Re Ranker _ Shivam Solanki.pdf 87.6KB   ✕

Ask a question about your PDF:

What score does Colbert uses?

ColBERT uses a MaxSim operator to calculate the score of the document by summing up all of the MaxSims.

Figure 8-3. *PDF Q&A*

We use Langchain and OpenAI in this app instead of HuggingFace to provide diverse tools and options for text processing and question-answering. Langchain is a specialized library for NLP tasks, offering specific features tailored to our needs, such as text splitting and question-answering chains. OpenAI provides a powerful LLM for accurate text generation (question-answering based on the context from PDF in this case). It's worth mentioning that we have already utilized HuggingFace in other applications, and in this case, we wanted to explore alternative approaches and leverage the strengths of Langchain and OpenAI.

PDF Summarization

PDF summarization using LLMs is a powerful technique that extracts crucial information and generates concise summaries from PDF documents. We leverage the Langchain library and the T5 model to achieve this, specifically the `MBZUAI/LaMini-Flan-T5-248M`[34] variant. The text is preprocessed and split into smaller chunks by uploading a PDF file through our user-friendly interface. The T5 model then generates a summary of the content. This technique saves time and effort by quickly extracting essential insights from lengthy PDFs, benefiting researchers, professionals, and students.

CHAPTER 8   BUILDING DEMO APPLICATIONS USING LLMS

Let's dive into the code now. We will use the following Streamlit application to understand this and reuse PDF summarizer for enterprises/business use cases:

```python
import streamlit as st
from langchain.text_splitter import RecursiveCharacterTextSplitter
from langchain.document_loaders import PyPDFLoader, DirectoryLoader
from langchain.chains.summarize import load_summarize_chain
from transformers import T5Tokenizer, T5ForConditionalGeneration
from transformers import pipeline
import torch
import base64

# Downloading the model for the first time
tokenizer = T5Tokenizer.from_pretrained("MBZUAI/LaMini-Flan-T5-248M")
base_model = T5ForConditionalGeneration.from_pretrained("MBZUAI/LaMini-Flan-T5-248M")

# File loader and preprocessing
def preprocess_file(file):
    loader = PyPDFLoader(file)
    pages = loader.load_and_split()
    text_splitter = RecursiveCharacterTextSplitter(chunk_size=200, chunk_overlap=50)
    texts = text_splitter.split_documents(pages)
    final_texts = ""
    for text in texts:
        final_texts = final_texts + text.page_content
    return final_texts

# LLM pipeline
def run_llm_pipeline(filepath):
    pipe_sum = pipeline(
        'summarization',
        model=base_model,
        tokenizer=tokenizer,
        max_length=200,
        min_length=50)
```

```python
        input_text = preprocess_file(filepath)
        result = pipe_sum(input_text)
        result = result[0]['summary_text']
        return result

@st.cache_data
# Function to display the PDF of a given file
def display_pdf(file):
    # Opening file from file path
    with open(file, "rb") as f:
        base64_pdf = base64.b64encode(f.read()).decode('utf-8')

    # Embedding PDF in HTML
    pdf_display = f'<iframe src="data:application/pdf;base64,{base64_pdf}" width="100%" height="600" type="application/pdf"></iframe>'

    # Displaying File
    st.markdown(pdf_display, unsafe_allow_html=True)

# Streamlit code
st.set_page_config(layout="wide")

def main():
    st.title("Document Summarization App using Language Model")

    uploaded_file = st.file_uploader("Upload your PDF file", type=['pdf'])

    if uploaded_file is not None:
        if st.button("Summarize"):
            col1, col2 = st.columns(2)
            filepath = "data/" + uploaded_file.name
            with open(filepath, "wb") as temp_file:
                temp_file.write(uploaded_file.read())
            with col1:
                st.info("Uploaded File")
                pdf_view = display_pdf(filepath)

            with col2:
                summary = run_llm_pipeline(filepath)
```

```
            st.info("Summarization Complete")
            st.success(summary)

if __name__ == "__main__":
    main()
```

Now it's time to dive into the code and understand each block separately.

1. First, we import the required libraries.

    ```
    import streamlit as st
    from langchain.text_splitter import RecursiveCharacterTextSplitter
    from langchain.document_loaders import PyPDFLoader, DirectoryLoader
    from langchain.chains.summarize import load_summarize_chain
    from transformers import T5Tokenizer, T5ForConditionalGeneration
    from transformers import pipeline
    import torch
    import base64
    ```

    The previous code block imports the necessary libraries and modules for our PDF summarization application. It includes Streamlit for the user interface, Langchain for text splitting and document loading, Transformers for language model handling, Torch for deep learning operations, and Base64 for encoding PDF files.

2. Next, we download and load the T5 language model for summarization.

    ```
    tokenizer = T5Tokenizer.from_pretrained("MBZUAI/LaMini-Flan-T5-248M")
    base_model = T5ForConditionalGeneration.from_pretrained("MBZUAI/LaMini-Flan-T5-248M")
    ```

    Here, we download and load the T5 language model for summarization using the MBZUAI/LaMini-Flan-T5-248M variant. We create a tokenizer and a base model for a conditional generation.

3. The next step is to process the PDF document as shown here:

```
def file_preprocessing(file):
    loader = PyPDFLoader(file)
    pages = loader.load_and_split()
    text_splitter = RecursiveCharacterTextSplitter(chunk_size=200,
    chunk_overlap=50)
    texts = text_splitter.split_documents(pages)
    final_texts = ""
    for text in texts:
        final_texts = final_texts + text.page_content
    return final_texts
```

The previous file_preprocessing function performs the preprocessing of the uploaded PDF file. It uses the PyPDFLoader to load and split the PDF into pages.

Next, we create an instance of the RecursiveCharacterTextSplitter class. This splitter is responsible for dividing the pages into smaller chunks of text. We set the chunk_size to 200 characters and the chunk_overlap to 50 characters, ensuring the chunks overlap for better context understanding.

We use the text splitter to split the pages into smaller texts, and the resulting texts are stored in the texts variable as a list. To combine all the text chunks into a single text, we initialize an empty string called final_texts. We iterate over each text in the texts list and concatenate its content with the final_texts string.

Finally, we return the final_texts, which represents the combined text content of the PDF file. This preprocessing step prepares the text for further processing, such as summarization or other natural language processing tasks.

4. Now that the PDF document is processed, we can summarize the content of the document using the LLM as shown here:

```
def llm_pipeline(filepath):
    pipe_sum = pipeline(
```

```
            'summarization',
            model = base_model,
            tokenizer = tokenizer,
            max_length = 200,
            min_length = 50)
    input_text = file_preprocessing(filepath)
    result = pipe_sum(input_text)
    result = result[0]['summary_text']
    return result
```

The previous llm_pipeline function performs the PDF summarization using the T5 language model. The function takes the filepath parameter, representing the PDF file's path.

First, we create a pipeline using the pipeline function from the Hugging Face library. This pipeline is designed explicitly for text summarization and is initialized with the summarization task. We provide the necessary components, such as the base_model and tokenizer, which were previously downloaded, along with the max_length and min_length parameters to control the length of the generated summaries.

Next, we call the file_preprocessing function, passing the filepath to it. This function processes the PDF file and returns the combined text content of the file. We then use the pipe_sum pipeline to summarize the input_text containing the preprocessed text. The result is stored in the result variable. Since the result is returned as a list of dictionaries, we access the summary_text key of the first element in the list to extract the actual summary text. This summary text is assigned to the result variable.

Finally, we return the result, which represents the summarized text generated by the LLM pipeline. This function encapsulates the entire text summarization process, making it easier to integrate into our application and retrieve the summarization results efficiently.

CHAPTER 8 BUILDING DEMO APPLICATIONS USING LLMS

5. We can also display the uploaded PDF using the utility function as shown here:

```
@st.cache_data
def displayPDF(file):
    with open(file, "rb") as f:
        base64_pdf = base64.b64encode(f.read()).decode('utf-8')
    pdf_display = F'<iframe src="data:application/pdf;base64,{base64_pdf}" width="100%" height="600" type="application/pdf"></iframe>'
    st.markdown(pdf_display, unsafe_allow_html=True)
```

This function `displayPDF` displays the uploaded PDF file in the Streamlit application. It reads the PDF file, encodes it using Base64, and embeds it as an HTML `iframe` for rendering in the Streamlit interface.

6. Finally, we can define the main function of the Streamlit application as shown here:

```
st.set_page_config(layout="wide")

def main():
    st.title("Document Summarization App using Language Model")
    uploaded_file = st.file_uploader("Upload your PDF file", type=['pdf'])
    if uploaded_file is not None:
        if st.button("Summarize"):
            col1, col2 = st.columns(2)
            filepath = "data/"+uploaded_file.name
            with open(filepath, "wb") as temp_file:
                temp_file.write(uploaded_file.read())
            with col1:
                st.info("Uploaded File")
                pdf_view = displayPDF(filepath)
            with col2:
                summary = llm_pipeline(filepath)
                st.info("Summarization Complete")
```

CHAPTER 8 BUILDING DEMO APPLICATIONS USING LLMS

```
            st.success(summary)

if __name__ == "__main__":
    main()
```

In the final code block, we configure the Streamlit page layout and define the main function for our application. The Streamlit interface includes a title and a file uploader for the PDF file. When the Summarize button is clicked, the file is saved temporarily, and the uploaded file's content is displayed in one column. At the same time, the summarization process is performed in the other column. The summary is then displayed in a success message.

You can execute the previous application by running the following command:

```
streamlit run PDF-Summary-Streamlit.py
```

You can view the UI in your favorite browser at `http://localhost:8501/`.

I have created a small PDF for this demo about the re-ranker, and you can summarize this PDF as shown in Figure 8-4. It is important to note that the model used in this app is small, so if you want to scale up this application, you need to use a larger model so that you can increase the token limit for processing and ultimately use a large PDF for summarization to convert this simple demo application into an enterprise-grade application.

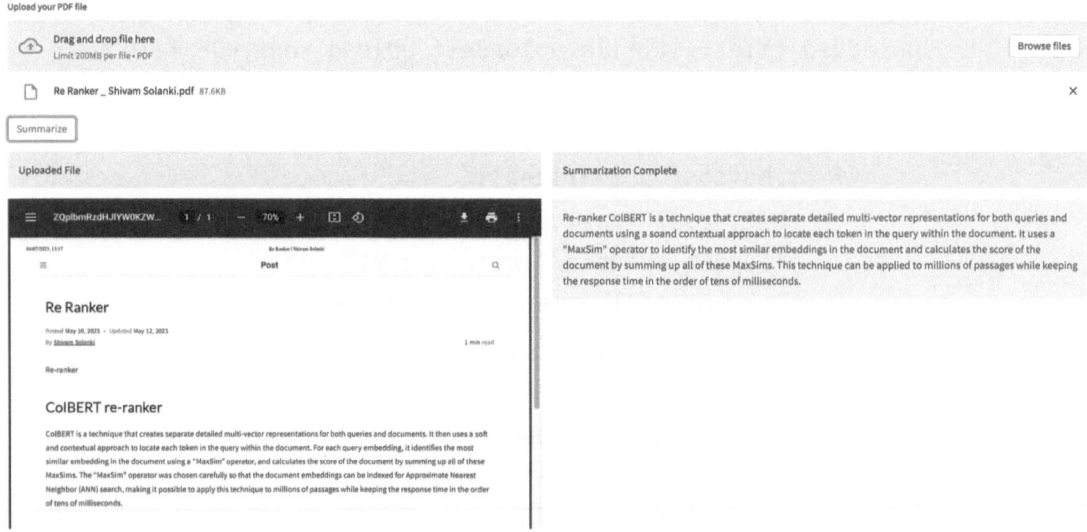

Figure 8-4. *PDF summarization*

This demo app works well for small PDFs. However, it wouldn't work for long PDFs as this model will run out of context length. So, we need to apply the following strategy for long PDFs:

- Use a larger model with a longer context length.
- Split documents into smaller chunks that can fit into the context window of the LLM.
- Summarize each chunk individually.
- Combine all the summaries.
- Extract the final summary of the individual summaries.

You could also use other strategies like creating a vector embedding and indexing each passage. Then, you can use a retriever to extract the most relevant passages for the summary and summarize those passages only. Therefore, it will depend on the use case and the data (PDF contents).

Extracting Insights from Video Transcripts

Extracting insights from video transcripts has become increasingly valuable in today's digital landscape. As video content continues dominating various platforms, businesses recognize the untapped potential of the information embedded within these videos. Video transcripts offer a wealth of textual data that can be leveraged to gain valuable insights and address critical business use cases.

By analyzing video transcripts, businesses can unlock a multitude of benefits. First, extracting insights from video transcripts enables the efficient and accurate indexing of video content. Instead of relying solely on manual tagging or limited metadata, businesses can utilize the text within transcripts to categorize, search, and retrieve specific video segments. This enhances content discoverability and empowers users to find relevant information quickly, improving user experience and engagement.

Furthermore, video transcripts provide a foundation for comprehensive content analysis. By applying natural language processing (NLP) techniques, businesses can extract critical concepts, sentiments, and thematic patterns from the transcripts. This analysis helps identify popular topics, customer sentiments, and emerging trends within the video content. These insights enable businesses to refine their marketing strategies, understand their target audience better, and tailor their offerings to meet evolving customer preferences.

Video transcripts also facilitate the creation of concise summaries and highlights. By automatically generating summaries of video content, businesses can distill lengthy videos into easily digestible snippets. These summaries can be repurposed for marketing materials, social media posts, or internal knowledge sharing, amplifying the reach and impact of the original video content. Additionally, extracting highlights from transcripts allows businesses to showcase the most compelling moments of a video, increasing engagement and maximizing the value derived from the range.

Another significant advantage of extracting insights from video transcripts is applying question-answering systems. By combining video transcripts with powerful language models, businesses can create intelligent systems that answer specific questions based on the video content. This opens opportunities for interactive and personalized experiences, as users can obtain instant answers or seek clarification on particular topics within the video. Such applications can be precious in e-learning, customer support, and knowledge-intensive industries.

In summary, extracting insights from video transcripts empowers businesses to unlock the hidden potential of their video content. It enables efficient indexing, comprehensive content analysis, concise summarization, and interactive question-answering capabilities. By leveraging the textual data embedded within video transcripts, businesses can enhance user experience, refine their strategies, and make informed decisions based on the valuable insights extracted.

In this section, we will learn how to extract insights from video transcripts using the open-source LLMs on Hugging Face and the Langchain and OpenAPI libraries. Let's start with the open-source models and Streamlit application.

Video Caption Summarization and Q&A

We will build a demo application that allows users to extract insights from video transcripts. The app generates a summary of the video transcript and answers the user's question based on the cleaned transcript using the Hugging Face text generation models. This is achieved using the following code:

```
import streamlit as st
import requests
import urllib.parse
from langchain.document_loaders import YoutubeLoader
```

```python
import json
import re

# HuggingFace Inference API endpoint
SUMMARIZATION_ENDPOINT = "https://api-inference.huggingface.co/models/tiiuae/falcon-7b-instruct"
QA_ENDPOINT = "https://api-inference.huggingface.co/models/gpt2-large"

# Streamlit app
def main():
    st.title("YouTube Caption Summarizer and Q&A")

    # Get user input
    url = st.text_input("Enter the YouTube URL:")
    question = st.text_input("Ask a question based on the caption:")

    if st.button("Summarize and Answer"):
        if url:
            # Extract video ID from YouTube URL
            video_id = extract_video_id(url)
            print("VIDEO_ID", video_id)

            # Get YouTube captions
            # captions = get_youtube_captions(video_id)
            transcript = get_youtube_captions(video_id)
            print("======TRANSCRIPT", transcript)

            # Transcribe the captions into text
            # transcript = transcribe_captions(captions)

            # Answer the question based on the summary
            answer = answer_question(transcript, question)
            print("**********", answer)

            # Generate a summary using the HuggingFace model
            summary = generate_summary(transcript)
            print("---------SUMMARY", summary)

            # Display the results
            st.subheader("Summary:")
            st.write(summary)
```

```python
            st.subheader("Answer:")
            st.write(answer)
        else:
            st.warning("Please enter a valid YouTube URL.")
# Function to extract video ID from YouTube URL
def extract_video_id(url):

    # url = "https://www.youtube.com/watch?v=bZQun8Y4L2A&ab_
    channel=MicrosoftDeveloper"

    # Parse the query string from the URL
    parsed_url = urllib.parse.urlparse(url)
    query_string = urllib.parse.parse_qs(parsed_url.query)

    # Extract the video ID from the query string
    video_id = query_string["v"][0]

    # print(video_id)
    return video_id

# Function to get YouTube captions
def get_youtube_captions(video_id):
    # Get the captions for the specified video ID using the
    YouTubeTranscriptApi
    loader = YoutubeLoader(video_id, language="en")
    summarization_docs = loader.load_and_split()
    summarization_text = summarization_docs[0].page_content
    summarization_text = summarization_text[0:2000]

    # Remove unwanted strings using regular expressions
    cleaned_text = re.sub(r"\[.*?\]", "", summarization_text)
    # Remove square brackets and their contents
    cleaned_text = re.sub(r"\(.*?\)", "", cleaned_text)
    # Remove parentheses and their contents
    cleaned_text = re.sub(r"\'s", "", cleaned_text)
    # Remove apostrophe followed by 's'
    cleaned_text = re.sub(r"\w+://\S+", "", cleaned_text)
    # Remove URLs
```

```python
    # Remove unwanted characters
    unwanted_chars = ["'", '"', ",", ".", "!", "?"]
    cleaned_text = ''.join(c for c in cleaned_text if c not in
    unwanted_chars)

    # Remove extra whitespace
    cleaned_text = re.sub(r"\s+", " ", cleaned_text).strip()

    return cleaned_text

# Function to generate summary using HuggingFace model
def generate_summary(context):
    # Set up the HuggingFace API request
    headers = {"Authorization": "Bearer XXXXXXX "}

    model_input = f"Summarize the following transcript. " + \
        "Transcript: "  + context

    json_data = {
            "inputs": model_input,
            "parameters": {'temperature': 0.5,
                        'max_new_tokens': 100,
                        'return_full_text': False,
                        },
        }

    # HuggingFace response
    response = requests.post(SUMMARIZATION_ENDPOINT, headers=headers,
    json=json_data)
    json_response = json.loads(response.content.decode("utf-8"))
    print("-----------------------------------", json_response)
    summary = json_response[0]['generated_text']
    return summary

# Function to answer a question based on the context
def answer_question(context, question):
    # Set up the HuggingFace API request
    headers = {"Authorization": "Bearer XXXXX "}
```

```
    # context = clean_words_joined
    model_input = f"Answer the question based on the context below. " + \
        "Context: " + context + \
        " Question: " + question

    json_data = {
            "inputs": model_input,
            "parameters": {'temperature': 0.5,
                           'max_new_tokens': 100,
                           'return_full_text': False,
                           },
        }

    # Send a request to the HuggingFace Inference API
    response = requests.post(QA_ENDPOINT, headers=headers, json=json_data)
    # print("====================", response.json())

    # Extract and return the answer from the API response
    answer = response.json()[0]["generated_text"]
    return answer

if __name__ == "__main__":
    main()
```

Now, let's dissect the code and understand each block.

1. We will begin by importing the required libraries:

   ```
   import streamlit as st
   import requests
   import urllib.parse
   from langchain.document_loaders import YoutubeLoader
   import json
   import re
   ```

   We import the necessary libraries for our application, including Streamlit for building the user interface, requests for making API requests, `urllib` for parsing URLs, and other libraries for specific functionalities.

CHAPTER 8   BUILDING DEMO APPLICATIONS USING LLMS

2. Then, we will set API endpoints:

   SUMMARIZATION_ENDPOINT = https://api-inference.huggingface.co/models/tiiuae/falcon-7b-instruct

   QA_ENDPOINT = https://api-inference.huggingface.co/models/gpt2-large[35]

   We define the endpoints for the HuggingFace Inference API, which we will use for summarization and question-answering tasks.

3. Now let's define the Streamlit app:

   ```
   def main():
       st.title("YouTube Caption Summarizer and Q&A")

       # Get user input
       url = st.text_input("Enter the YouTube URL:")
       question = st.text_input("Ask a question based on the caption:")
   ```

   Here we defined the main function of our Streamlit app. It sets the app's title and prompts the user to enter a YouTube URL and a question based on the video caption.

4. Next, we define the button click event of the Streamlit app as shown here:

   ```
           if st.button("Summarize and Answer"):
               if url:
                   # Extract video ID from YouTube URL
                   video_id = extract_video_id(url)
                   print("VIDEO_ID", video_id)

                   # Get YouTube captions
                   transcript = get_youtube_captions(video_id)
                   print("======TRANSCRIPT", transcript)
   ```

```
            # Answer the question based on the summary
            answer = answer_question(transcript, question)
            print("**********", answer)

            # Generate a summary using the HuggingFace model
            summary = generate_summary(transcript)
            print("---------SUMMARY", summary)

            # Display the results
            st.subheader("Summary:")
            st.write(summary)

            st.subheader("Answer:")
            st.write(answer)
    else:
        st.warning("Please enter a valid YouTube URL.")
```

When the Summarize and Answer button is clicked in the previous code block, we check if a valid YouTube URL is provided. If so, we extract the video ID, get the YouTube captions, generate a summary of the transcript using the HuggingFace model, and answer the user's question based on the transcript. Finally, we display the generated summary and answer.

5. Now we will create some utility functions starting with extracting video ID from the YouTube URL as shown here:

```
def extract_video_id(url):
    parsed_url = urllib.parse.urlparse(url)
    query_string = urllib.parse.parse_qs(parsed_url.query)
    video_id = query_string["v"][0]
    return video_id
```

The previous function extract_video_id takes a YouTube URL as input, parses it to extract the video ID from the query string, and returns the video ID.

6. The following utility function is to get YouTube Captions/transcripts as described here:

```python
def get_youtube_captions(video_id):
    loader = YoutubeLoader(video_id, language="en")
    summarization_docs = loader.load_and_split()
    summarization_text = summarization_docs[0].page_content
    summarization_text = summarization_text[0:2000]
    cleaned_text = re.sub(r"\[.*?\]", "", summarization_text)
    cleaned_text = re.sub(r"\(.*?\)", "", cleaned_text)
    cleaned_text = re.sub(r"\'s", "", cleaned_text)
    cleaned_text = re.sub(r"\w+://\S+", "", cleaned_text)
    unwanted_chars = ["'", '"', ",", ".", "!", "?"]
    cleaned_text = ''.join(c for c in cleaned_text if c not in unwanted_chars)
    cleaned_text = re.sub(r"\s+", " ", cleaned_text).strip()
    return cleaned_text
```

The get_youtube_captions function takes a video ID as input and uses the YoutubeLoader from the langchain.document_loaders library to load the captions for the specified YouTube video. It cleans the text by removing unwanted strings, characters, URLs, and extra whitespace before returning the cleaned text.

7. Finally, we define a function to generate a summary using the HuggingFace model:

```python
def generate_summary(context):
    headers = {"Authorization": "Bearer XXXXXXX"}
    model_input = f"Summarize the following transcript. " + \
        "Transcript: " + context
    json_data = {
        "inputs": model_input,
        "parameters": {'temperature': 0.5,
                       'max_new_tokens': 100,
                       'return_full_text': False,
                      },
    }
```

CHAPTER 8   BUILDING DEMO APPLICATIONS USING LLMS

```
    response = requests.post(SUMMARIZATION_ENDPOINT,
    headers=headers, json=json_data)
    json_response = json.loads(response.content.
    decode("utf-8"))
    summary = json_response[0]['generated_text']
    return summary
```

The generate_summary function takes the cleaned transcript as input, sets up the API request to the Hugging Face Summarization API with the necessary headers and parameters, sends the request, and retrieves the generated summary from the API response.

8. Similarly, we define a function to answer a question based on context.

```
def answer_question(context, question):
    headers = {"Authorization": "Bearer XXXXX "}
    model_input = f"Answer the question based on the context 
    below. " + \
        "Context: "  + context + \
        " Question: " + question
    json_data = {
        "inputs": model_input,
        "parameters": {'temperature': 0.5,
                       'max_new_tokens': 100,
                       'return_full_text': False,
                      },
    }
    response = requests.post(QA_ENDPOINT, headers=headers, 
    json=json_data)
    answer = response.json()[0]["generated_text"]
    return answer
```

The previous answer_question function takes the cleaned transcript and the user's question as input, sets up the API request to the Hugging Face Text Generation API, sends the request, and retrieves the generated answer from the API response.

CHAPTER 8 BUILDING DEMO APPLICATIONS USING LLMS

9. Finally, we define the main execution of the app as shown here:

```
if __name__ == "__main__":
    main()
```

The previous code ensures that the `main()` function is executed when the script is run directly but not when imported as a module.

You can execute the previous application by running the following command:

`streamlit run Video-chatbot.py`

You can view the UI in your favorite browser on `http://localhost:8501/`, as shown in Figure 8-5.

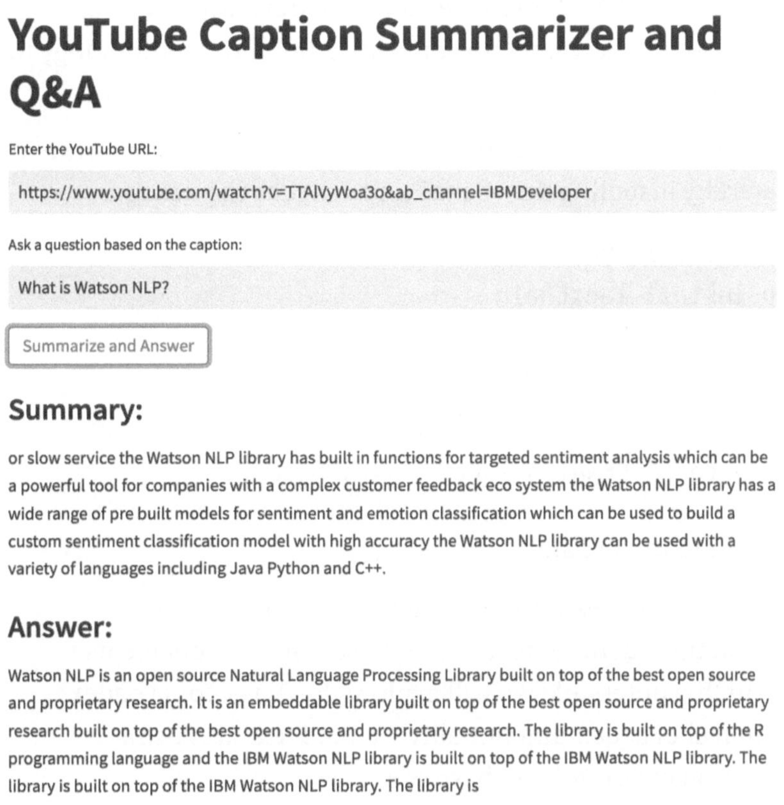

YouTube Caption Summarizer and Q&A

Enter the YouTube URL:

https://www.youtube.com/watch?v=TTAIVyWoa3o&ab_channel=IBMDeveloper

Ask a question based on the caption:

What is Watson NLP?

[Summarize and Answer]

Summary:

or slow service the Watson NLP library has built in functions for targeted sentiment analysis which can be a powerful tool for companies with a complex customer feedback eco system the Watson NLP library has a wide range of pre built models for sentiment and emotion classification which can be used to build a custom sentiment classification model with high accuracy the Watson NLP library can be used with a variety of languages including Java Python and C++.

Answer:

Watson NLP is an open source Natural Language Processing Library built on top of the best open source and proprietary research. It is an embeddable library built on top of the best open source and proprietary research built on top of the best open source and proprietary research. The library is built on top of the R programming language and the IBM Watson NLP library is built on top of the IBM Watson NLP library. The library is built on top of the IBM Watson NLP library. The library is

Figure 8-5. Video chatbot

CHAPTER 8   BUILDING DEMO APPLICATIONS USING LLMS

Following this code structure, we created a demo application allowing users to enter a YouTube URL and a question. Then, we generate a summary of the video transcript and answer the query using the Hugging Face models.

Video Transcript Analysis Using Langchain and OpenAPI

In this section, we will create a Jupyter notebook demonstrating the process of analyzing video transcripts using LLMs like GPT. The Langchain package is used for data loading, processing, and answering questions. The notebook covers various aspects such as installing the required packages, loading text files using `DirectoryLoader`, transcribing YouTube videos using YoutubeLoader, data processing with embeddings and chaining methods, question-answering using `RetrievalQA`, and post-processing techniques. Additionally, the notebook includes an example of transcript summarization using an external API. This notebook will equip you with a different methodology to perform the same task as in the "Video Caption Summarization and Q&A" section.

1. First, we need to install and set up the development environment. We start by installing the openai and langchain packages.

   ```
   !pip install openai
   !pip install langchain
   ```

   Next, we set the OpenAI API key:

   ```
   import os
   os.environ["OPENAI_API_KEY"] = "XXXXXXX"
   ```

2. Then, we load the data.

   Langchain provides various loaders that allow us to load and read unstructured text or files based on our specific use case. In this notebook, we will explore the `DirectoryLoader` and `YoutubeLoader` functionality. To load text files using `DirectoryLoader`, follow these steps:

   ```
   from langchain.document_loaders import DirectoryLoader
   text_loader = DirectoryLoader("", glob="*.txt")
   text_documents = text_loader.load_and_split()
   ```

The `DirectoryLoader` is initialized with the path to the directory containing the documents and the glob pattern to filter the files. In this example, we are loading all the text files (*.txt) located in the same directory as the script. The `load_and_split()` function starts the loading process.

For transcribing YouTube videos, we can use the `YoutubeLoader` module, which utilizes the `youtube_transcript_api` package.

```
from langchain.document_loaders import YoutubeLoader
from youtube_transcript_api import YouTubeTranscriptApi
yt_loader = YoutubeLoader(video_id="1rRt9uzWtqU", language="en")
transcripts = yt_loader.load_and_split()
YouTubeTranscriptApi.get_transcript("1rRt9uzWtqU")
```

The previous code snippet accepts the video ID and an optional language parameter (default: en) as inputs.

3. It's time for data processing.

   LLMs like GPT have limitations on the number of tokens they can handle, especially with larger documents. There are three common approaches to address these limitations.

   a. Using embeddings or a vector space engine

   b. Experimenting with different chaining methods, such as map-reduce or refine

   c. Combining both strategies

   The following example demonstrates the combination of embeddings with the "stuff" chaining method, consolidating all documents into a single prompt.

   To begin, we process our transcripts, referred to as `docs`, by ingesting them into a vector space using the `OpenAIEmbeddings` module. The resulting embeddings are stored in an in-memory embeddings database called Chroma.

   ```
   !pip install chromadb
   from langchain.embeddings.openai import OpenAIEmbeddings
   ```

```
from langchain.vectorstores import Chroma
embedding_model = OpenAIEmbeddings()
document_store = Chroma.from_documents(transcripts,
embedding_model)
```

Next, we specify the model_name we want to use for data analysis. In this case, we select gpt-3.5-turbo. You can find a complete list of available models in the documentation. The temperature parameter determines the sampling temperature, where higher values result in more random outputs, while lower values make the answers more focused and deterministic.

```
from langchain.chains import RetrievalQA
from langchain.chat_models import ChatOpenAI

language_model = ChatOpenAI(model_name="gpt-3.5-turbo",
temperature=0.9)

question_answer = RetrievalQA.from_chain_type(llm=language_model,
chain_type="stuff", retriever=document_store.as_retriever())
```

4. The next step is question-answering.
   We can now ask the model questions about our documents with the model and data prepared.

   ```
   query = "Who scored more goals in one Champions League season?"
   question_answer.run(query)
   ```

   ```
   query = "Who scored more goals in one season between Ronaldo and Haaland?"
   question_answer.run(query)
   ```

5. Then comes post-processing.
   In some cases, incomplete answers may be generated where the response abruptly ends after a few words. The token limitation of the model typically causes this. If the prompt given to the model is lengthy, it may not have sufficient remaining tokens to generate a complete answer. One approach to address this issue is to switch to a different chain type, such as refine.

CHAPTER 8   BUILDING DEMO APPLICATIONS USING LLMS

```
llm = ChatOpenAI(model_name="gpt-3.5-turbo", temperature=0.2)
question_answer = RetrievalQA.from_chain_type(llm=llm, chain_
type="refine", retriever=document_store.as_retriever())
```

However, it's worth noting that using a different chain type may result in less concrete results. Another way to handle these issues is to rephrase the question and make it more substantial.

6. Next comes the transcript summarization.
   To summarize a transcript, we can use various techniques. Here, we demonstrate using the Hugging Face API and the T5-based model google/flan-t5-xxl[36] to generate a summary for a given document.

```
import requests
import json
truncated_doc = str(summarization_docs[0])[0:1000]
API_URL = "https://api-inference.huggingface.co/models/google/
flan-t5-xxl"
headers = {"Authorization": "Bearer XXXXX "}

def query(payload):
    response = requests.post(API_URL, headers=headers,
    json=payload)
    return response.json()

model_input = f"Provide a summary for the following document:" +
truncated_doc

# Define the JSON data for the Hugging Face API
json_data = {
    "inputs": model_input,
    "parameters": {'temperature': 0.5, 'max_new_tokens': 300},
}

# Send a POST request to the Hugging Face API to get the model's
response
response = requests.post(API_URL, headers=headers, json=json_data)
json_response = json.loads(response.content.decode("utf-8"))
```

```
model_output = json_response[0]['generated_text']
model_output
```

The Langchain package offers a convenient way to analyze LLMs like GPT for text documents or transcripts with just a few lines of code. It's important to note that the package is still in its early stages, so updates and code changes can be expected. The provided code snippets in this notebook may be subject to modification.

When working with LLMs in your projects, it's essential to prioritize proper data cleaning, optimize token usage, and adhere to best practices such as setting budget limits or alarms to ensure optimal performance and resource management.

This equips you with two frameworks for extracting insights from the video (YouTube in this case) transcripts.

Summary

In this chapter, we explored various use cases of LLMs by leveraging open-source frameworks and the OpenAI API. We covered tasks such as question-answering and summarization for website analysis, PDF documents, and video transcripts. Refer to Table 8-1 for a summary of all the tasks covered.

Table 8-1. Summary of Tasks Covered

Use Case	Task	Model	Model Use	Library	Open Source?
Website analysis	Question-answering	BLOOM 165B	Hugging Face API	Hugging Face	✓
Website analysis	Summarization	Falcon-7B	Hugging Face API	HuggingFace	✓
PDF	Question-answering	OpenAI Embedding	OpenAI API	HuggingFace	✗
PDF	Summarization	Lamini-Flan-T5	Local	HuggingFace	✓

(continued)

Table 8-1. (*continued*)

Use Case	Task	Model	Model Use	Library	Open Source?
Video transcript	Summarization	Falcon-7B-instruct	Hugging Face API	HuggingFace	✓
Video transcript	Question-answering	GPT2-Large	Hugging Face API	HuggingFace	✓
Video transcript	Question-answering	GPT3.5 Turbo	OpenAI API	Langchain	✗
Video transcript	Summarization	FLAN-T5-XXL	Hugging Face API	HuggingFace	✓

For website analysis, we utilized the BLOOM 176B model for question-answering and the Falcon-7B model for summarization, both accessed through the Hugging Face API. These open-source frameworks allowed us to work efficiently with LLMs and achieve impressive results in understanding website content.

In the case of PDF documents, we experimented with the OpenAI Embedding model for question-answering using the OpenAI API. Additionally, we used the Lamini-Flan-T5 model, available locally through Hugging Face, for summarization. These approaches enabled us to quickly and accurately extract insights from PDF files.

For video transcript analysis, we employed the Falcon-7B-instruct model for summarization and the GPT2-Large model for question-answering, both accessed via the Hugging Face API. These open-source frameworks provided reliable solutions for understanding video content.

Finally, we explored the capabilities of the GPT3.5 Turbo model for question-answering using the OpenAI API in conjunction with the Langchain library. Additionally, we utilized the FLAN-T5-XXL model for summarization through the Hugging Face API. These experiments showcased the potential of integrating LLMs with open-source and API-based solutions for various use cases.

By utilizing these diverse models and frameworks, we demonstrated the flexibility and power of LLMs in solving real-world problems and building practical applications. Whether through open-source libraries or API services, LLMs offer a wide range of natural language understanding and generation possibilities, making them an invaluable asset for any data scientist and AI developer.

CHAPTER 9

Building Enterprise-Grade Applications Using LLMs

In the preceding chapter, you explored the incredible capabilities of large language models (LLMs) and their practical application in crafting demo applications. You discovered how LLMs can make sense of website content through data scraping, question-answering, and summarization. Additionally, you learned to uncover insights from PDF documents and video transcripts, delving into video caption summarization and Q&A.

Now, we will take a significant leap forward, venturing into the world of enterprise-grade applications powered by LLMs. In this chapter, we shift gears and immerse ourselves in the complexities of developing applications that transcend the boundaries of simple demos. Enterprise-grade applications demand a higher level of sophistication, scalability, and adaptability to meet the exacting standards of the business world. Our journey together will explore the art of harnessing LLMs to develop robust solutions that address the diverse needs of enterprises.

We understand that building enterprise-grade applications requires a deep understanding of domain-specific challenges and a commitment to delivering excellence. To equip you with the necessary expertise, we will unravel advanced techniques and best practices that empower you to meet the unique demands of the corporate landscape.

As we progress through this chapter, you will gain insights into extracting meaningful insights from unstructured data with question-answering and into crafting compelling user interfaces that enhance user experiences.

By the time you reach the end of this chapter, you will possess the knowledge and skills to develop sophisticated enterprise-grade applications using LLMs. These applications will empower businesses to make informed decisions, automate processes, and unlock hidden value from vast amounts of data.

Let's embark together on this transformative journey of building enterprise-grade applications using LLMs. We will unlock the full potential of LLMs, harnessing their power to revolutionize enterprise solutions and shape the future of artificial intelligence in the corporate world.

Retrieval-Augmented Question-Answering Chatbot

Imagine the excitement that swept through the online world when LLMs like ChatGPT burst onto the scene. These web-based chatbots possessed an astonishing ability to understand and generate text, leaving people, myself included, genuinely amazed.

Eager to explore their capabilities, many of us logged in to put these LLMs to the test. We asked them to compose haikus, draft motivational letters, or craft email responses. We quickly discovered that LLMs excelled in generating imaginative content and tackling everyday language tasks and other natural language processing challenges.

As the hype around LLMs grew, people started contemplating ways to integrate them into their applications. However, it soon became apparent that merely developing a simple wrapper around an LLM API wouldn't guarantee success. To truly make an impact, offering something beyond what the LLMs already provided was essential.

One major hurdle faced by LLMs is what we call the *knowledge cutoff*. This means that LLMs need to gain awareness of any events or information after training. So, if you were to ask ChatGPT (based on GPT-3.5) about an event in 2023, it wouldn't be able to give you an accurate response.

These limitations, though significant, didn't dampen the enthusiasm for LLMs. Instead, they fueled the quest to enhance their capabilities and find innovative ways to leverage their strengths while overcoming their cutoff knowledge challenge.

The same problem will occur if you ask an LLM about an event that isn't included in its training dataset. The LLM's knowledge is limited to the information it was exposed to during its training, which means it has a knowledge cutoff date. This date is significant when it comes to publicly available information. However, it's important to note that the LLM does not know private or confidential data that might exist even before the knowledge cutoff.

Let's consider a scenario where a company wants to utilize a customized LLM to answer specific questions about their confidential information. In this case, the LLM's training data would not include that private information, making it unable to provide accurate responses.

On the other hand, the publicly available information that the LLM is aware of might already need to be updated. The world constantly evolves, and new events and discoveries occur daily. The LLM's training might have yet to encompass the latest updates, which could impact the accuracy and relevance of its responses to current events.

These limitations highlight the importance of understanding the context and scope of the LLM's knowledge when seeking answers or utilizing it for specific applications. It's crucial to evaluate the LLM's training data, consider the relevance of the information, and recognize the potential gaps in its understanding of both public and private domains.

LLMs pose another challenge: their ability to generate text that sounds remarkably realistic, even if it could be less accurate. Spotting invalid information can be pretty tricky, particularly regarding missing data. In such cases, there's a high likelihood that the LLM will conjure up an answer that sounds convincing but is ultimately incorrect rather than acknowledging its lack of foundational facts. This phenomenon is known as *hallucination* in the LLM realm.

To illustrate, let's explore the realm of research and court citations. Verifying the accuracy of such sources is of utmost importance. At the time of writing this chapter, a lawyer found themselves in trouble because they unthinkingly trusted the court citations generated by ChatGPT. Without conducting the necessary due diligence, the lawyer relied solely on the sources provided by the LLM. Unfortunately, this led to unforeseen complications, highlighting the potential pitfalls of unquestioningly accepting information generated by LLMs.

This real-world example is a stark reminder of the need to exercise caution and critical thinking when relying on LLMs. While LLMs excel in crafting text that appears genuine, it's vital to independently verify the accuracy and validity of the information they produce. Especially in areas like legal research, where precision is paramount, unthinkingly accepting LLM-generated citations can lead to severe consequences. Conducting thorough fact-checking and corroborating information from multiple reliable sources is essential to ensure accuracy and avoid potential pitfalls.

The retrieval augmented approach takes a different approach, utilizing the power of LLMs to generate answers based on relevant documents from your data source. Figure 9-1 shows the simplified version of the retrieval-augmented generation (RAG) pipeline.

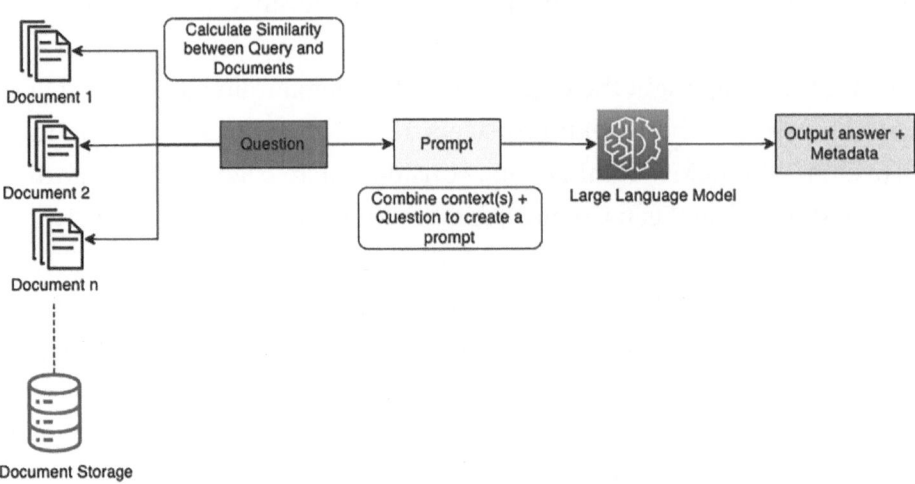

Figure 9-1. Simplified RAG pipeline

When a user asks a question, the first step is to calculate the similarity between the query and documents in your document database. Then you combine the extracted document as context with the question to create your prompt. Finally, you pass the prompt to the LLM to generate an answer that is truthful and based on the augmented knowledge from your database. We will dive into the detailed components and implementation steps of a RAG pipeline using an LLM in the following sections of this chapter.

Rather than relying solely on the internal knowledge of the LLM, this method leverages the LLM's capability to augment information from the provided documents. This way, you can access specific and contextual information beyond what the LLM has been trained on.

To illustrate, let's consider the ChatGPT browser plugins as an example of a retrieval-augmented approach to LLM applications. When you enable a browsing plugin in the ChatGPT interface, it allows the LLM to search the Internet for real-time and up-to-date information, which it can then incorporate into its final answer.

With the browsing plugin enabled, ChatGPT can answer questions like who won the Oscars for different categories in 2023. However, it's important to note that ChatGPT's internal knowledge is limited to information up until 2021. ChatGPT utilizes the browser plugin to access external information and provide answers based on current data to overcome this knowledge cutoff. These browser plugins serve as integrated augmentation mechanisms within the OpenAI platform, enhancing the capabilities of ChatGPT and enabling it to deliver answers enriched with real-time information.

By adopting a retrieval-augmented approach, you can harness the strengths of LLMs while leveraging external resources to ensure the most accurate and up-to-date responses. It opens exciting possibilities for accessing dynamic information and keeping pace with the ever-evolving world.

Real-World Use Cases of Retrieval Augmentation Generation

Let's review some of the real-world use cases of RAG use cases before diving into the architecture.

- **Up-to-Date Customer Support:** Customer support systems can benefit from retrieval augmentation to access the most up-to-date product information and user manuals. This helps the LLMs give customers accurate answers to product queries. For instance, a tech company could use this technology to inform customers about the latest software updates or troubleshooting steps.

- **Financial Reporting and Analysis:** Financial analysts use retrieval-augmented generation to access real-time market data, regulatory updates, and company financials. This enables them to generate up-to-date reports, forecasts, and analyses that reflect the current financial landscape and make better-informed decisions.

- **Healthcare Information Synthesis:** Healthcare professionals can enhance their understanding and knowledge by utilizing LLMs, augmented with patient records, clinical guidelines, and recent medical research. This synthesized information can be valuable in aiding diagnosis, planning treatments, and keeping up-to-date with the latest medical advancements while maintaining patient privacy.

- **Legal Research:** Lawyers and legal researchers use RAG to access and interpret vast repositories of legal documents. This enables them to conduct thorough research by referencing current laws, case precedents, and legal articles, streamlining the legal research process.

- **Content Creation and Curation:** In media and content creation, retrieval augmentation enables LLMs to incorporate relevant facts, statistics, and references, guaranteeing accurate and well-supported content. This is particularly useful in journalism, where factual accuracy is critical.

- **Technical Documentation:** Technical writers use LLMs with retrieval augmentation to automatically update documentation by pulling in the latest technical specs and product updates. This ensures that manuals, help guides, and FAQs remain accurate as products evolve.

- **Supply Chain Management:** LLMs can predict supply chain disruptions and respond proactively by augmenting with live data feeds from logistics platforms, inventory systems, and market conditions.

RAG Architecture

Figure 9-2 shows the main components of a RAG architecture.

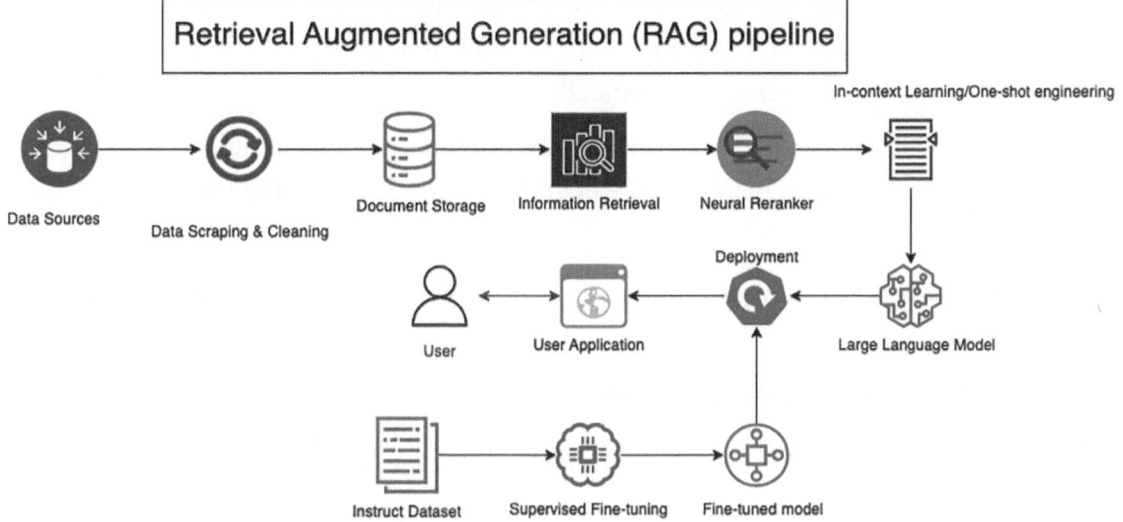

Figure 9-2. *RAG architecture*

These components serve as the building blocks of the RAG pipeline. Also, these components are interchangeable LEGO pieces, and you have the ability to choose from a variety of options available. For example, the information retrieval could be done using Elasticsearch, Solr, Watson Discovery, or any other retriever. Here is a brief description of the main components:

- **Document Store:** All the documents are first scraped and cleaned to store in this document storage.

- **Information Retrieval:** A retriever is used to search for the most relevant document based on the user query.

- **Reranker:** The reranker component finds the most relevant document for the user query.

- **Language Model:** The large language model generates/extracts the answer from the document provided by the retriever.

- **User Application:** This acts as the interface between the end user and the pipeline, allowing the user to enter their question and have the answer generated automatically.

- **Fine-tuning (Optional):** You can also implement a fine-tuning pipeline using an instruct dataset and feedback, as described in Chapter 7.

We will look into each component in detail in the following section of this chapter.

Creating a Knowledge Base

Creating a knowledge base is essential for integrating large language models to build enterprise-grade applications. With the increasing amount of data generated daily, it is becoming increasingly difficult for businesses to keep track of everything. A knowledge base is a central repository of information that can help businesses organize, manage, and share their knowledge effectively. You can leverage this knowledge to help the LLM to avoid hallucination, as you learned in Chapter 7. Figure 9-3 shows the full architecture diagram for the RAG pipeline. However, in this section, we will focus on the highlighted section in the rectangle.

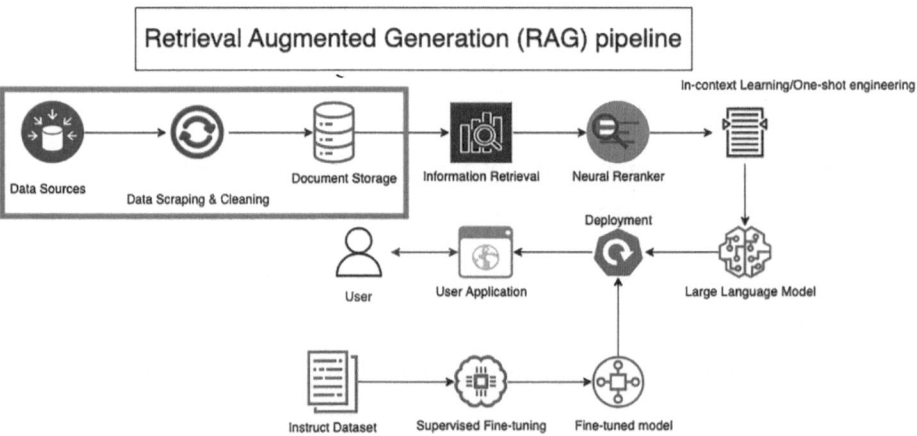

Figure 9-3. RAG knowledge base

Step 1: Collecting Data

Let's begin by collecting data for building the knowledge base. You will create a corpus for user manuals. You can obtain these manuals from http://www.manualsonline.com/.

You can either write your scraping script or download these scraped manuals from this Google Drive: https://drive.google.com/drive/folders/1-gX1DlmVodP6OVRJC3W BRZoGgxPuJvvt.

Once you have downloaded the data, you can extract the one zip file that you have downloaded from Google Drive to get 16 more zip files. Please extract all the zip files in the same directory so you can index them all at once in the following steps.

Step 2: Setting Up a Retriever

We will use Solr as a retriever in this chapter. The Solr retriever acts as a knowledgeable assistant, swiftly retrieving relevant information to answer your queries precisely and accurately. However, you can use any other retriever like Elasticsearch, Apache Lucene, Amazon CloudSearch, etc.

Step 2.1: Downloading a Solr Binary File

Download the binary release for Solr from https://solr.apache.org/downloads.html.

Step 2.2: Install Solr

Extract the Solr distribution archive to your local home directory. Ensure that the command matches with the Solr version you are using.

```
cd ~/
tar zxf solr-9.2.1.tgz
```

Step 2.3: Start the Solr Cloud Service Locally

Start Solr locally in interactive mode.

```
./bin/solr start -e cloud
```

The script prompts you to specify the number of Solr nodes for your local cluster. It is set to 2 by default, but you can modify this value to meet your specific needs. This interactive feature enables easy customization of your local Solr cluster setup.

After starting up all nodes in the cluster, the script prompts you for the name of the collection to create, where you can enter `Manuals` as the name of your collection.

Then the script will prompt you for the number of shards to distribute the collection across and the number of replicas to create for each shard. You can leave the default of 2 in both prompts.

Lastly, the script will ask you to specify the configuration directory name for your collection. You have three options to choose from: `basic_configs`, `data_driven_schema_configs`, or `sample_techproducts_configs`.

You can select `sample_techproducts_configs` as your configuration for the Manuals collection unless you want to create your config with a schema from scratch.

To observe the deployment status of your collection across the cluster, navigate to the Solr Admin UI's cloud panel. You can access it by visiting http://localhost:8983/solr/#/~cloud.

Step 2.4: Index the Documents

The purpose of indexing is to organize and structure the data, making it searchable and allowing for faster and more accurate querying within the Solr search engine. Indexing documents in Solr involves adding content to the Solr index, enabling efficient searching and retrieval of information.

You can index all 55,000 manuals using the following command:

```
./bin/post -c Manuals /Users/shivamsolanki/Desktop/Generative-AI-Book/Chapter-9/EManuals_Corpus/*
```

Please ensure you provide the correct directory with all the zip files. This process will take around three hours, so you can take a break and return later to check the status for completion. You can observe a similar message.

```
124476 files indexed.
COMMITting Solr index changes to http://localhost:8983/solr/Manuals/update...
Time spent: 3:03:03.075
```

You can also observe the indexed documents on the Solr UI by copying the URL from your terminal and opening it in the browser, as shown in Figure 9-4.

CHAPTER 9   BUILDING ENTERPRISE-GRADE APPLICATIONS USING LLMS

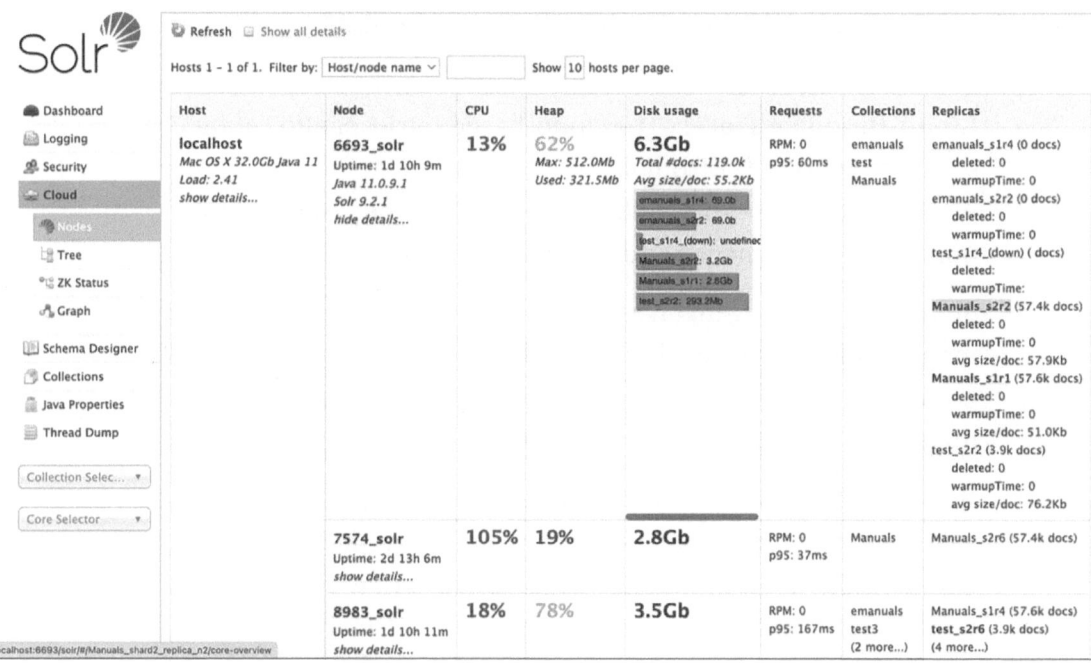

Figure 9-4. Solr Indexing UI

Scaling Up Indexing for Enterprise-Grade Knowledge Bases

The steps outlined earlier explain how to set up Solr in a local development environment. However, this method has constraints when dealing with large amounts of data. Indexing tens of thousands of documents is feasible, but enterprise knowledge bases can consist of millions or even billions of documents. Running Solr on a single machine may lead to challenges in handling such massive volumes of data.

These are the limitations of single-machine Solr:

- **Indexing Time:** Processing and indexing large datasets on a single machine can take days or weeks.

- **Performance Bottlenecks:** Memory and processing power limitations on a single machine can lead to slow query responses, impacting user experience.

- **Scalability:** Adding more data requires significant hardware upgrades, becoming less practical as the knowledge base grows.

To overcome these limitations and handle enterprise-grade data volumes, consider these scaling-up strategies:

- **Cloud-Based Retriever:** The cloud-based retriever allows you to scale your cluster horizontally by adding more nodes. Each node in the cluster shares the indexing and serving load, significantly improving performance and scalability.

- **Sharding:** Sharding involves dividing your data into smaller, more manageable chunks. Each shard is then replicated across multiple nodes, distributing the indexing and querying load across the cluster, improving efficiency and performance.

- **Replication Factor:** Replication refers to creating copies of each shard across multiple nodes. This ensures redundancy and fault tolerance. If one node fails, another replica can serve the data, minimizing downtime and maintaining data availability.

These are some additional considerations:

- **Hardware Resources:** Scaling a retriever requires adequate hardware resources for each node in the cluster. Consider factors like CPU cores, memory, and storage capacity to handle the anticipated data volume and query load.

- **Monitoring and Optimization:** Continuously monitor your cluster's performance and resource utilization. Implement automated scaling strategies and optimize Solr configurations to ensure optimal performance as your knowledge base grows.

By leveraging cloud-based retriever, sharding, and replication strategies, you can build a robust and scalable knowledge base infrastructure that efficiently handles massive data volumes for your enterprise-grade RAG applications.

Setting Up a Retrieval System

Now that we have created a knowledge base, we will set up the information retrieval component, as shown in Figure 9-5.

CHAPTER 9   BUILDING ENTERPRISE-GRADE APPLICATIONS USING LLMS

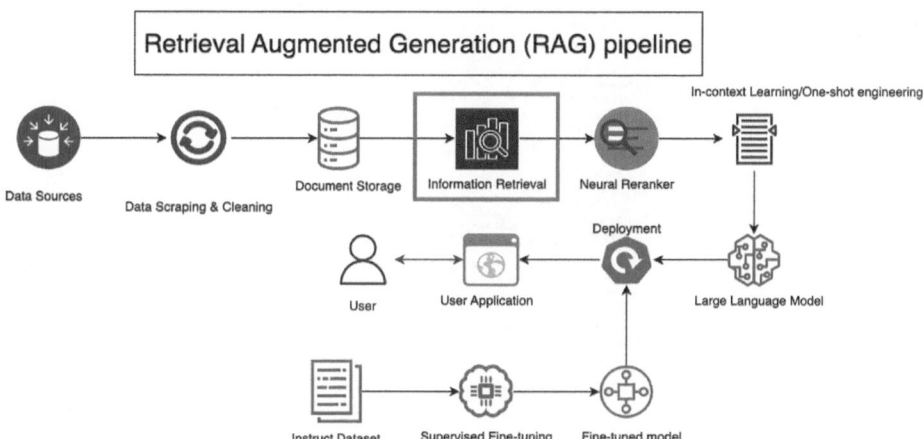

Figure 9-5. *RAG information retrieval*

When you perform a search in Solr, it uses a clever mechanism to find the most relevant documents for your query. First, Solr analyzes your search query and compares it with the indexed data to identify the documents that match your search terms. Then, it ranks these documents based on various factors like how frequently the search terms appear in the document. This ensures that the most relevant results are displayed to you, making your search experience faster and more accurate.

You can search for a single term using a query like this:

```
curl "http://localhost:8983/solr/Manuals/select?q=Nokia"
```

Or you can also do a query directly from the Solr UI, as shown in Figure 9-6.

CHAPTER 9   BUILDING ENTERPRISE-GRADE APPLICATIONS USING LLMS

Figure 9-6. *Solr Query UI example*

We collected and created a knowledge base for manuals in the previous step. So, when you search for Nokia, you will observe 1,463 documents containing mention of Nokia. Alternatively, you can review some documents in the zip file and search for another product.

Now it's time to start building our demo app for retrieval-augmented question-answering chatbot. Let's start with the retriever part of the app. First define a function called solr_retriever() with user input for querying.

```
import requests
import re
import pandas as pd

def solr_retriever(question):
    solr_url = f'http://localhost:7574/solr/Manuals/select?q={question}&q.op=AND&wt=json'
    response = requests.get(solr_url)
    query_results = response.json()
    
    total_documents = query_results['response']['numFound']
    print(f"{total_documents} documents found.")
    
    results_list = []
```

```python
    if total_documents > 0:
        total_documents = min(total_documents, 10)
        for i in range(total_documents):
            content_unicode = query_results['response']['docs'][i]
            ['content'][0]
            content_decoded = content_unicode.encode("ascii", "ignore").
            decode()
            keyword = "{: shortdesc} "
            cleaned_text = skip_unwanted_characters(content_decoded,
            keyword)
            pattern = r'\{\s*:\s*[\w#-]+\s*\}|\{\s*:\s*\w+\s*\}|\n\s*\n'
            cleaned_text = re.sub(pattern, '', cleaned_text)
            cleaned_text = preprocess_text(cleaned_text)

            document_info = {
                "document": {
                    "rank": i,
                    "document_id": query_results['response']['docs'][i]
                    ['id'][0],
                    "text": cleaned_text[1000:3000],
                },
            }
            results_list.append(document_info)

        results_to_display = [result['document'] for result in
        results_list]
        df = pd.DataFrame.from_records(results_to_display, columns=['rank',
        'document_id', 'text'])
        df.dropna(inplace=True)
print('=========================================================
===========')
print(f'QUERY: {question}')
return results_list
```

Block 1: URL Construction and Request

```
solr_url=f'http://localhost:7574/solr/Manuals/select?q={question}&q.
op=AND&wt=json'
response = requests.get(solr_url)
query_results = response.json()
```

In this block, a URL is constructed to make a request to a Solr server. The URL includes the provided question as a query parameter. The `requests.get()` function sends a GET request to the Solr server using the constructed URL. The response from the server is then converted to JSON format and stored in the `query_results` variable.

Block 2: Document Information Extraction

```
total_documents = query_results['response']['numFound']
print(f"{total_documents} documents found.")
```

This block retrieves the total number of documents found in the Solr search results from the `query_results`. It is accessed using the appropriate keys in the JSON structure. The total number of documents is then printed.

Block 3: Search Result Processing

```
results_list = []
if total_documents > 0:
    total_documents = min(total_documents, 10)
    for i in range(total_documents):
        content_unicode = query_results['response']['docs'][i]
        ['content'][0]
        content_decoded = content_unicode.encode("ascii", "ignore").
        decode()
        keyword = "{: shortdesc} "
        cleaned_text = extract_desired_text(content_decoded, keyword)
        pattern = r'\{\s*:\s*[\w#-]+\s*\}|\{\s*:\s*\w+\s*\}|\n\s*\n'
        cleaned_text = re.sub(pattern, '', cleaned_text)
        cleaned_text = clean_text(cleaned_text)

        document_info = {
```

```
            "document": {
                "rank": i,
                "document_id": query_results['response']['docs'][i]
                ['id'][0],
                "text": cleaned_text[1000:3000],
            },
        }
        results_list.append(document_info)
```

This block processes the search results if there are documents found. It limits the total number of documents to a maximum of 10 (or the actual number if it's less). It then iterates over the document indices using a `for` loop.

For each document, it retrieves the content in Unicode format, encodes it to remove non-ASCII characters, and decodes it back to a string. The `extract_desired_text` function is called to extract the desired text from the content based on a keyword.

A regular expression pattern is defined to remove specific patterns from the cleaned text. The `re.sub` function substitutes the pattern with an empty string. The `clean_text` function is called to clean the text further.

The relevant information for each document, including the rank, the document ID, and a portion of the cleaned text, is stored in a dictionary `document_info`. This dictionary is then appended to the `results_list`.

Block 4: Results Display

```
results_to_display = [result['document'] for result in results_list]
df = pd.DataFrame.from_records(results_to_display, columns=['rank',
'document_id', 'text'])
df.dropna(inplace=True)
```

This block prepares the results for display by extracting the document key from each item in `results_list` and storing them in `results_to_display`. A Pandas DataFrame called `df` is created from these records with specified column names. Any rows with missing values are dropped from the DataFrame.

Block 5: Output and Return

```
print(f'QUERY: {question}')
return results_list
```

CHAPTER 9   BUILDING ENTERPRISE-GRADE APPLICATIONS USING LLMS

This block prints a separator line and the original query. Finally, the `results_list` is returned as the output of the `solr_retriever` function.

Therefore, in summary, the `solr_retriever` function retrieves and processes search results from a Solr server based on a provided question. It extracts relevant information from each document, cleans the text, and stores the results in a list. The processed results are displayed and returned.

Neural Reranker

A neural reranker is a component used in the retrieval augmented approach of information retrieval systems. It is designed to enhance the ranking of search results generated by a base retrieval system, such as a traditional search engine. The neural reranker utilizes advanced machine learning techniques, often based on neural networks, to re-evaluate and rerank the initial search results, improving their relevance and ensuring that the most relevant results are presented to the user. This helps to enhance the overall search experience and increase the accuracy of search result rankings. Figure 9-7 shows the reranker component in the RAG pipeline.

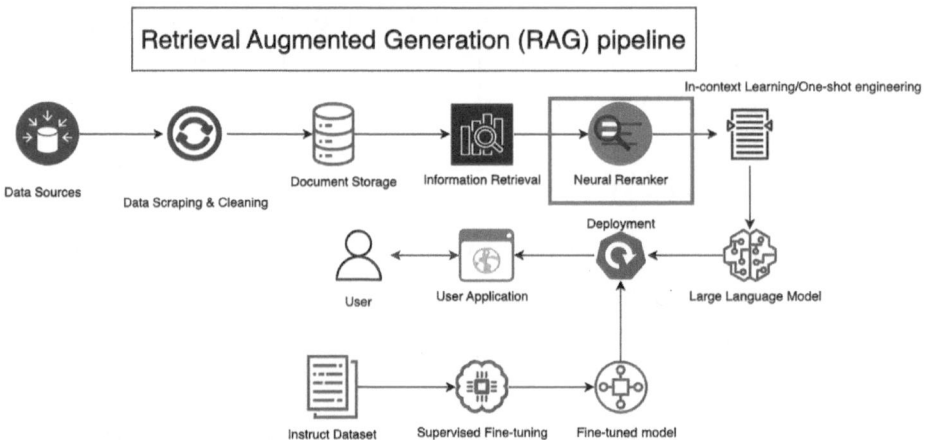

Figure 9-7. *RAG reranker*

Figure 9-8 shows an example of the importance of reranker. All the product documentation was first indexed, and then a simple question was asked: "What is IBM Cloud Pak for Data?" The response from the retriever and reranker is in the left and right columns, respectively.

Before	After
Palantir for IBM Cloud Pak for Data enables building no-/low-code line of business applications using data, machine learning, and optimization from IBM Cloud Pak for Data. Ontology managers can define business-oriented data models integrating data from IBM Cloud Pak for Data. Application builders can use Palantir tools to create applications using these data models. Additionally, applications can integrate machine learning models from IBM Cloud Pak for Data to infuse predictions, as well as decision optimization result data from IBM Cloud Pak for Data to determine optimized actions based on data and predictions.\nThis blog post explains how to create AI-infused apps using Palantir ontology and application building tools together with IBM Cloud Pak for Data model deployments and data and AI catalog. ....[continued]	Collect, Govern and Analyze Hadoop Data using IBM Cloud Pak for Data - Written by Basem Elasioty and Gregor Meyer \n IBM Cloud Pak for Data is a cloud-native solution that enables data scientists, data engineers and business experts to collaborate defining, developing, validating and deploying analytic products. It provides the needed tools for data science projects including connectivity to various data sources, governing and organizing data, training and evaluating AI and ML models and finally deploying these models and infuse them into business processes. ....[continued]

Figure 9-8. *Reranker before and after*

In Figure 9-8, you can observe that before implementing the reranker, the first document identified by the retriever contained information about the IBM Cloud Pak for Data product but was not directly relevant to answering the question. However, after implementing the reranker, it identified the document that contains the definition of IBM Cloud Pak for Data and is more relevant to this question.

DrDecr Reranker

The ColBERT-based DrDecr model is trained using a supervised learning approach. This means that the model is trained on a dataset of labeled data. The labeled data consists of pairs of queries and documents, where the documents are ranked according to their relevance to the queries.

The model is trained to predict the ranking of a document given a query. The model does this by first encoding the query and the document using a transformer encoder. The transformer encoder is a neural network that is specifically designed for natural language processing tasks.

Once the query and document have been encoded, the model uses a scoring function to predict the ranking of the document, as shown in Figure 9-9. The scoring function is a neural network that is trained to predict the probability that a document is relevant to a query.

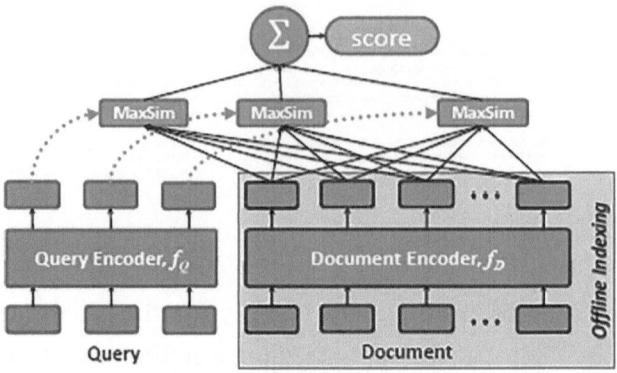

Figure 9-9. *Colbert reranking (source: https://paperswithcode.com/paper/ colbert-efficient-and-effective-passage)*

The model is trained using a loss function that measures the difference between the predicted ranking and the ground truth ranking. The loss function is minimized using an optimization algorithm, such as stochastic gradient descent.

Let's implement the DrDecr[37] reranker to get the most relevant document after getting results from Solr.

```
def drdecr_reranker(question, max_reranked_documents=10):
    reranker = ColBERTReranker(model=model_name_or_path)
    reranker.load()

    search_results = solr_retriever(question)
    if len(search_results) > 0:
        reranked_results=reranker.predict(queries=[question],
        documents=[search_results], max_num_documents=max_reranked_
        documents)

        print(reranked_results)

        reranked_results_to_display = [result['document'] for result in
        reranked_results[0]]
        df = pd.DataFrame.from_records(reranked_results_to_display,
        columns=['rank', 'document_id', 'text'])
        print('=========================================================
        =============')
        print(f'QUERY: {question}')
```

```
        return df['text'][0]
else:
    return "0 documents found", "None"
```

The previous function can be described as follows:

1. The drdecr_reranker function takes a question as input and an optional parameter for the maximum number of reranked documents.

2. An instance of the ColBERTReranker class is created and assigned to the variable reranker. This class is responsible for reranking the search results using a specific model.

3. The load method is called on the reranker object to load the model.

4. The solr_retriever function is called to retrieve the search results for the given question. The results are stored in the search_results variable.

5. If search results are available (i.e., the length of search_results is greater than zero), the reranker's predict method is called. The original question, search results, and a maximum number of reranked documents are passed as arguments.

6. The reranked results are printed for debugging or informational purposes.

7. The reranked results are then processed to extract relevant information, such as the rank, document ID, and text. These results are stored in the reranked_results_to_display list.

8. A Pandas DataFrame, df, is created from the reranked_results_to_display list with columns named rank, document_id, and text.

9. A separator line is printed to separate the reranked results from the original query. The original query is printed.

CHAPTER 9   BUILDING ENTERPRISE-GRADE APPLICATIONS USING LLMS

10. Finally, the first document's text is returned as the output if reranked results are available. Otherwise, a message indicating no documents were found is returned along with the value None.

Overall, this code block implements the DrDecr reranking algorithm by utilizing the `ColBERTReranker` class. It loads the reranker model, retrieves search results, reranks them, and returns the reranked text of the first document if available.

Generative LLM

We have learned about in-context learning using LLMs in Chapter 7. Let's refresh our learning about in-context learning.

In-context learning using LLMs refers to the ability of these models to acquire new knowledge or adapt their responses based on specific contexts or prompts. It involves updating the model with additional data or examples related to a particular domain or task. Figure 9-10 shows the in-context learning component of the RAG pipeline.

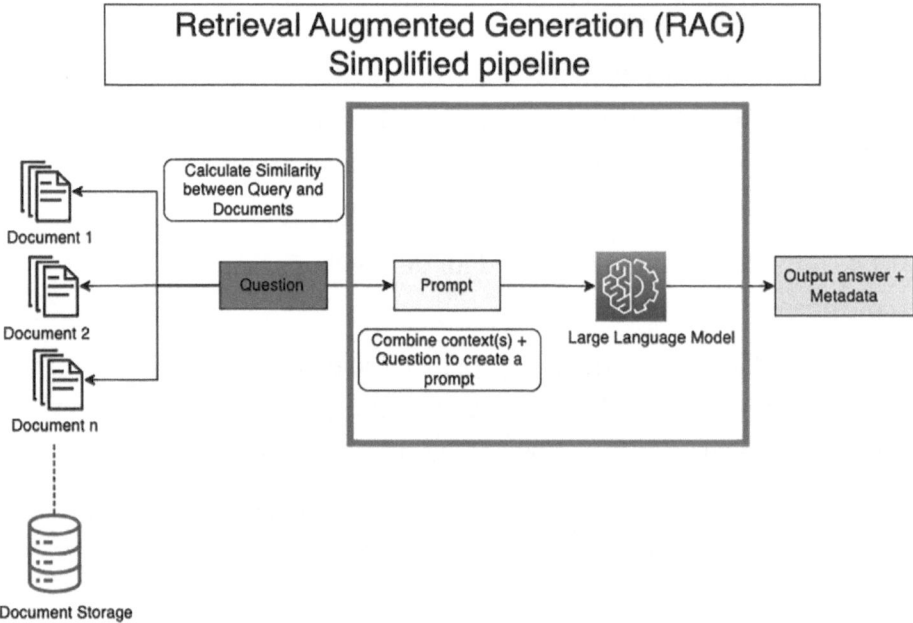

Figure 9-10. In-context learning

CHAPTER 9   BUILDING ENTERPRISE-GRADE APPLICATIONS USING LLMS

LLMs like GPT-3 have been pre-trained on vast amounts of text from the Internet, which gives them a broad understanding of language and knowledge. However, they might need more domain expertise or up-to-date information. In-context learning allows these models to specialize and refine their responses within a given context.

In-context learning helps improve the accuracy, relevance, and safety of the model's responses in a particular domain. It allows the model to adapt to various tasks, such as providing medical advice, answering legal queries, or assisting in customer support. Large language models can provide more tailored and appropriate responses by incorporating context-specific information, making them more valuable and reliable in real-world applications.

We will use the concept of In-context learning to build a chatbot that can answer custom queries from the context provided by the retriever-reranker pipeline created in the previous steps.

Let's take an example passage from our document corpus and ask a question based on the passage.

```
output = query({
    "inputs": "Answer the question based on the context below. \
    Context: HIGHLIGHTS Here are just some of the things you can do
    with the X700 mobile phone. Send colour pictures and video clips to
    friends and family via Multimedia messaging and e-mail messaging. Send
    pictures, video, electronic business cards, and more using Bluetooth ,
    Infrared , or USB connections to other phones or devices.. Swap
    content between appropriate devices with the included miniSD Card
    and adapter. Surf the Internet using your XHTML configured wireless
    Internet browser. Add frames, text, and pop art to VGA size images.
    Capture video clips . Play 40 tone polyphonic ringtones . No more
    wired headset! Use a Bluetooth ® wireless headset. Synchronise your
    Contacts and Calendar with your other networked devices . See who is
    calling with picture Caller Line ID on the X700 external display.
    Download games , images , themes, sounds , videos , and Java ™ and
    Series 60 applications . View all your media files in one place , the
    Media Gallery. Search for your files ; including games, images, install
    packages, and video and sound clips. Receive, edit , and view Microsoft
    ® Word, Excel, and PowerPoint ® files . Beam documents to printers
    and projectors (with appropriate accessory). Speak a name or number
```

CHAPTER 9 BUILDING ENTERPRISE-GRADE APPLICATIONS USING LLMS

```
        to dial the number using Voice Commands. Say "open Messages" or "open
        Camera" and your phone will go directly to the application using Voice
        Commands. 1 \
        Question: What are the capabilities of Panasonic X700?",
        "parameters": {'temperature': 0.5,
                    'max_new_tokens': 200,
                      'return_full_text': False,},
})
output
```

The generated answer is shown here and in Figure 9-11.

```
Panasonic X700 has built-in Bluetooth , Infrared , USB and miniSD Card
ports. It is capable of sending pictures, video, electronic business cards,
and more using Bluetooth , Infrared , or USB connections to other phones
or devices. It can also swap content between appropriate devices with the
included miniSD Card and adapter.
```

In-Context Learning Example

Context

HIGHLIGHTS Here are just some of the things you can do with the X700 mobile phone. Send colour pictures and video clips to friends and family via Multimedia messaging and e-mail messaging. Send pictures, video, electronic business cards, and more using Bluetooth , Infrared , or USB connections to other phones or devices.. Swap content between appropriate devices with the included miniSD Card and adapter. Surf the Internet using your XHTML configured wireless Internet browser. Add frames, text, and pop art to VGA size images. Capture video clips . Play 40 tone polyphonic ringtones . No more wired headset! Use a Bluetooth ® wireless headset. Synchronise your Contacts and Calendar with your other networked devices . See who is calling with picture Caller Line ID on the X700 external display. Download games , images , themes, sounds , videos , and Java ™ and Series 60 applications . View all your media files in one place , the Media Gallery. Search for your files ; including games, images, install packages, and video and sound clips. Receive, edit , and view Microsoft ® Word, Excel, and PowerPoint ® files . Beam documents to printers and projectors (with appropriate accessory). Speak a name or number to dial the number using Voice Commands. Say "open Messages" or "open Camera" and your phone will go directly to the application using Voice Commands. 1

Question

What are the capabilities of Panasonic X700?

Answer

Panasonic X700 has built-in Bluetooth , Infrared , USB and miniSD Card ports. It is capable of sending pictures, video, electronic business cards, and more using Bluetooth , Infrared , or USB connections to other phones or devices. It can also swap content between appropriate devices with the included miniSD Card and adapter.

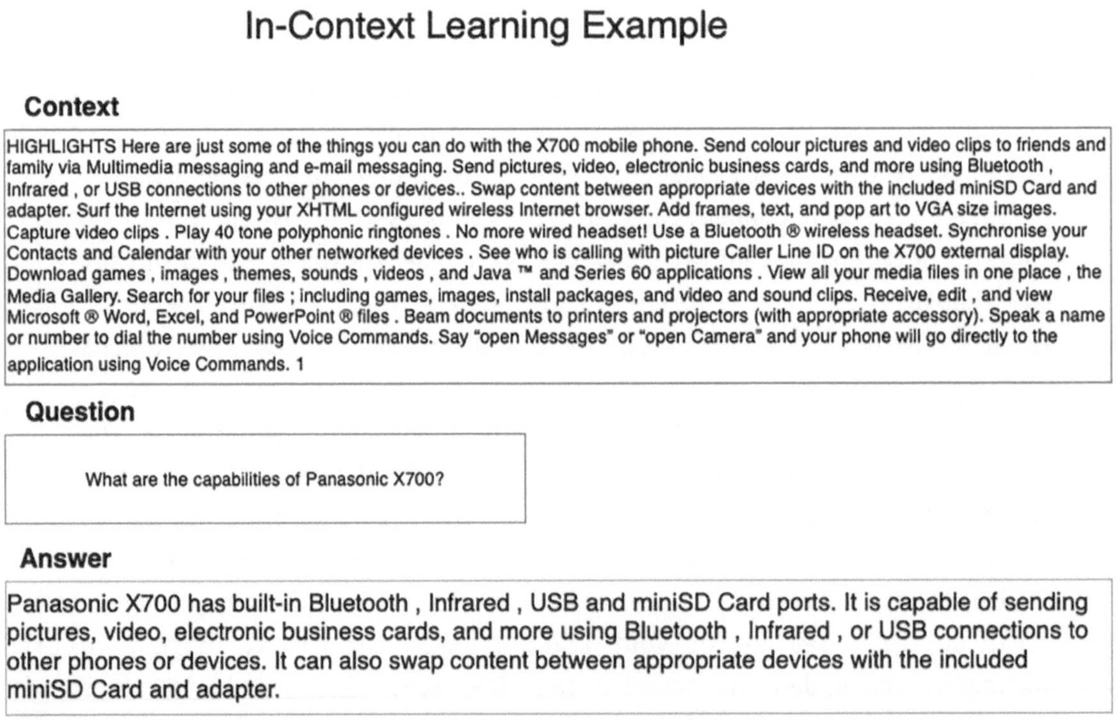

Figure 9-11. In-context learning example

You can observe that you can leverage this type of large language model to create a chatbot that can answer customer queries. Now, all you need to do is fit the retriever-reranker components with the Generative LLM component and create a UI for customer interaction.

Mitigating Hallucination Risks in In-Context Learning

In-context learning is an integral feature of generative LLMs, allowing them to adapt to new information and contexts rapidly. However, a known challenge in applying these models, particularly in generating responses based on retrieved content, is the risk of hallucination where the model generates plausible but factually incorrect or unverifiable information. You can use the following strategies to mitigate hallucination.

Data Veracity Checks:

Incorporate additional verification layers to check the factual accuracy of the data being retrieved and used for in-context learning.

- Cross-reference retrieved information with verified databases or trusted sources. For example, add the URL of the documentation used to generate the answer in the response.

- Apply data provenance techniques to trace the origin of the information.

Controlled Generation:

Adjust generative parameters to constrain the model's creative liberties, thus reducing the likelihood of hallucination.

- Calibrate the `temperature` parameter to lower the randomness of the generated text.

- Limit `max_new_tokens` to control the verbosity of the model and keep outputs concise and on topic.

Post-Generation Validation:

After generating responses, apply a post-processing step where outputs are checked for accuracy against known facts or external databases.

- Employ human-in-the-loop validation for critical use cases to ensure the reliability of responses.

Domain-Adaptive Training

Fine-tune your LLM on domain-specific datasets with previously identified and corrected hallucinations.

- Collect instances where the model has produced hallucinations and use these as negative examples in training.
- Continuously update the training set with examples to cover emerging information and topics.

User Feedback for Continuous Learning:

Incorporate user feedback mechanisms to identify and correct hallucinations, using this data to inform ongoing model training.

- Design feedback loops that allow users to flag incorrect responses.
- Regularly review flagged content to refine the model and update the knowledge base.

Implementing these strategies can reduce the risks of hallucinations in LLMs. This ensures that the learning process produces reliable and precise outcomes. This approach improves the usefulness of generative language models in real-world scenarios and builds trust in AI-powered systems among users.

User Interface

We will use the Dash framework to build the user interface in this exercise. However, similar to the previous exercises, you can choose any framework and think of the individual components as the LEGO pieces, which can be combined to build the overall application.

Let's begin by importing the required libraries. You can run the command `pip3 install -r requirements.txt` to install any missing libraries before importing them.

```
import os
import time
from textwrap import dedent
import dash
from dash import html
from dash import dcc
import dash_bootstrap_components as dbc
from dash.dependencies import Input, Output, State
```

```
from PIL import Image
import json
import re
import requests
from bs4 import BeautifulSoup
import plotly.express as px
import pandas as pd
import openai
from primeqa.components.reranker.colbert_reranker import ColBERTReranker
```

Then we will set up our env with the API keys required to make the API calls to the LLMs. You can create an .env file with the following:export HF_API_KEY="Bearer hf_XXXXXXXXX"

```
export OPENAI_API_KEY="sk-XXXXXXXXX"
```

And in the main app file (rag_dash_app.py), you can call the env variables, as shown here:

```
openai.api_key = os.getenv("OPENAI_API_KEY")
hf_key = os.getenv("HF_API_KEY")
```

Now let's define the UI components one by one.

```
def Header(name, app):
    title = html.H1(name, style={"margin-top": 5})
    logo = html.Img(
        src=app.get_asset_url("apress-logo.png"), style={"float": "right",
        "height": 60}
    )
    return dbc.Row([dbc.Col(title, md=8), dbc.Col(logo, md=4)])
```

The Header function is a helper function that creates a header component for the web application. It takes two parameters: name and app.

Inside the function, it creates an html.H1 element for the title, using the provided name parameter and applying a style that sets the top margin to 5 pixels.

It also creates an html.Img element for the logo, using the app.get_asset_url method to get the URL of the logo image file. The style of the logo is set to float right and has a height of 60 pixels.

CHAPTER 9 BUILDING ENTERPRISE-GRADE APPLICATIONS USING LLMS

Finally, the function returns a dbc.Row component that contains two columns: one for the title (dbc.Col(title, md=8)) and another for the logo (dbc.Col(logo, md=4)). The md=8 and md=4 parameters specify the column widths for medium-sized screens.

The next step is to define a text box that will take user queries as input.

```python
def textbox(text, box="AI", name="Knowa"):
    text = text.replace(f"{name}:", "").replace("You:", "")
    style = {
        "max-width": "60%",
        "width": "max-content",
        "padding": "5px 10px",
        "border-radius": 25,
        "margin-bottom": 20,
    }

    if box == "user":
        style["margin-left"] = "auto"
        style["margin-right"] = 0

        return dbc.Card(text, style=style, body=True, color="primary",
        inverse=True)

    elif box == "AI":
        style["margin-left"] = 0
        style["margin-right"] = "auto"

        thumbnail = html.Img(
            src=app.get_asset_url("apress-logo.png"),
            style={
                "border-radius": 50,
                "height": 36,
                "margin-right": 5,
                "float": "left",
            },
        )
        textbox = dbc.Card(text, style=style, body=True, color="light",
        inverse=False)
```

```
        return html.Div([thumbnail, textbox])

    else:
        raise ValueError("Incorrect option for `box`.")
```

The `textbox` function creates a text box component for a chat interface. It takes three parameters: `text`, `box`, and `name`.

First, the function modifies the `text` variable by removing any occurrences of the `name` parameter followed by a colon (`:`) and the word "You:". This is done using the `replace` method.

Next, a `style` dictionary is created to define the appearance of the text box. It sets properties such as the maximum width, width based on content, padding, border radius, and bottom margin.

The function then checks the value of the `box` parameter to determine the type of text box to create.

If `box` is set to `user`, it modifies the `style` dictionary by setting the left margin to `auto` and the right margin to 0. This creates a text box for user messages. The function returns a `dbc.Card` component with the modified `text` and `style` properties. The card is styled with a primary color and an inverse color scheme.

If `box` is set to `AI`, it modifies the `style` dictionary by setting the left margin to 0 and the right margin to `auto`. This creates a text box for AI responses. Additionally, the function creates an `html.Img` element for a thumbnail image, using the `app.get_asset_url` method to get the image's URL. The thumbnail image is styled with a border radius, a height of 36 pixels, a right margin of 5 pixels, and floated to the left. The function returns an `html.Div` element containing the thumbnail and a `dbc.Card` component with the modified `text` and `style` properties. The card is styled with a light color and a regular color scheme.

If none of these conditions is met, the function raises a `ValueError` with the message "Incorrect option for box."

The `textbox` function is versatile for creating text box components with different styles based on the `box` parameter.

Finally, we will define the layout of the Dash app as shown here:

```
app.layout = dbc.Container(
    fluid=False,
    children=[
        Header("Generative Q&A", app),
```

CHAPTER 9   BUILDING ENTERPRISE-GRADE APPLICATIONS USING LLMS

```
            dbc.Row(html.H6("Enterprise Knowledge Bank", style={'textAlign':
            'left'}),
                        className="me-auto",
                        align='left',
                        justify='let'
                  ),
        html.Hr(),
        dcc.Tabs(id="tabs-example-graph", value='tab-1-example-graph',
        children=[
        dcc.Tab(label='AI powered QA', children=[
                                        dcc.Store(id="store-
                                        conversation", data=""),
                                        conversation,
                                        controls,
                                        dbc.Spinner(html.
                                        Div(id="loading-component")),
            ]),
        # dcc.Tab(label='KPA-Slack channel analysis', children=[
        #                                 slack_channel_dropdown,
        #                                 html.Br(),
        #                                 html.H4("Top 10 Key Points
                                        from the selected slack
                                        channel"),
        #                                 html.Ol(id='slack-kpa-
                                        list_new'),
        #                                 slack_kpa_figure,
        # ]),
            ]),
        html.Br(),
        html.Footer(children="Please note that this content is made
        available with the Generative AI book. \
                        The content may include systems & methods
                        pending patent with USPTO and protected under US
                        Patent Laws. \
```

Copyright - Apress Publication")
```
    ],
)
```

The previous code sets up the layout for a web application using the Dash and Bootstrap components.

1. The `app.layout` variable is assigned to a `dbc.Container` component, which serves as the main container for the application's content. The `fluid` parameter is set to `False`, indicating that the container should have a fixed width.

2. The `children` parameter defines several child components inside the container.

3. The `Header` function is called to create a header section with the title `Generative Q&A` and a logo. The `Header` function is passed the `app` object as a parameter.

4. Next, a `dbc.Row` component contains an `html.H6` element with the text `Enterprise Knowledge Bank`. The `className` parameter is set to `me-auto` to align the row to the right, and the `align` and `justify` parameters are set to `left` to align the text to the left.

5. An `html.Hr` element is added to create a horizontal line.

6. The `dcc.Tabs` component creates a tabbed interface. It has an `id` of `tabs-example-graph` and a `value` of `tab-1-example-graph`.

7. Inside the tabs is a `dcc.Tab` component labeled `AI-powered QA`. It contains several child components, including a `dcc.Store` component, a `conversation` component, a `controls` component, and a `dbc.Spinner` component wrapped around an `html.Div` element. These components are used for AI-powered question-answering and conversation handling.

8. After the tabs, there is an `html.Br` element for adding a line break.

9. Finally, there is an `html.Footer` element with text content explaining that the application's content is made available with the *Generative AI* book. It also mentions that the content may include pending patent systems and methods protected under U.S. patent laws.

Overall, this code sets up the layout structure for a web application with a header, content sections, tabs, and a footer, using various Dash and Bootstrap components.

Now that we have set up all the UI components, let's set up three callback functions for updating the display, clearing the input, and running the chatbot.

Update Display Callback Function:

Let's define the first callback function to update the display.

```
@app.callback(
    Output("display-conversation", "children"),
    [Input("store-conversation", "data")]
)
def update_display(chat_history):
    conversation_parts = chat_history.split("<split>")[:-1]
    conversation_components = [
        textbox(x, box="user") if i % 2 == 0 else textbox(x, box="AI")
        for i, x in enumerate(conversation_parts)
    ]
    return conversation_components
```

The code defines a callback function, `update_display`, for a Dash application. The callback is triggered when the data in the `store-conversation` component changes.

The function takes the chat history as input and splits it into separate conversation parts using the `<split>` delimiter. Then, it creates conversation components by iterating over the conversation parts and calling the `textbox` function with appropriate parameters based on the index of the conversation part. If the index is even, it creates a user textbox component. Otherwise, it creates an AI textbox component.

Finally, the function returns the list of conversation components, which will be displayed as the children of the `display-conversation` component.

In summary, the code updates the display of the conversation by generating and rendering the appropriate textbox components based on the chat history.

Clear Output Callback Function:

The second callback function is to clear the textbox once Submit is clicked so the user can reenter the query.

```
@app.callback(
    Output("user-input", "value"),
    [Input("submit", "n_clicks"), Input("user-input", "n_submit")],
)
def clear_input(n_clicks, n_submit):
    return "" if n_clicks or n_submit else dash.no_update
```

The previous code sets up a callback in a Dash application that updates the value of the user-input component. If there were any clicks on the submit button or form submission, it clears the input value by returning an empty string. Otherwise, it keeps the current value unchanged by returning dash.no_update.

Run Chatbot Callback function:

Let's create our final and most crucial callback function for running the chatbot.

```
@app.callback(
    [Output("store-conversation", "data"), Output("loading-component",
    "children")],
    [Input("submit", "n_clicks"), Input("user-input", "n_submit")],
    [State("user-input", "value"), State("store-conversation", "data")],
)
def run_chatbot(n_clicks, n_submit, user_input, chat_history):
    if n_clicks == 0 and n_submit is None:
        return "", None
    if user_input is None or user_input == "":
        return chat_history, None

    context = drdecr_reranker(user_input)
    chat_history += f"Answer the question based on the context below. " + \
        "Context: "  + context + \
        " Question: " + user_input

    model_input = chat_history.replace("<split>", "\n")

    # HuggingFace LLM model
    model_output = call_huggingface_api(model_input)
```

```
model_output = format_model_output(model_output)

# OpenAI LLM model - Uncomment to use OpenAI model
# model_output = call_openai_api(model_input)

model_output = f"{model_output}<split>"

return model_output, None
```

This code defines a callback function, `run_chatbot`, for our Dash application. The callback function is triggered by the `n_clicks` event of the `submit` button and the `n_submit` event of the `user-input` field. The function performs the following steps:

1. It checks if the button has not been clicked (`n_clicks == 0`) and no submit event has occurred (`n_submit is None`). If this condition is true, it returns empty values (`""`, `None`).

2. It checks if the user input is empty or None (`user_input is None or user_input == ""`). If this condition is true, it returns the current `chat_history` and None for the loading component.

3. It generates the context based on the `user_input` using the `drdecr_reranker` function.

4. It appends the context, user input, and additional text to the `chat_history` string.

5. It replaces the `<split>` tokens in the `chat_history` with newlines (`\n`) to format it as the `model_input` string.

6. It calls the `call_huggingface_api` function to interact with the Hugging Face language model API using the `model_input` as input. Alternatively, you call the `call_openai_api` function to interact with the OpenAI LLM API.

7. It formats the returned `model_output` using the `format_model_output` function.

8. It concatenates the `model_output` with the `<split>` token to separate it from the previous conversation.

9. It returns a tuple containing the `model_output` and None for the loading component. These values will be assigned to the specified `Output` components in the Dash application.

Overall, this callback function handles user input, updates the conversation history, interacts with a language model API, formats the model's output, and provides the updated conversation data and loading component for the Dash application.

You can find the complete code on GitHub under Chapter 9, in a file called rag_dash_app.py. You can execute the Python script through the terminal/cmd prompt or the Python IDE of your choice. Once you execute this script, you can access your App at http://127.0.0.1:8051/.

Let's try our RAG application with a sample question: "How are button numbers assigned in Flicker Flasher game?" The manual for this game is indexed in Solr, so the retriever should provide the right context to the LLM for generating an answer. You can observe that our RAG application has returned a correct answer for this question, as shown in Figure 9-12. The generated answer is, "Answer: The button numbers are assigned as follows: Tommy character button number 1 Angelica character button number 2 Chuckie character button number 3 Lil character button number 4 Phil character button number 5 Reptar character button number 6."

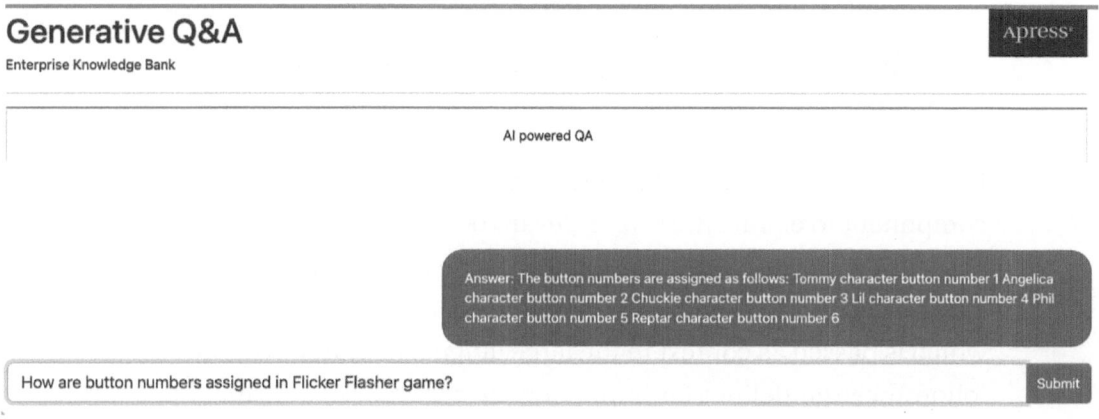

Figure 9-12. RAG UI

By combining retrieval-based methods with generative language modeling, we have achieved a more robust and efficient question-answering system. The pipeline's ability to access and process a vast knowledge base allows it to address a wide range of user inquiries accurately and efficiently.

CHAPTER 9   BUILDING ENTERPRISE-GRADE APPLICATIONS USING LLMS

Suggested Improvements in the RAG Pipeline for Generative Q&A

You have created a basic version of an enterprise-grade application using the LLM. Now you can work on improving different components to achieve the desired results, as shown here:

1. **Data Cleaning:** When creating a knowledge base for your enterprise, data cleaning is the most important step, as the accuracy of the retriever, the reranker, and, finally, the generated output depends on the efficacy of data cleaning steps. You can follow these guidelines for data cleaning:

    - Use regular expressions to remove nontextual elements like HTML tags, special characters, and stop words.

    - Use techniques such as stemming and lemmatization to normalize your text data further.

    - Employ bias detection techniques and consider data augmentation strategies to create a more balanced dataset.

2. **Neural Retriever:** You can replace the information retriever component to extract the most relevant documents/passages based on your use case. The accuracy of the generated output/answer depends on the accuracy of the retrieved document, which is passed as context to the large language model. You can follow these guidelines for improving the accuracy of information retrieval:

    - Use the BM25 algorithm to create a similarity-based index and perform a similarity/fuzzy search on this index.

    - Use a sparse encoder model like Elastic Learned Sparse EncodeR (ELSER) to create an index in Elasticsearch and perform a semantic search using this index.

    - Use a sentence transformer to create a dense vector index and perform vector search using KNN queries on this index.

- Use hybrid queries by leveraging multiple fields in the index to perform semantic + similarity-based search.

- Run multiple experiments to get benchmark results on your dataset and identify the best retriever for your use case.

3. **Reranker:** You can also try different deep learning rerankers available or train your own reranker model if you have a labeled dataset of queries and documents to get the optimal results from the reranker component. You can follow these guidelines for improving the reranker component of the RAG pipeline:

    - Train your own reranker model if you have a labeled dataset of queries and documents.

    - Use the retriever ranking function like reciprocal rank fusion from Elasticsearch.

    - Compare the results of basic retriever + reranker with the result of semantic search and/or hybrid search query and choose the one that gives a higher accuracy.

4. **Generative LLM:** You should choose the LLM based on the business requirements or the use case.

    - **Accuracy vs. Conciseness:** For concise and accurate answers, consider using a model like FLAN-T5-XXL, known for its factual language understanding.

    - **Long and Descriptive Answers:** If you require lengthy and detailed responses, explore larger models like BLOOM with more than 100B parameters. These models can generate more comprehensive and creative text formats.

    - **Infrastructure Considerations:** Be mindful of your computational resources. If deploying the LLM on-premises, factor in GPU requirements for generating answers. Cloud-based solutions might be an option for resource-intensive models.

5. **Fine-Tuning LLM:** As we learned in Chapter 7, you can also fine-tune a pre-trained LLM using an instruct dataset to tune the model for the specific task of question answering.

6. **User Application:** You have created a simple UI using the Dash framework, which has some restrictions. You can choose to develop the UI fully using other frameworks, such as React.

Therefore, this sample enterprise-grade application based on the RAG pipeline for LLM equips you with the tools and concepts to guide you through the development life cycle of the enterprise-grade applications based on LLM, and the sky is the limit.

Summary

In this chapter, we dived deep into the fascinating world of retrieval-augmented question-answering chatbots powered by LLMs. Our journey began by creating a knowledge base, a crucial foundation for any intelligent chatbot.

Step-by-step, we explored the intricacies of knowledge base creation, starting with data collection, the lifeblood of any intelligent system. Next, we set up a retriever, a key component responsible for fetching relevant information from the knowledge base. By understanding the significance of retrieval systems, we laid the groundwork for building powerful and efficient chatbots.

But we didn't stop there. Our exploration led us to the realm of neural rerankers, where we fine-tuned our chatbot's responses, ensuring the highest possible relevance and accuracy in answers. The integration of neural rerankers bolstered the chatbot's capability to engage in meaningful and contextually relevant conversations with users.

As we delved deeper, we unlocked the potential of generative LLMs. These language models enabled our chatbots to generate human-like responses, elevating the user experience to a whole new level. Harnessing the power of generative LLMs, we witnessed the transformation of our chatbots into intelligent conversational agents.

User interface design became a crucial aspect of our journey as we learned how to create intuitive and user-friendly interfaces for our chatbots. The user interface played a pivotal role in enhancing the user experience and ensuring seamless interactions with our intelligent systems.

Finally, we explored avenues for improvement in the RAG pipeline. By identifying areas for enhancement, we fine-tuned our chatbots, making them more efficient, accurate, and adaptable to a wide range of real-world scenarios.

Throughout this chapter, we ventured into the world of intelligent chatbots, armed with the knowledge and tools to build sophisticated and powerful applications. We embraced the challenges, learned from the intricacies, and emerged with a profound understanding of retrieval-augmented question-answering chatbots, ready to shape the future of conversational AI.

But the journey doesn't end here. The field of conversational AI is constantly evolving. Here's a glimpse into what the future holds:

- **Personalization:** Imagine chatbots that can tailor their responses to individual users based on past interactions and user profiles. Advancements in natural language understanding and personalization techniques will enable chatbots to deliver a more relevant and engaging user experience.

- **Multimodality:** The future of chatbots goes beyond text. Integration with speech recognition and natural language generation will create chatbots that can seamlessly converse through voice, further blurring the line between humans and machines.

- **Explainability and Transparency:** As chatbots become more sophisticated, it will be crucial to ensure their decisions and responses are explainable and transparent. Techniques like attention mechanisms and interpretable AI will shed light on the reasoning behind the chatbot's outputs, fostering trust and user confidence.

- **Ethical Considerations:** As chatbots become ubiquitous, addressing potential biases and ensuring responsible development will be paramount. Careful data selection, bias mitigation techniques, and a focus on ethical alignment will be essential to ensure chatbots serve humanity for good.

By staying informed about these advancements and ethical considerations, you can continue to build and refine your RAG chatbots, shaping the future of conversational AI and its impact on various industries. Remember, the ultimate goal is to create chatbots that are intelligent but also helpful, informative, and trustworthy companions in our digital interactions.

CHAPTER 9   BUILDING ENTERPRISE-GRADE APPLICATIONS USING LLMS

Conclusion: Generative AI Journey

This book has provided you with a thorough examination of Generative AI, exploring its many applications across different fields. We started with an introduction to Generative AI in Chapter 1, setting the stage for a more detailed exploration of its particular features.

Chapters 2 and 3 showcased the captivating realm of creative content generation. We explored the magic of transforming text descriptions into captivating images (Chapter 2) and of breathing life into scripts through text-to-video generation (Chapter 3). Chapter 4 further expanded our horizons, exploring how Generative AI bridges the gap between text and audio, opening possibilities for innovative applications.

Recognizing the importance of language as a core component of Generative AI, Chapter 5 introduced you to the fascinating world of LLMs. Chapters 6 and 7 delved deeper into the capabilities of LLMs, showcasing their proficiency in various natural language generation tasks, from creative writing to text summarization. We also explored advanced techniques like prompting to unlock the full potential of LLMs.

Building upon this knowledge, Chapters 8 and 9 ventured into the practical application of LLMs. Chapter 8 focused on creating engaging demo applications, while Chapter 9 explored the development of robust enterprise-grade applications powered by LLMs. These chapters taught you to leverage LLMs for real-world problem-solving and process automation.

The Generative AI Revolution: A World of Opportunity

As you conclude this journey, Generative AI is no longer a theoretical concept but a rapidly evolving field with immense potential. Here's how you can become part of this revolution:

- **Identify Use Cases in Your Domain:** Consider how text-to-image generation, text-to-video creation, or natural language generation with LLMs could benefit your field or organization.

- **Experiment with Creative Applications:** Do not hesitate to try these generative techniques to produce new content or streamline workflows.

- **Explore Open-Source Tools:** Numerous open-source libraries, such as Transformers and Hugging Face, offer user-friendly tools for exploring Generative AI functionalities.

- **Stay Updated with Advancements:** The field of Generative AI is constantly evolving. Stay informed about new research, projects, and best practices.

The Future of Generative AI: What Lies Ahead

The future of Generative AI holds many exciting possibilities. Progress in areas such as personalization, multimodality (voice interaction), explainability, and ethical considerations will shape this technology. By actively engaging with this evolving landscape, you can play a role in the responsible and impactful development of Generative AI.

Remember, Generative AI is not just a technological marvel; it's a tool with the potential to transform how we create, interact, and access information. By understanding these capabilities, you can leverage its power to foster innovation, enhance creative expression, and contribute to a future shaped by human-machine collaboration.

References

[1] Radford, A., Kim, J. W., Hallacy, C., Ramesh, A., Goh, G., Agarwal, S., Sastry, G., Askell, A., Mishkin, P., Clark, J., Krueger, G., & Sutskever, I. (2021). *Learning Transferable Visual Models From Natural Language Supervision.*

[2] Podell, D., English, Z., Lacey, K., Blattmann, A., Dockhorn, T., Müller, J., Penna, J., & Rombach, R. (2023). *SDXL: Improving Latent Diffusion Models for High-Resolution Image Synthesis.*

[3] Rombach, R., Blattmann, A., Lorenz, D., Esser, P., & Ommer, B. (2022). High-Resolution Image Synthesis With Latent Diffusion Models. *Proceedings of the IEEE/CVF Conference on Computer Vision and Pattern Recognition (CVPR), 10684–10695.*

[4] Piedrafita, M. (2022). *Nouns auto-captioned.* Retrieved from https://huggingface.co/datasets/m1guelpf/nouns/

[5] Luo, Z., Chen, D., Zhang, Y., Huang, Y., Wang, L., Shen, Y., Zhao, D., Zhou, J., & Tan, T. (2023). *VideoFusion: Decomposed Diffusion Models for High-Quality Video Generation.*

[6] Li, J., Li, D., Savarese, S., & Hoi, S. (2023). *BLIP-2: Bootstrapping Language-Image Pre-training with Frozen Image Encoders and Large Language Models.*

[7] Chen, K., Du, X., Zhu, B., Ma, Z., Berg-Kirkpatrick, T., & Dubnov, S. (2022). *HTS-AT: A Hierarchical Token-Semantic Audio Transformer for Sound Classification and Detection.*

Wu, Y., Chen, K., Zhang, T., Hui, Y., Nezhurina, M., Berg-Kirkpatrick, T., & Dubnov, S. (2024). *Large-scale Contrastive Language-Audio Pretraining with Feature Fusion and Keyword-to-Caption Augmentation.*

REFERENCES

[8] Radford, A., Kim, J. W., Xu, T., Brockman, G., McLeavey, C., & Sutskever, I. (2022). *Robust Speech Recognition via Large-Scale Weak Supervision*.

[9] Ardila, R., Branson, M., Davis, K., Henretty, M., Kohler, M., Meyer, J., Morais, R., Saunders, L., Tyers, F. M., & Weber, G. (2020). *Common Voice: A Massively-Multilingual Speech Corpus*.

[10] Ao, J., Wang, R., Zhou, L., Wang, C., Ren, S., Wu, Y., Liu, S., Ko, T., Li, Q., Zhang, Y., Wei, Z., Qian, Y., Li, J., & Wei, F. (2022). *SpeechT5: Unified-Modal Encoder-Decoder Pre-Training for Spoken Language Processing*.

[11] Ravanelli, M., Parcollet, T., Plantinga, P., Rouhe, A., Cornell, S., Lugosch, L., Subakan, C., Dawalatabad, N., Heba, A., Zhong, J., Chou, J.-C., Yeh, S.-L., Fu, S.-W., Liao, C.-F., Rastorgueva, E., Grondin, F., Aris, W., Na, H., Gao, Y., ... Bengio, Y. (2021). *SpeechBrain: A General-Purpose Speech Toolkit*.

[12] Ao, J., Wang, R., Zhou, L., Wang, C., Ren, S., Wu, Y., Liu, S., Ko, T., Li, Q., Zhang, Y., Wei, Z., Qian, Y., Li, J., & Wei, F. (2022). *SpeechT5: Unified-Modal Encoder-Decoder Pre-Training for Spoken Language Processing*.

[13] Ao, J., Wang, R., Zhou, L., Wang, C., Ren, S., Wu, Y., Liu, S., Ko, T., Li, Q., Zhang, Y., Wei, Z., Qian, Y., Li, J., & Wei, F. (2022). *SpeechT5: Unified-Modal Encoder-Decoder Pre-Training for Spoken Language Processing. Proceedings of the 60th Annual Meeting of the Association for Computational Linguistics (Volume 1: Long Papers)*, 5723–5738.

[14] Vaswani, A., Shazeer, N., Parmar, N., Uszkoreit, J., Jones, L., Gomez, A. N., Kaiser, L., & Polosukhin, I. (2023). *Attention Is All You Need*.

[15] Devlin, J., Chang, M.-W., Lee, K., & Toutanova, K. (2018). BERT: Pre-training of Deep Bidirectional Transformers for Language Understanding. *CoRR, abs/1810.04805*. http://arxiv.org/abs/1810.04805

REFERENCES

[16] Devlin, J., Chang, M.-W., Lee, K., & Toutanova, K. (2018). BERT: Pre-training of Deep Bidirectional Transformers for Language Understanding. *CoRR, abs/1810.04805*. http://arxiv.org/abs/1810.04805

[17] Devlin, J., Chang, M.-W., Lee, K., & Toutanova, K. (2018). BERT: Pre-training of Deep Bidirectional Transformers for Language Understanding. *CoRR, abs/1810.04805*. http://arxiv.org/abs/1810.04805

[18] Gurulingappa, H., Rajput, A. M., Roberts, A., Fluck, J., Hofmann-Apitius, M., & Toldo, L. (2012). Development of a benchmark corpus to support the automatic extraction of drug-related adverse effects from medical case reports. *Journal of Biomedical Informatics, 45*(5), 885–892. https://doi.org/10.1016/j.jbi.2012.04.008

[19] Sanh, V., Debut, L., Chaumond, J., & Wolf, T. (2019). DistilBERT, a distilled version of BERT: smaller, faster, cheaper and lighter. *ArXiv, abs/1910.01108*.

[20] Radford, A., Wu, J., Child, R., Luan, D., Amodei, D., & Sutskever, I. (2019). *Language Models are Unsupervised Multitask Learners.*

[21] Lewis, M., Liu, Y., Goyal, N., Ghazvininejad, M., Mohamed, A., Levy, O., Stoyanov, V., & Zettlemoyer, L. (2019). BART: Denoising Sequence-to-Sequence Pre-training for Natural Language Generation, Translation, and Comprehension. *CoRR, abs/1910.13461*. http://arxiv.org/abs/1910.13461

[22] Yu, H., Yang, Z., Pelrine, K., Godbout, J. F., & Rabbany, R. (2023). *Open, Closed, or Small Language Models for Text Classification?*

[23] Dang, N. C., Moreno-García, M. N., & De la Prieta, F. (2020). Sentiment Analysis Based on Deep Learning: A Comparative Study. *Electronics, 9*(3). https://doi.org/10.3390/electronics9030483

REFERENCES

[24] Penedo, G., Malartic, Q., Hesslow, D., Cojocaru, R., Cappelli, A., Alobeidli, H., Pannier, B., Almazrouei, E., & Launay, J. (2023). The RefinedWeb dataset for Falcon LLM: outperforming curated corpora with web data, and web data only. *ArXiv Preprint ArXiv:2306.01116.* https://arxiv.org/abs/2306.01116

[25] Hu, E. J., Shen, Y., Wallis, P., Allen-Zhu, Z., Li, Y., Wang, S., & Chen, W. (2021). LoRA: Low-Rank Adaptation of Large Language Models. *CoRR, abs/2106.09685.* https://arxiv.org/abs/2106.09685

[26] Hu, E. J., Shen, Y., Wallis, P., Allen-Zhu, Z., Li, Y., Wang, S., & Chen, W. (2021). LoRA: Low-Rank Adaptation of Large Language Models. *CoRR, abs/2106.09685.* https://arxiv.org/abs/2106.09685

[27] Jin, Q., Dhingra, B., Liu, Z., Cohen, W., & Lu, X. (2019). PubMedQA: A Dataset for Biomedical Research Question Answering. *Proceedings of the 2019 Conference on Empirical Methods in Natural Language Processing and the 9th International Joint Conference on Natural Language Processing (EMNLP-IJCNLP)*, 2567–2577.

[28] Chung, H. W., Hou, L., Longpre, S., Zoph, B., Tay, Y., Fedus, W., Li, E., Wang, X., Dehghani, M., Brahma, S., Webson, A., Gu, S. S., Dai, Z., Suzgun, M., Chen, X., Chowdhery, A., Narang, S., Mishra, G., Yu, A., ... Wei, J. (2022). *Scaling Instruction-Finetuned Language Models.* arXiv. https://doi.org/10.48550/ARXIV.2210.11416

[29] Narayan, S., Cohen, S. B., & Lapata, M. (2018). Don't Give Me the Details, Just the Summary! Topic-Aware Convolutional Neural Networks for Extreme Summarization. *ArXiv, abs/1808.08745.*

[30] Raffel, C., Shazeer, N., Roberts, A., Lee, K., Narang, S., Matena, M., Zhou, Y., Li, W., & Liu, P. J. (2020). Exploring the Limits of Transfer Learning with a Unified Text-to-Text Transformer. *Journal of Machine Learning Research, 21*(140), 1–67. http://jmlr.org/papers/v21/20-074.html

References

[31] Zhang, X., Zhao, J., & LeCun, Y. (2015). Character-Level Convolutional Networks for Text Classification. *ArXiv:1509.01626 [Cs]*.

[32] Radford, A., Wu, J., Child, R., Luan, D., Amodei, D., & Sutskever, I. (2019). *Language Models are Unsupervised Multitask Learners*.

[33] Penedo, G., Malartic, Q., Hesslow, D., Cojocaru, R., Cappelli, A., Alobeidli, H., Pannier, B., Almazrouei, E., & Launay, J. (2023). The RefinedWeb dataset for Falcon LLM: outperforming curated corpora with web data, and web data only. *ArXiv Preprint ArXiv:2306.01116*. https://arxiv.org/abs/2306.01116

[34] Wu, M., Waheed, A., Zhang, C., Abdul-Mageed, M., & Aji, A. F. (2024). LaMini-LM: A Diverse Herd of Distilled Models from Large-Scale Instructions.

[35] Radford, A., Wu, J., Child, R., Luan, D., Amodei, D., Sutskever, I., & others. (2019). Language models are unsupervised multitask learners. *OpenAI Blog*, *1*(8), 9.

[36] Chung, H. W., Hou, L., Longpre, S., Zoph, B., Tay, Y., Fedus, W., Li, E., Wang, X., Dehghani, M., Brahma, S., Webson, A., Gu, S. S., Dai, Z., Suzgun, M., Chen, X., Chowdhery, A., Narang, S., Mishra, G., Yu, A., ... Wei, J. (2022). *Scaling Instruction-Finetuned Language Models*. arXiv. https://doi.org/10.48550/ARXIV.2210.11416

[37] Li, Y., Franz, M., Sultan, M. A., Iyer, B., Lee, Y.-S., & Sil, A. (2021). *Learning Cross-Lingual IR from an English Retriever*. https://doi.org/10.48550/ARXIV.2112.08185

Index

A

Abstractive summarization
   decoder-only model, 311
      data pre-processing, 312–314
      inference/evaluation, 320–322
      libraries and data loading, 312
      quantization technique, 314
      training model, 314–319
   encoder-decoder model, 299
      data preprocessing, 300–303
      evaluation, 309–311
      evaluation/benchmark setup, 303–305
      libraries/data loading, 299, 300
      sequence-to-sequence model, 305
      tokenization parameters, 302
      training model, 305–309
   RLHF, 298
   text document, 298
Adverse drug events (ADEs), 190, 191, 202
Artificial intelligence (Generative AI), 1
Audio-to-text generation
   challenges, 118
   CLAP model
      architecture, 121
      implementation, 120
      inference, 123–125
      infrastructure, 121
      libraries/data loading, 122, 123
   conversion technologies, 130
   CTC, 125
   digital representation, 115
   evolution, 113, 114
   frequency spectrum, 117
   fundamentals, 115
   inherent challenges, 119, 120
   sampling representation, 115, 116
   spectrogram, 118
   speech recognition technologies, 113
   waveform, 116, 117
Automatic speech recognition (ASR), 129

B

Bidirectional and Auto-Regressive Transformers (BART)
   encoder-decoder models, 223
Bidirectional Encoder Representations from Transformers (BERT), 229
   bidirectional context approach, 184
   encoder models, 184
   fine-tuning process
      accuracy method, 195
      ADE dataset, 191
      benefits, 191
      classification, 197
      classification dataset, 192
      classification task, 201, 202
      data pre-processing, 193–195
      DistilBERT model, 196
      evaluation process, 197–200
      flowchart, 189, 190
      libraries/data loading, 192
      prediction process, 200

INDEX

Bidirectional Encoder Representations from Transformers (BERT) (*cont.*)
    sentence classification, 192
    sentence-level tasks, 188
    training model, 195–197
    training parameters, 190
  masked language model, 184
  NER, 202
  next-sentence prediction, 185
  pre-training objective, 184–187
  RLHF generation, 337
Bootstrapped Language Image Pretraining 2 (BLIP-2), 97

C

Colab environment
  clone repository, 27
  content, 26
  download file, 29
  drive, 26
  libraries, 25
  pro premium subscription, 23
  runtime type, 25
  upload files, 28
Connectionist Temporal Classification (CTC)
  architectures, 126–128
  seq2seq architectures, 125
  seq2seq models, 128–130
Contrastive language-audio pretraining (CLAP)
  audio-to-text generation, 120–125
Contrastive language image pretraining (CLIP), 39
  data preprocessing, 47, 48
  fundamental architecture, 44, 45
  implementation, 46
  libraries/data loading, 47
  model inference stage, 48
  pre-training architecture, 45
  textual descriptions/visual imagery, 44
  vintage leather backpack, 45
Convolutional neural networks (CNNs), 6
Cyan, Magenta, Yellow, Key/Black (CMYK) model, 42

D

Deep learning, 3
Diffusion models
  generative AI models, 10
  model training
    cvtImg function, 54
    data preparation, 53
    forward method, 60
    generate_images function, 65
    generative model, 53, 64
    gradual improvement, 66
    noisy image, 67
    NumPy arrays, 54
    stages, 55
    TimeStepUNet class, 60, 61
    traindata_loader, 65
    U-Net architecture, 55–58, 60
    untrained model, 62–64
    visualization, 67
  text-to-image models
    complex data, 49
    data preprocessing, 51, 52
    implementation, 50
    libraries, 50, 51
    transformations, 52
Discriminative *vs.* generative AI models, 8, 9

INDEX

E

Elastic Learned Sparse EncodeR (ELSER), 436
Encoder-decoder models
    BART model, 223
    challenges/directions, 226
    definition, 222
    evolution, 224
    real-world applications, 225
    source code, 224
    strategies, 223
    training/fine-tuning, 225
    transformer model, 225
Enterprise-grade applications
    RAG, 402

F

Fine-tuning LLMs
    abstractive summarization, 298
    advantage, 258, 259
    benefits, 258
    catastrophic forgetting, 262
    computational resources, 259
    parameter efficiency, 262–264
    pre-trained model, 258
    Question Answering (Q&A), 263
    RLHF, 324
    sentiment analysis
        customer reviews, 260
        disadvantages, 261
        steps, 260, 261
    SFT memory requirements, 322–324

G

Generative adversarial networks (GANs), 3, 4, 38, 82
    neural networks, 7
    generator/discriminator, 10, 11
Generative AI models
    data science, 18–20
    deep learning technologies, 3
    development environment
        Colab, 23–29
        Hugging Face account, 30–32
        OpenAI resources, 32, 33
        troubleshooting issues, 33–35
    diffusion models, 10
    discriminative models, 8, 9
    diverse domains
        audio, 21
        overview, 20
        problem-solving, 22
        text generation, 22
        visual domain, 20
    ethical/technical challenges
        capabilities, 17
        data privacy and security, 17
        DeepMind approach, 17, 18
        innovation/responsibility, 15, 16
    evolution timeline, 4, 5
    flourishing ecosystem, 2
    generative adversarial networks, 10, 11
    industries/reshape, 1
    neural networks, 6, 7
    PixelRNNs, 12, 13
    real-world application/advantages, 13–15
    restricted Boltzmann machines, 11
    text-to-image, 3, 37
    text-to-video, 81
    types/techniques, 9
    variational autoencoders (VAEs), 11

INDEX

Generative Pre-trained Transformer
(GPT), 7, 229
  decoder-only models
    autoregressive modeling, 221
    opportunities, 222
    real-world implications/technical
      applications, 222
    sequence generation, 221
    source code, 220
    training phase, 220
  large language models, 173

H, I, J, K

Huggingface token, 31

L

Large language models (LLMs), 4, 39, 171,
    297, 349, 401
  auto-encoding models, 183
  benchmarks, 179
  BERT, 184
  creative text generation, 179
  decoder-only models, 219–222
  definition, 173
  encoder-decoder models, 222–226
  enterprise-grade, 401
  fine-tuning, 258
  generalization, 178
  growth, 175
  in-context learning, 422–426
  key directions, 226, 227
  language transformer models
    architecture, 179, 180, 182
    attention layers, 180, 182
    comparison, 183
    unique training process, 181
  language understanding, 178
  limitations/ethical concerns, 174
  natural language generation, 241–250
  NLP, 230
  prompt completion, 178
  prompting techniques, 250–257
  research and innovation, 179
  search engine, 173
  training/adoption
    fine-tuning phase, 176, 177
    high-level overview, 175
    life cycle, 175
    pre-training, 176
  transfer learning, 178
  unsupervised learning, 173
  website content, 349
Latent Dirichlet Allocation (LDA), 240
Long Short-Term Memory (LSTM), 225
Low-Rank Adaptation (LoRA)
  abstractive summarization, 317
  advantages, 264
  computational benefits, 264
  data preprocessing
    components, 267
    data entry, 267
    loading dataset, 268
    preprocessing function, 275–277
    PubMedQA, 267
    tokenizer, 272–274
    train-test split, 268
    transform dataset, 269–272
  development environment, 266
  evaluation process
    DataFrame, 293–295
    loading model, 288, 289
    metrics dataframe, 294
    output comparison, 291
    pre-trained model, 291–293

question-answering tasks, 288
  test dataset, 289–291
 fine-tuning, 263–266
 model training/fine-tuning
  checkpoint size, 288
  DataCollator, 282, 283
  hyperparameters, 283–285
  loading model, 278, 279
  preparation, 279–282
  quantization, 277, 278
  saving model, 286–288
  training model, 286
 QLoRA, 265

M

Masked language model (MLM), 184

N

Named Entity Recognition (NER)
 dataframe, 206
 data preparation, 204–211
 displaCy library, 217
 entities, 205, 207
 evaluation process, 214–219
 fine-tuning, 202, 214
  entity-level recognition, 202
 libraries/data loading, 202–204
 output process, 218
 tagging function, 210
 training model, 211–214
 visualization, 216
Natural language generation (NLG), 229, 232
 creative writing tasks, 241, 242
 dialogue generation
  chatbots/virtual assistants, 248
  practical deployment, 249, 250
  technical integration, 248
  text generation, 247, 248
  user experience considerations, 249
 fine-tuning, 242, 243
 genre, 243
 hallucination, 245, 246
 summarization, 244, 245
Natural language processing (NLP), 2, 4, 38, 82, 229, 324
 classification tasks, 230
 entity extraction
  entities, 237
  model/pipeline/tokenizer, 236
  real-world scenarios, 237, 238
 Falcon/LLaMA, 230
 prompt engineering/prompt tuning, 230
 RoBERTa, 231
 sentiment analysis
  classification, 234
  NLG tasks, 232
  principles, 236
  prompt engineering, 234–236
  text generation pipeline, 233
  training procedure, 231
 seq2seq model, 128
 societal impact, 18
 topic modeling task, 239, 240
 traditional modeling methods, 240
 video transcripts, 383
Neural networks
 architectures, 6, 7
Next-sentence prediction (NSP), 185

O

OpenAI access account, 32, 33

INDEX

P

Parameter efficient fine-tuning (PEFT), 262–264, 297, 316
Peak Signal-to-Noise Ratio (PSNR), 111
Pixel Recurrent Neural Networks (PixelRNNs), 12, 13
Portable document format (PDF), 398
    insights/gaining, 368
    question-answering, 369–375
    summarization, 369, 375–383
Prompting techniques
    chain-of-thought (CoT)
        approach, 253
        complex reasoning, 254
        implementation, 254
        interpretability, 254
    few-shot prompting
        approach, 252
        code generation tasks, 252
        industry projects, 252
        patterns, 251
        personalization, 253
        zero-shot prompting, 251
    fine-tuning
        application, 257
        decision flowchart, 255
        decision-making framework, 257
        limitations, 255
        prompt engineering, 256
        trade-offs, 256
Proximal Policy Optimization (PPO), 347
    algorithm, 338
    environment, 330
    monitoring, 340
    optimization, 343
    positive/coherent continuations, 330
    PPOTrainer, 336
    results, 344
    sentiment analysis, 332
    training, 331, 342

Q

Quantization Low-Rank Adaptation (QLoRA), 265–268

R

Recall-Oriented Understudy for Gisting Evaluation (ROUGE), 303
Recurrent neural networks (RNNs), 6, 225
Reinforcement learning from human feedback (RLHF), 177, 298
    approaches, 347
    benefits, 326
    challenges and considerations, 326
    controlled review generation
        BERT classifier, 337
        configure environment, 331–333
        controlled reward generation, 330
        dataset loading, 333–335
        dependencies, 331
        environment, 330
        GPT2 model, 336
        optimization process, 336
        PPOTrainer, 336
        reward tuning, 330
        text generation, 337
    data collection, 329
    definition, 325
    evaluation/ensuring reward model, 345–347
    evaluation process, 341–345
    mitigating bias, 327
    positive reviews
        training loop, 338–341

PPO monitoring, 340
reward model
    implementation, 328–330
save model, 345
stages, 325
Restricted Boltzmann machines
    (RBMs), 4, 11
Retrieval-augmented generation (RAG)
    architecture, 406–408
    colbert reranking, 420
    components, 407
    generative Q&A
        components, 436
        data cleaning, 436
        deep learning rerankers, 437
        fine-tuning, 438
        generative LLM, 437
        information retriever, 436
    in-context learning
        components, 422, 423
        mitigating hallucination, 425, 426
        passage, 423, 424
        retriever-reranker components,
            424, 425
    information retrieval Solr
        component, 412, 413
        document information
            extraction, 416
        output/return, 418
        Query UI, 414
        results display, 417
        search result processing, 416, 417
        solr_retriever() function, 414
        URL construction/request, 416
    knowledge base
        creation, 408
        data collection, 408
        retriever, 409–411

scaling-up strategies, 412
single-machine, 411, 412
neural reranker
    component, 418
    DrDecr model, 419–422
    product documentation, 418, 419
Question-Answering chatbot
    hallucination, 403
    knowledge cutoff, 402
    limitations, 403
    simplified pipeline, 403
    web-based chatbots, 402
real-world application, 405, 406
retriever, 409
    cloud service locally, 409, 410
    documents, 410
    indexing UI, 411
    installation, 409
    Solr binary file, 409
user interface, 426–435

S

Sequence-to-sequence models, 125,
    128–130, 222
Speech to Text (STT) model
    audio-to-text, 130
    evaluation process, 134–136
    features, 131
    fine-tuning
        data processing, 139, 140
        evaluation/inference, 148, 149
        Hill Mari language, 145
        libraries/data loading, 137, 138
        training model, 140–147
        Whisper small model, 137
    inference model, 133, 134
    libraries/data loading, 131, 132

INDEX

Speech to Text (STT) model (*cont.*)
   requirements/data scopes, 130
   seq2seq, 128
   word error rate (WER), 131, 133
Stochastic Differential Equation editing (SDEdit), 68
Structural similarity index (SSIM), 111
Supervised fine-tuning (SFT), 297
   consumption, 324
   memory requirements, 322, 323
   task/domain tuning, 323
   token lengths, 323
   types, 323

T

Text-to-audio generation, 113
Text-to-image generation
   CLIP model, 39, 44–50
   diffusion models, 39, 49–67
   education/research, 37
   fine-tuning
      libraries/loading data, 71–73
      model inference, 76, 77
      model training phase, 73–75
      stable diffusion model, 71
      troubleshooting, 77, 78
   fundamentals, 39
   image data
      digital images, 40
      file formats, 43
      foundational elements, 40
      pixels/color models, 42
      resolution/quality, 42
      vector/raster images, 41
      visual representation, 42
   pre-trained model
      fine-tuning, 71–78

      image generation, 70
      libraries, 69
      model inference, 69–71
      SDEdit, 68
   real-life applications, 37, 38
   stable diffusion model, 39
Text-to-speech (TTS)
   architecture, 149, 150
   data pre-processing, 153–156
   fine-tuning
      components, 168
      data pre-processing, 158–163
      inference, 169, 170
      libraries/data loading, 157
      model training process, 163–169
      speaker embeddings, 165
      SpeechT5, 157
      troubleshooting, 170
   inference, 156, 157
   libraries/data loading, 152, 153
   mean opinion score (MOS), 151
   pre-trained SpeechT5 model, 151
   speech generation process, 150
   SpeechT5, 151
   STT, 129
Text-to-video generation
   augmented reality/virtual reality, 112
   fine-tuning
      BLIP-2 architecture, 97, 98
      customization, 96
      data loading/preprocessing tasks, 100–102
      inference, 107–111
      Large File Storage (LFS), 103
      libraries, 99, 100
      training model, 103–107
      vision transformer (ViT), 98
   historical development, 82

industries, 81
innovative tool, 82
transitions, 83
VideoBERT, 83
video data, 84
Transformers, 7

U

User interface (UI)
    extract_combined_passage, 365, 366
    Hugging Face API, 364, 365
    libraries, 363
    RAG model
        application, 435
        callback functions, 432–435
        components, 427
        Dash app, 429
        Dash/Bootstrap components, 431
        libraries, 426
        main app file, 427
        parameters, 427
        step-step process, 434
        textbox function, 429
        user queries, 428, 429
    step-step process, 366–368
    Streamlit app, 360–363

V

Variational autoencoders (VAEs), 11, 75
Video data
    annotation/quality, 90
    audio components, 86
    color depth/bitrate, 86
    comprehensive overview, 84
    compression/storage, 87, 88
    contextual and semantic scenarios, 89
    ethical/privacy concerns, 90
    formats/codecs, 84
    frame rate, 85
    integration, 89
    metadata, 87
    pre-trained model, 94
        libraries, 94
        model inference, 95, 96
    resolution/aspect ratio, 86
    technical/computational/creative challenges, 88–91
    temporal dimension, 88
    textual data
        content analysis/personalization, 91, 92
        metadata/annotations, 91
        semantic metadata tagging, 92, 93
        transcripts/subtitles, 91
Video transcripts
    benefits, 383
    caption summarization/Q&A, 384–394
    chatbot, 393
    extracting insights, 383, 384
    Langchain package/OpenAPI
        Chroma, 395
        data processing, 395
        DirectoryLoader, 394
        methodology, 394
        notebook, 394
        post-processing, 396
        transcript summarization, 397
        YoutubeLoader, 395
    significant advantage, 384
    Streamlit app, 389
    summarization, 398
    YouTube Captions/transcripts, 390

INDEX

W, X, Y, Z

Website content
    concepts, 349
    data scraping, 351–353
    in-context learning, 353
    prompts, 353
    question-answering
        pattern, 353–357
    summarization, 357–360
    user interface/application, 360–368